Tourism Theories, Concepts and Models

Bob McKercher and Bruce Prideaux

(G) Goodfellow Publishers Ltd

(G) Published by Goodfellow Publishers Limited,
26 Home Close, Wolvercote, Oxford OX2 8PS
http://www.goodfellowpublishers.com

British Library Cataloguing in Publication Data: a catalogue record for this title is available from the British Library.

Library of Congress Catalog Card Number: on file.

ISBN: 978-1-911635-35-2

DOI: 10.23912/9781911635352-4280.

Design and typesetting by P.K. McBride, www.macbride.org.uk

Cover design by Cylinder

Printed by Lightning Source

Contents

Acknowledgements

This book has been 30 years in the making, since the two authors began their journeys as tourism academics. And, so, there are almost too many people to thank by name for making this project possible. However, many need to be acknowledged directly and indirectly.

We started at a time when tourism scholarship was undergoing unprecedented growth. Most of the new programmes were situated in business schools or commerce departments, while most of the tourism scholars came from non-business, social science backgrounds. Our knowledge was based on the first generation of scholars who began to explore this area of study beginning in the 1970s. Their work provided the foundation for how we think about tourism the way we think about it today. The works of those pioneering scholars form the basis of this book. We, and indeed, all of tourism academics, owe a debt of gratitude to them.

In addition, we have to thank the many thousands of students we have taught over the years for challenging us, forcing us to rethink much of what was accepted to be true about tourism, yet had little real supporting evidence, and for being sounding boards for us to bounce ideas on them.

At a more prosaic level, special thanks need to go to Professors Chris Cooper and Alan Fyall along with Drs. Vincent Tung, Aaron Yankholmes and Patrick L'Espoir Decosta for providing valuable feedback on various drafts of this manuscript. And, of course we have to thank the people at Goodfellows Publishers for their support and encouragement.

While many people can share rightfully in any success the book may have, the two authors take full responsibility for any of its deficiencies.

Preface

What is tourism? Simple question. Difficult to answer. How you define tourism lies within the eyes of the beholder. We all know it exists. We all know or think we know what is definitely tourism, who tourists are and what the tourism industry is. We also all know, or think we know, what is definitely not tourism, is not a tourist and what business sectors fall outside of the tourism industry. But tourism is neither absolute nor exclusive. Instead, it is classified by vague and fuzzy boundaries. That is what makes it such an exciting field of study, on the one hand, and such an exasperating phenomenon to study on the other hand.

The situation is complicated further by the traditional disciplinary silos that have influenced tourism studies, where academics have worked narrowly, within their own disciplines, often with little or no input from other disciplines. As a result, discrete sets of rules, guidelines and regulations have evolved. They begin at the most basic level about how each discipline even defines tourism or the tourist. Indeed, it seems fantastical that no single agreed upon definition of tourism has been developed, in spite of 70 years or more of academic enquiry. Maybe it is not so surprising, given the vast array of disciplines and fields of study that have explored this phenomenon.

And, the situation is only going to get more confusing as we move inexorably to the trend of increasing adjectification of tourism. It began in the late 1980s with the rise of the anti-tourism or alternative tourism movement, accelerated rapidly in the 1990s with the emergence of ecotourism and the alleged shattering of the mass market into a series of special interest tourism markets until it seems today 'tourism' no longer exists. Instead, we have to deal with golf tourism, ski tourism, film tourism, dark tourism, and almost any other descriptor. As we will discuss in the book, the lack of common terminology is something that affects all aspects of tourism studies. Not only do we not agree on a definition, but the ideas of a destination, product, attraction and even sustainability are beset by confusion, conflation and value judgements.

As a result, much of what we think we know about tourism is well-known but not necessarily known well. Most topics are covered in tourism texts or cited frequently in the academic literature. But, the core ideas that drive how we think about tourism are often condensed into a short paragraph, a few sentences, or simply included as one of a series of references to illustrate a point. Knowledge for many comes from these brief summations and not from the foundational works themselves. We are hoping that through this text, people can reflect more on how we think about tourism and what epistemological and ontological processes we have followed to think the way we think.

This observation led to the genesis of the need for a book of this type. It has a rather ambitious aim of trying to pull together some of the vast array of literature conceptualizing the tourism phenomenon into a single body of work. The title recognizes that tourism is not and can never be a singular, unified field of study, for the frameworks and models have been developed to explore the discrete component parts of tourism come from diverse sources. This book examines the core theories, models, frameworks and concepts that lead to our understanding of each of the component elements of tourism in an attempt to develop a common, cross disciplinary understanding. It then seeks to show the linkages, forward and backward, between each of these elements and the other constituent components of the tourism system.

This book becomes increasingly important as the nature of tourism studies shifts from a discipline-focused approach that uses tourism as a means to explore the core disciplines and/or uses core disciplines to explain tourism, to the emergence of the dominant business studies/training orientation adopted increasingly in higher educational institutions. The emergence of stand-alone business tourism schools/departments or the emergence of stand-alone tourism specialisms within business schools provides an enhanced range of career opportunities for students. But these benefits come at the risk of students receiving a selective, cursory overview of the role social science research has played in our understanding of tourism, our beliefs about how tourism works and, importantly, our grounding in an understanding of how the social sciences explain much of the business of tourism. As a result, we are at risk of producing technically competent individuals who often know very little about tourism as a social, human phenomenon. *Tourism Theories, Concepts and Models* hopes to fill the void that exists in our collective knowledge base. It explains why we think about tourism the way we do through a series of detailed chapters that review key concepts, models and theories.

The book is divided into five sections. Part 1 sets the stage for the rest of the book. It begins with a review of what theory is and how models and concepts fit into this broad paradigm. In particular, the first chapter looks at how the term 'theory' has been misused in much academia and identifies seven different types of 'theory' used in tourism. A discussion of the challenge of defining tourism follows with a review of various models that describe the structure of tourism.

Part 2 examines the key features that explain how tourism works. Successful tourism is a function of both attractions, access, politics and various supply and demand functions. This section begins with a detailed discussion of what attractions are and how they drive tourism. Because the term 'attraction' is one of the most misused in tourism academia, a conceptual taxonomy of attractions is developed as a means of adding some structure to the topic. The issue of access is discussed next looking at the valuation of time and how both market access and physical access influence tourism flows. The following chapter looks at international tourism and the role the World Trade Organisation, and other global bodies, play in facilitating or inhibiting travel. Finally, various supply/demand and economic models are discussed.

Part 3 looks at the evolution of tourism. A critical review of Butler's and Plog's lifecycle models is undertaken highlighting the conceptual strengths and limitations of each one. It also discusses unrecognized subtleties of the models. An alternate lifecycle model using complexity theory is then discussed. We argue it is a more reliable way to document destination evolution as its dynamic nature allows for multiple factors to play a role in influencing change.

Part 4 focusses on sociological and psychological investigation of the tourism. Here various foundation works by Erik Cohen, Jafar Jafari, Graham Dann, Philip Pearce, John Crompton and others are discussed. This part of the book begins with a discussion of the evolution of motivation theory from Maslow's hierarchy through push-pull theory to the current state of the art of the travel career pattern. The following chapters look at sociological and anthropological aspect of tourism before we embark on a discussion of tourism and the individual and various barriers and opportunities to travel

Part 5 examines various issues relating to tourism impacts, sustainability, climate change and matters relating to the development of sustainable planning. It begins with a historic review of impacts framed within a conflict paradigm and then argues that most impacts are a function of place change. The various historical and contemporary concepts of sustainability and sustainable tourism are discussed next, before moving onto the most pressing global issue of climate change. This section concludes with a review of various planning models and how they must fit within a sustainability context.

The Epilogue presents a table that shows the interlinkages between and among the more than 50 theories, concepts and models discussed throughout the book. Here, the authors call for tourism academics to appreciate the multidisciplinary nature of tourism studies

We hope you enjoy the book and that it makes you think about what tourism is and how it works.

Bob and Bruce.

About the authors

Prof Bob McKercher has been a tourism academic since 1990. Prior to that he worked in the Canadian tourism industry in a variety of advocacy and operational roles. He has broad ranging interest in tourism. He is a Fellow of the International Academic for the Study of Tourism, the Council for Australian University Tourism and Hospitality Education and the International Academy of Culture, Tourism and Hospitality Research.

Prof Bruce Prideaux commenced his academic career in 1992 at the University of Queensland where he later obtained a PhD. He began his professional life as a high school teacher and later moved to the public service where he worked as a transport analyst. His pre academia wanderings through Africa, Europe and Asia played a large part in developing his interest in tourism as both a traveller and later as a scholar. He has a wide range of research interests including sustainability issues, climate change, transition to carbon-neutrality, drive tourism, protected area tourism, rural tourism, remote area tourism and crisis management

To our families for putting up with us.

Part 1 – Setting the Stage

The first three chapters set the stage for the study of how our understanding of tourism has evolved over the past 50 years. Students of tourism need to be aware of how terms are conflated, misused and misunderstood. So, in these first chapters, we start to cut through the mist of what tourism is and how it is structured. Most importantly, it is essential to be able to appreciate how each discipline defines tourism and uses it for its own narrow needs.

No text examining tourism theories, concepts and models would be complete without first defining its terms and conditions. Chapter 1 sets the stage by exploring the issues of what is a theory, concept and model. Special reference is made to work by Stephen Smith and his colleagues about how the term theory is used in seven different ways to explore this phenomenon we call tourism. The chapter then moves on to a discussion of the differences between theory, concept and models.

Chapter 2 begins the exploration of tourism by examining the contribution that different disciplines and fields of study have made to the development of this field (or is it a discipline?). The chapter then enters into the almost impossible debate about trying to define tourism and concludes that any comprehensive definition must include both heuristic and technical definitions, which in many ways are mutually exclusive.

Chapter 3 looks at the structure of tourism and reviews the major models that have been used to identify the component parts of this activity. In particular, it examines the idea of a destination and examines how it is comprised of three types of space.

1 Theory, Concepts and Models

By the end of this chapter, the reader will be able to:

- Know what theory is and distinguish between good and bad theory
- Be able to identify and analyse seven types of theory used in tourism
- Define the differences between theory, concepts and models

Introduction

Recently, a number of academics have begun to think about theory in tourism in much more detail. As a relatively young field, there have been many complaints that tourism lacks distinct theory that has held back its development. However, Stergiou and Airey (2018:165) note tourism theory does not exist independently of reality; rather it has pragmatic relevance, facilitating researchers and others to make sense of and explain the so-called real world in the form of data or other phenomena.

We as academics seem focussed on producing theory and developing new ideas as 'theory'. Yet, most of us know very little about what theory is and what constitutes good theory. A key challenge, and one that must be addressed right from the outset, is the meaning of theory, in general terms and specifically how it applies to tourism. 'Theory' is one of most misused terms today, for it is often construed as an expression of an unproven and often unprovable opinion. In reality, fact and theory go hand in hand for each is necessary for the other to be true (Cooper and Schindler. 2014).

The terms 'theory', 'concept' and 'model' are often used interchangeably in tourism, when in fact they mean quite different things (Sarokin, 2017). While interrelated, they are also quite different. This chapter discusses each idea.

Theory

Theory in a social science context is an ill-defined notion, meaning different things to different disciplines and discourses (Stergiou and Airey, 2018). In fact, there are almost as many (mis)definitions of theory as there are users. Part of the problem is that theory means something very different in lay language than it does in science (Ghose, 2013). To most people, theory is just an idea, when in fact it is much more. As Feldman (2004:565) notes when talking about the management discipline, "theory is often in the eye of the beholder. What is theory to one reader may not be theory to another" Abend (2008) says sociologists use theory in seven different ways, while information systems researchers also struggle with the term (Lee, 2014). It is not surprising that such a diverse field of study as tourism has even more challenges in describing theory, leading Smith and Lee (2010) to conclude there is demonstrated substantially and increasingly inconsistent use of the word.

So, let's start at the beginning. What is theory? Cooper and Shindler (2014) define theory as the systematically interrelated concepts, definition and propositions that are advanced to define and predict phenomena. Within a tourism context, Dann, Nash and Pearce (1988: 4) defined theory as the body of logically interconnected propositions that provide an interpretive basis for understanding phenomena. It is a well-defined scientific principal that is supported by convincing experimental and observational evidence that has strong explanatory power and helps scientists understand a phenomena and make predictions about future events (Sarokin, 2017). Sutton and Staw (1995) add it must answer the question "why?", while Gorelick (2011) feels theory involves the formation of hypotheses that can be tested empirically and rejected or not rejected. Perhaps Stewart, Harte and Sambrook (2011) summarise it best when they feel theory is an explanation that offers accounts of how and why things are as they are.

Clearly, theory involves more than just unproven ideas or suppositions. Summarizing the literature, Bricker, Donohoe, Becerra and Nickerson (2015: 3) see theory as serving as a tool which guides inquiry and analysis. It guides inquiry in the form of shaping the development of hypotheses and it shapes research analysis by aiming to explain observations through concepts. It is accepted that theories must be refutable, testable or falsifiable. They should also be verifiable through observation.

Wacker (1998) states theory must have four basic criteria. It must have:

- Conceptual definitions where existence and properties are asserted in order to account for what is observable, that define who or what is included and specifically excluded. Moreover it should use common and clear language that is understandable by all.
- Domain limitations which specify the exact setting where the theory can be applied.
- Relationship-building qualities which logically assemble a causal relationship between items and state how variables are related or unrelated to other variables.
- The ability to predict relationships and offer internally consistent predictions in the real world.

As Wacker (1998: 363) notes "theories carefully outline the precise definitions in a specific domain to explain why and how relationships are logically tied so that the theory gives specific predictions." The goal of 'good' theory is a clear explanation of how and why specific relationships lead to specific events. Sutton and Stack (1995: 378) add:

> "theory is about the connections among phenomena, a story about why acts, events, structure, and thoughts occur. Theory emphasizes the nature of causal relationships, identifying what comes first as well as the timing of such events. Strong theory, in our view, delves into underlying processes so as to understand the systematic reasons for a particular occurrence or nonoccurrence."

Wacker (1998), Ritchie et al (2008) and Smith et al. (2013) state 'good' scientific theory must have the following characteristics. It must demonstrate:

- **Uniqueness** where one theory must be differentiated from another and if two theories are identical, then they are really one theory;
- **Accuracy** within its domain;
- **Parsimony**, whereby if two theories are similar in most respects, the one making fewer assumptions and requiring fewer definitions probably is better;
- **Conservativeness**, in that it cannot be replaced until a new, competing theory is shown to be superior;
- **Generalizability**, where the more areas a theory can be applied to, the better;
- **Fecundity**, where theory can be used to expand the area of enquiry into new directions, and it should have broad scope with explanations that go beyond those it was initially used to explain;

- **Internal consistency** and consistency with other accepted theories that are related to the same phenomena;
- **Abstraction**, and;
- **Independence** in time and space.

In short, theory should be able to explain existing phenomena and also to predict new phenomena. Theories, therefore, are sets of statements, some of which state laws, while others are singular factual or existential claims. Bramwell (2015: 205, 206) illustrates that theoretical activity seeks to find general accounts or explanations of the world. In doing so, it can increase understanding through a systemised structure to thinking that involves relatively abstract ideas, categories, groupings or interpretations. He then adds theoretical thinking is necessary because the world is complex, and by simplifying complex processes we can identify things that are more important and look for underlying causes or relationships that affect how things work. Looking at the literature, Bramwell (2015: 206) argues empirical investigation is often bypassed so that attention can be focussed on mental processes, generalisation and explanation. In doing so, it can direct future empirical research.

As such, it plays a key in driving and/or shaping research. Again Bricker et al. (2015) argue theories prevent research errors, assist in making sense of observed patterns and shape and direct research efforts. Udo-Akang (2012) feels theory and research are interrelated. While research contributes to the explanatory power of theory, theory provides the framework to develop the research process. Wacker (1998) suggests three other relationships. First, theory provides the framework for analysis that provides structure where differences of opinions exist and, importantly provides commonality of language. Second, it represents an efficient method for field development by reducing errors in problem solving by building on existing knowledge. Third, it establishes clear explanations for the pragmatic world, which give the basis for empirical support of rules, advances knowledge and guides future research.

It has proven difficult to define theory in a clear and concise manner, with the conflation of the terms 'paradigm', 'theory' and 'model' all too frequent in social science research (Khan 2002). Instead, it is often easier to identify what is not theory. Sutton and Staw (1995) address many of the issues that seem to be confusing to many tourism academics. They identify five cases where things have been identified as 'theory' when they are not. They are summarised in Table 1.1.

Table 1.1: What theory is *not*.

What is *not* theory	Why
References	References provide the basis for grounding work conceptually. They are sometimes used like a smoke screen to hide the absence of theory where people use a string of references to hide the fact that we really do not understand the phenomenon in question. References are valuable for they need to explicate which concepts and causal arguments are adopted from cited sources and how they are linked to the theory being developed or tested.
Data	Empirical evidence plays an important role in confirming, revising, or discrediting existing theory and in guiding the development of new theory. Theory and data each play a distinct role in behavioural science research. Data describe which empirical patterns were observed and theory explains why empirical patterns were observed or are expected to be observed.
Lists of variables or constructs	This is especially relevant given the vast amount of structural equation modelling (SEM) work being produced. Theory must explain why variables or constructs come about or why they are connected. Comparative tests of variables should not be confused with comparative tests of theory, because a predicted relationship must be explained to provide theory; simply listing a set of antecedent or casual ordering does not make a theoretical argument.
Diagrams	Diagrams show the nature of relationships but are not causal by design. Instead, the authors feel diagrams and figures should be considered as stage props rather than the performance itself. While boxes and arrows can add order to a conception by explicitly delineating patterns and causal connections, they rarely explain why the proposed connections will be observed.
Hypotheses or predictions	Hypotheses can be an important part of a well-crafted conceptual argument but are not theory in and of themselves. Hypotheses are concise statements about what is expected to occur, not why it is expected to occur.

Source: From Sutton and Staw (1995)

■ Use and misuse of theory in tourism

Renowned academic Steven Smith has explored the use and misuse of the term theory in tourism research. His first foray into this area (Smith and Lee, 2010) determined that there has been a substantial increase in the misuse of the term over the past few years, perhaps as a result of tourism research being embedded in non-traditional disciplines, which means that academics may not have the deep grounding that their predecessors have had in the past.

Later he and a number of colleagues (Smith et al., 2013) identified seven ways the term 'theory' is used or misused in tourism. Table 1.2 identifies them, defines their key features and their constituent elements.

Table 1.2: Seven ways theory has been used in tourism research.

Key feature	**Element**
1. Theory of the form used in natural sciences	
Falsifiable hypotheses that have been tested multiple times with positive results Single verifiable 'truth'	Based on the belief that there is a knowable, objective reality that transcends the researcher's own opinions or biases. Normally only one theory can exist to explain a given phenomenon. If two or more competing theories exist, one (or more) will eventually be proven wrong. Can be independently tested by other researchers. Understand their theories may ultimately be proven to be incomplete or incorrect. Limited to explanatory and predictive models of some phenomena supported by repeated tests, logically linked to other concepts and theories that provide an integrated and coherent understanding of some aspects of reality, and produces significant falsifiable predictions. Repeated tests with positive results cannot conclusively prove a belief is correct. One contrary result can prove an assumption, model, or theory is wrong.
2. Theory of the form often used in social sciences	
Two or more valid competing theories Falsifiable hypotheses that have been tested multiple times with positive results	Usually associated with complex social science phenomena. Results of any test of a hypothesis may show only equivocal support for the theory. Multiple theories may exist with equal validity within the scope of available evidence because such theories address complicated phenomena for which data and understanding may be incomplete. The failure of a Type 2 theory to support a hypothesis is, by itself not usually seen as a sufficient reason to reject the theory, at least not until after repeated failures.
3. Theory is equated with statistical analysis	
No a priori theory – instead test an ad hoc model. (If testing an a prior theory, then it is likely an empirical test of Type 1 or 2 theory)	Statistical models that are formulated and presented as theory but without a tested a priori conceptual mode. Falsifiable in that an independent researcher can check the results or even replicate the study to determine whether the results are reliable using the data in question.
4. Theory is an untested/untestable verbal or graphic model	
Models with no empirical, testable hypotheses	Similar to Type 3 but not falsifiable by an independent observer. Verbal or graphic models that are expressed in such a form that independent, objective testing is not done or may not even be feasible.

5. Epistemology or a research design presented as theory	
Self-fulfilling prophecy See the world through the 'theory' and shape studies to confirm it	Epistemology presents a worldview and identifies which questions and data are appropriate for scholarly enquiry and which are not. Examples: Postcolonial theory, conflict theory, and feminist theory. Epistemologies-cum-theory specify what questions are legitimate to ask, what data may be collected, how data may be collected, and how data are to be analyzed and interpreted. Does not produce objectively testable hypotheses. Rather, it produces interpretations that are based on the assumptions of the epistemology. Self-fulfilling prophecy type of 'theory' tends to be self-perpetuating in that those who hold a particular epistemological perspective tend to view all the evidence they collect in terms of their chosen intellectual filter rather than allowing for the possibility of contrary findings. Tendency to interpret any results as confirming one's predetermined conclusions. Boundary between theories of the fourth type (untested models) and the fifth type (epistemology) can blur in casual reading. Theories of the fourth type are expressed in terms of a specific model or concept, whereas theories of the fifth type are formally articulated epistemologies described as theories.
6. Grounded theory	
Grounded theory - method	Inductive methodological approach of data collection, analysis and interpretation utilizing a systematic process of iteration and comparison for coding transcripts and other data from which theories emerge.
7. Theory as a casual term or used as an analogy; 'Other'	
Analogy – idea cannot be tested Conclude ' theoretical' insights without offering theory	All other uses of the term, primarily the use of the word in a casual sense such as speculation. In this group of theories, findings may be described as offering theoretical insights but these insights are not based on either a falsifiable hypothesis or an a priori model. Borrowing from one field as an analogy to tourism – i.e. chaos theory. Theory as analogy.

Source: Smith et al., 2013.

Essentially, there are two main themes to this study. The first is that the term 'theory' is used very loosely in much of our literature. Indeed, they suggest that apart from the first and second types, the term 'theory' is misused, a synonym for different research methods, or as an untested set of opinions.

The second theme, and one that is more important, relates to theory of the first and second type. In the hard sciences, only one theory can exist. It must be testable and can be only rejected or not rejected. The scientific method is both cautious and skeptical in nature. If a theory can be rejected, that means another competing theory may be more appropriate. Theories can only be 'accepted' if they are tested repeatedly and not rejected. But the key feature is that only one theory can exist at any one time and that when competing theories purport to explain the same phenomenon, only one can work. This is the essence of the debate between the theory of evolution and the theory of intelligent design. The two are incompatible and, therefore, one or both will be proven wrong. The body of evidence supporting the validity of the theory of evolution is overwhelming, while there is no compelling scientific body of evidence supporting the theory of intelligent design.

But, they also identify theory of the second type in the social sciences where different disciplines may look at the same phenomenon and develop discipline-specific theories to explain that phenomenon. Here, competing theories may be valid. For example, in tourism, lifecycle evolution theories have been proposed from the social science disciplines of human geography, social psychology and complexity. Each one seeks to explain the evolution of destinations from its own narrow, disciplinary perspective and each is valid within its own domain. None of them necessarily explains fully the phenomenon of destination evolution.

The challenge with something like tourism is that it has been explored from dozens of different disciplinary perspectives, with each one driven by its own ideology, boundaries, methods and frameworks. As a result, multiple complementary theories have emerged to try to explain tourism. Many, or most, of these theories fall into what Smith et al. (2013) would describe as theory of the third to seventh types. The key challenge with tourism, then, is to understand the perspective or disciplinary focus used to describe the phenomenon and, by the same token, to ensure that one does not mistakenly apply one discipline's theoretical constructs to explain another discipline's findings unless they are compatible.

Does tourism have theory? Good question. The short answer is it depends how one defines theory. Dann and Cohen (1991: 167) assert that no all-embracing theory of tourism exists because "tourism, like any other field of human endeavor, is a target field," simply providing fodder for the theoretical approaches of other disciplines.

Smith and Lee (2010) feel we are beginning to see the nascent development of theory in tourism. They feel most tourism theories borrow heavily from theories developed in other disciplines, as a reflection of a true multi-disciplinary field of study. Examples include the modification of the theories of planned behaviour and the theories of reasoned action. Type 3 theory is typified by structural equation modelling which may eventually lead to Types 1 or 2 theories if the model can be tested in multiple locations, falsified or supported. Most other types of tourism theory fall into Types 4 to 5 and need substantial development before they can be elevated into Types 1 or 2.

Ritchie et al. (2008) adopt a somewhat more positive attitude, where they see the beginnings of the development of core tourism theory as distinct from theory developed in other disciplines and applied to tourism. To them:

> "part of the problem that creates controversy over tourism as a field of inquiry or a discipline may be viewed as centering on defining what should be considered core tourism theory versus what is theory from other disciplines or foundational theory."

They feel these differences can be viewed as lying along a continuum with three key distinguishing characteristics used to separate core from foundational tourism theory. The first issue is generalizability, where theory that is core to tourism is specific to tourism and not readily generalizable to other disciplines, whereas foundational tourism theory aims to develop generalizations beyond tourism. The second issue relates to theory that is being built upon or critiqued. Here they argue theory that is core to tourism builds on or critiques existing core tourism theory, while foundational tourism theory builds on or critiques theory from its own discipline. Finally, the last issue concerns whether or not the theory is contained within a single traditional discipline. Here core tourism theory is inherently multi-disciplinary while foundational theory is confined to one discipline. This last point is critical, for core tourism theory integrates different disciplinary perspectives to explain tourism and especially to fill in the gaps where one discipline's perspective is lacking.

Concepts

Concepts are abstractions of ideas (Sarokin 2017). A concept can be broad or narrow and can be based on experience or imaginary. They are not theories, models or paradigms. Instead, they are mental boxes into which things are thrown that are thought to have things in common (Bosch, 2011). The level of abstraction can vary depending on the generalizability of the topic. Something like tourism is a very abstract concept. We know what it is, but as you will find out in the next chapter, it is virtually impossible to define in absolute terms. Other concepts, like paying for goods, are more concrete and observable, but payment is still an abstract concept at some level. Bosch (2011) identifies four key features when defining concepts:

- Keep the definition clear – which can be hard when we use words to describe words.
- Make the definition appropriate – use a definition that is consistent with what is generally meant by the idea.
- Avoid defining concepts with related concepts – try not to define tourism as a concept by describing tourists, another concept.
- Avoid circular arguments.

Models

The term 'model' is also problematic in that it too is used in different ways by different groups of scientists (Ghose, 2013). Essentially, in the social sciences, models are often visual representations that help build or explain theory (Gay and Weaver, 2014). Models, as a human constructs that help us better understand real world systems (anon, 2019), present a simplified set of relationships that hopefully show some causal direction. They are felt to have some input, some process function and some output (anon, 2019). As Bricker et al. (2015) note, models are useful to simplify theory and demonstrate its relevance and applicability to real life.

Within a tourism context, Getz (1986:22) identified two types of models based on their nature and purpose. Theoretical models describe aspects of the tourism system, while process models describe sequences aimed at achieving particular outcomes. Theoretical models can be further subdivided into:

- Descriptive models which define components of the tourism system,
- Explanatory models which show how a system or subsystem works with or without the specification of a causal relationship,
- Predictive models that enable forecasting.

Process models can also be subdivided into:

- Subjective types based on a particular dogma or idiosyncratic format,
- Problem solving that follows a sequence,
- Systems theory.

While Getz (1986) was concerned about the use of these models in a planning context, his classification has a broader application to the use of models in a tourism context.

Leiper (2004) discusses four benefits of models. A model represents a simplified version of reality that tries to capture the essence of something. In doing so, it can be used as a visual representation of a theoretical construct that can aid in analysis of a problem. Third, models should work at different scales, depending on fit for purpose. A good tourism flow model, for example, should be applicable at a scale ranging from travel between a hotel and an attraction, as well as to understand global tourism flows. Finally, a model can serve the needs of different disciplines.

Cooper and Schindler (2014) explain that the difference between a model and a theory is that theory's role is explanation, while a model's role is representation.

Typologies

Typologies do not employ the analytical approach taken by models and are used primarily for classification and comparison (Mo, Howard and Havitz, 1993; Pearce 1981). As a consequence, empirical testing of typologies is rare (Mo et al., 1993). Emphasis is placed on observing factors such as psychological, social, cultural and geographic relationships. Based on the results of these observations, classifications are developed. Primarily descriptive in nature, typologies provide a framework for description, understanding and interpretation but have limited predictive capability.

Conclusion

Theory is a difficult concept to understand fully, for there are almost as many theories about theory as there are definitions. But, one can envision theory, concept and model as existing in a nested manner (Figure 1.1), whereby theories represent simplified operationalization of a concept and a concept represents an abstraction of a theory. Theory in turn represents a broad logically set of interconnected propositions that provide a basis for understanding a phenomenon.

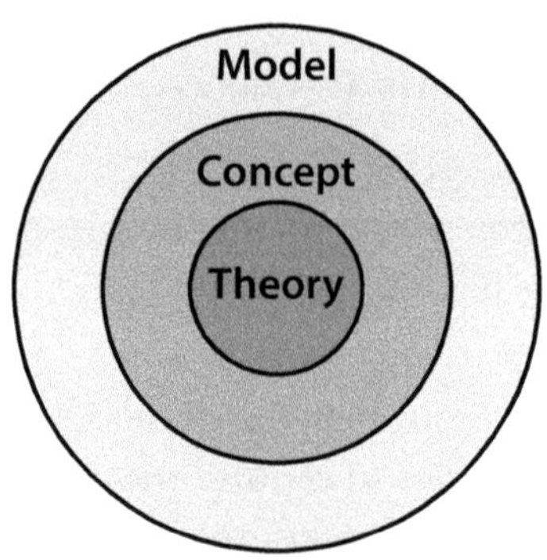

Figure 1.1: Nesting of theory, concept and model.

■ Discussion points / Exercises

1. Review five papers on a related theme and determine what type of Smith's theories have been used in these works, and indeed, if any theory has been used?
2. Discuss whether you feel tourism scholars understand the concept of Type 1 or 2 theory or are more likely to use it as Types 3 to 7.

Points to consider for revision

- Theories, concepts and models are interrelated.
- A theory is a well-defined scientific principal that is supported by convincing evidence and has strong explanatory powers.
- Theory is often misused in tourism research, to such an extent that seven different, increasingly less powerful uses have been identified.
- Concepts are abstractions of ideas.
- Models are visual representation of issues, sometimes use to build theory and sometimes used to explain phenomena.

Suggested additional readings

Smith, S., Xiao, H., Nunkoo, R. & E. Tukamushaba (2013) Theory in Hospitality, Tourism and Leisure Studies. *Journal of Hospitality Marketing and Management*, 22, 875-894.

Wacker, J. G. (1998). A definition of theory: research guidelines for different theory-building research methods in operations management. *Journal of Operations Management*, 16(4), 361-385.

References

Abend, G. (2008) The meaning of theory. *Sociological Theory*, 26(2), 173-199.

Anon (2019) What is a Model. https://serc.carleton.edu/introgeo/models/WhatIsAModel.html <downloaded Oct 24, 2019>.

Bosch R. (2011) Concepts, variables and measurement. http://polisci.usca.edu/apls301/Text/Chapter%204.%20Concepts,%20Variables,%20and%2Measurement.htm <downloaded Dec 2, 2019>.

Bramwell, B. (2015) Theoretical activity in sustainable tourism research. *Annals of Tourism Research*, 54, 204-218.

Bricker, K., Donohoe, H., Becerra, L. & Nickerson, N. (2015) Theoretical perspectives on tourism – An introduction. In Bricker, K & Donohoe, H. (eds) *Demystifying Theories in Tourism Research.* CABI: Boston. pp 1-6.

Cooper, D. & Schindler, P (2014) *Business Research Methods.* New York: McGraw Hill.

Dann, G. & Cohen, E. (1991) Sociology and tourism. *Annals of Tourism Research*, 18,155-169.

Dann, G., Nash, D. & Pearce, P. (1988) Methodology in tourism research, *Annals of Tourism Research*, 15, 1-28.

Feldman, D. (2004) What are we talking about when we talk about theory? *Journal of Management*, 30(5), 565-567.

Gay, B. & Weaver, S. (2011) Theory building and paradigms: A primer on the nuances of theory construction. *American International Journal of Contemporary Research*, 1(2), 24-32

Ghose, T. (2013) "Just a theory": 7 misused science words. *Scientific American*, April. https://www.scientificamerican.com/article/just-a-theory-7-misused-science-words/ <downloaded Oct 24, 2019>.

Gorelick, R (2011) What is theory? *Ideas in Ecology and Evolution*, 4, 1-10. https://rootgorelick.files.wordpress.com/2015/06/gorelick-2011-ideas-ecol-evol-4-1.pdf <downloaded Oct 28, 2019>.

Khan, H. (2002) On paradigms, theories and models. http://www.cirje.e.u-tokyo.ac.jp/research/dp/2002/2002cf156.pdf <downloaded Oct. 25, 2019>.

Lee, A. (2014) Theory is king? But first, what is theory. *Journal of Information Technology*, 29(4), 350-352.

Leiper, N. (2004). *Tourism Management (3rd ed).* French's Forest: Pearson Education Australia.

Mo, C., Howard, D. R. & Havitz, M. E. (1993). Testing an international tourist role typology, *Annals of Tourism Research*, 20, 319-335.

Pearce, P. (1981). *Tourism Development*, Longmans, London

Ritchie, J. Sheehan, L. & Timur, S. (2008) Tourism sciences or tourism studies? Implications for the design and content of tourism programming. *Teoros*, 27(1) http://teoros.revues.org/1621 <downloaded Dec 2, 2019>.

Sarokin, D. (2017) Differences between concepts, theories and paradigms. *Sciencing.com.* https://sciencing.com/differences-between-concepts-theories-paradigms-8415723.html <downloaded Oct 24, 2019>.

Smith, S. & Lee, H. (2010) A typology of 'theory' in Tourism. In Pearce, D. & Butler, R. (eds) *Tourism Research: A 20-20 Vision.* Goodfellow Publishing: Oxford. pp 28-40.

Smith, S., Xiao, H., Nunkoo, R. & E. Tukamushaba (2013) Theory in hospitality, tourism and leisure studies. *Journal of Hospitality Marketing and Management*, 22, 875-894.

Stergiou, D. & Airey, D. (2018) Understanding tourism theory. *Tourism Review*, 73(2), 156-168.

Stewart, J., Harte, V. & Sambrook, S. (2011) What is theory? *Journal of European Industrial Training*, 35(3), 221–229. https://www.emerald.com/insight/content/doi/10.1108/03090591111120386/full/pdf?title=what-is-theory <downloaded Oct 28, 2019>.

Sutton, R. I. & Staw, B. M. (1995) What theory is not. *Administrative Science Quarterly*, 371-384.

Udo-Akang, D. (2012) Theoretical constructs, concepts and applications. *American International Journal of Contemporary Research*, 2(9), 89-97.

Wacker, J. G. (1998) A definition of theory: research guidelines for different theory-building research methods in operations management. *Journal of Operations Management*, 16(4), 361-385.

2 What is Tourism?

By the end of this chapter, the reader will be able to:

- Identify and analyse the force field of tourism knowledge.
- Understand the development of myths about tourism and assess their validity.
- Identify the components of tourism.
- Distinguish between and critique heuristic and technical definitions of tourism.
- Explain the multidisciplinary nature of tourism.
- Demonstrate why pleasure tourism is an extreme form of leisure by explaining its evolutionary roots.

Introduction

Tourism is not absolute. That is what makes it such an exciting field of study on the one hand, and such an exasperating phenomenon to study on the other hand. Life is much easier when things are framed in black-and-white terms: good versus bad; healthy versus unhealthy; left versus right; right versus wrong. Increasingly, as well, people like to have the world presented to them in a simplistic manner where they are faced with apparently easy choices. Life would be easy, if only the world was simple. Yet, in reality the world is a rich tapestry of colours. What on the surface appears to be a simple yes/no choice, in reality becomes a very complex situation when one scratches below the surface.

Tourism is no different. We can all come up with absolute examples of what is and what is not tourism and who is and who is not a tourist. Visitors from China who spend five days on a package tour sightseeing and shopping in London are clearly tourists. Their activities and their resultant expenditure are also clearly tourism related. By contrast, if someone goes shopping for groceries in his or her home community, the person is clearly not a tourist. Other absolute categories such as, business travellers, people visiting friends and relatives, special interest tourists, can also be defined. But after that, what constitutes tourism enters a grey area. What if you have a second home and spend every weekend there? Are you a tourist? What if you are studying abroad for 11 months? Are you a tourist? What if you have family living someplace where you grew up and you visit them regularly? Are you a tourist? The answer depends on how tourism and the tourist are defined.

By the same token, identifying all businesses that represent the tourism sector and the people who work in this sector is challenging (Leiper, 1999). People talk about theme parks as being tourist attractions. Yet, a large portion of their visitors are local residents. Most so-called tourist retail shops, restaurants and even travel providers also rely heavily on local traffic to keep them viable. Even the hotel sector is reliant to a certain extent on the local market for functions, weddings, meetings, special events, food and beverage, and even accommodation.

Who works in the tourism industry? Again, we talk about tourism employees, but as Leiper (1999) mentions there is a huge grey area. You could probably say that hotel, airline, travel agent, theme park and tour operations staff work in the tourism industry, although certainly not all of their income comes from tourists. But, what about immigration officers, doctors, accountants, lawyers, insurance agents, the police and many others? Most people would not think they are part of the tourism sector, but they will deal with tourists directly on a daily basis or serve the tourism industry indirectly on a regular basis.

The same level of imprecision applies to the study of tourism concepts and theories. They are often presented as absolutes. But, as Smith et al. (2013) document in the previous chapter, the term 'theory' has been used very loosely. Even the word 'tourism' means quite different things depending on the disciplinary perspective. This chapter will examine some of the inherent complexities in tourism and explain the resulting difficulties that have arisen through the academic inquiry of this activity. Appreciating the complexity will also help appreciate how and why tourism scholarship has evolved the way it has.

Tourism as a complex topic of enquiry

Our understanding of tourism, and the resulting theories, principles, concepts and models developed to explain and/or explore this phenomenon are a reflection of this complexity. How can we understand tourism as an absolute entity when we cannot even agree on whether it is a discipline, a field of study or a post disciplinary subject of enquiry (Coles, Hall and Duval, 2006; Hall et al., 2014; Tribe, 1997) or if it is even an industry (Lew, 2012)? Tribe's paper on the 'Truth about Tourism' is an invaluable starting point, for he asks (2006:360):

> Is it possible to tell the truth about tourism? Surely an elaborate and expensive knowledge production system (universities, research departments, journals, conferences) exists just for that purpose. But could it be that the thousands of research articles that have been published have failed to uncover the truth?

Tourism has been the subject of academic inquiry for well over 70 years. A huge body of literature has developed over that time. More than 3,000 universities worldwide teach tourism and more than 400 academic journals publish refereed, scholarly research on this topic. The best guess is that somewhere in the neighbourhood of 3,000 academic papers are published each year and that the volume of published material is doubling every five years (McKercher and Tung, 2015). Many things promoted as being core truths about tourism are accepted as fact, even though the basic premises behind them either have no sound evidence or have been disproven repeatedly (McKercher and Prideaux, 2014). But, they are still being repeated.

Also, in spite of the 70 years of research, some key issues in tourism remain understudied, or have been examined at a superficial level. What do we really know about destination choice, market segmentation, life cycles? How much good work has been conducted on gender and tourism? Why is so little work being conducted on drive tourism when it is the most common form of tourism? Indeed, one of the ironies is that early researchers working in an information void, often produced the most insightful research. Today, when we have access to more data than ever before, more sophisticated research instruments and the much larger research community, much of the research being produced is not very exciting. It seems that numbers and adherence to certain statistical techniques get in the way of good ideas.

Tribe (2006) illustrates the root cause and the resultant challenge for tourism for he talks about how the force field of tourism provides a lens that distorts reality. The following section summarises his work. The treatise he argues is informative, for he states no research is value- or bias- free. Instead, he sees a gap between the fact of tourism and the phenomena of the collective 'truth' or knowledge held about the subject (Figure 2.1). The gap is caused by a five component 'force field' that knowingly or unknowingly distorts knowledge from fact. The five factors are:

- The person doing the research
- The rules guiding research
- The position of the individual
- The ends for which the research is conducted, and
- The ideology that drives research.

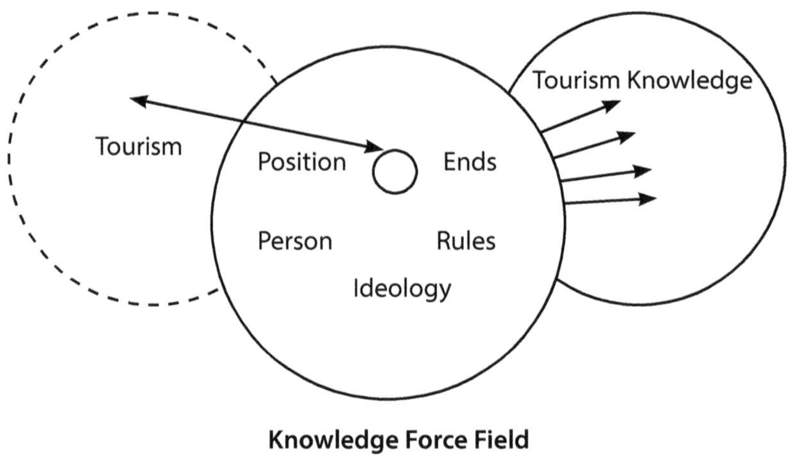

Figure 2.1: Force field of tourism knowledge. *Source*: Tribe, 2006.

Person

All research is influenced directly or indirectly by the researcher. And as much as we may try to be unbiased, it is impossible to separate the individual from the research process. At a minimum, any research must be filtered through the researcher's own cultural bias and also influenced by personal tastes and interests. Geographers see the world one way, while psychologists see it another way. How we see it will influence how we, individually and collectively, investigate the phenomenon of tourism. The individual's own perception of the merits of tourism, and possibly his or her own positive or negative experiences, as either a tourist or a resident of a destination, will influence whether they research tourism looking for positive or negative outcomes.

Sarniak (2015) identifies nine cases where the person, either the subject of the research or the researcher, may bias the findings, as shown in Table 2.1. Respondents often want to satisfy researchers and give what they feel are the 'right' answers to questions, even if they do not necessarily feel that way. This issue is especially present in sensitive research topics where a high degree of social desirability bias will influence opinions. In addition, a degree of respondent fatigue may set in whereby the participant simply gets bored and starts to give a similar type of answer to subtly different questions.

Researchers are also not bias free. Most of the research published in English language media is produced by North Americans, Australians or English-speaking Europeans. This Eurocentrism has to influence their perceptions of the world and their perceptions of how tourism functions. Self-knowledge and its counterpart global awareness also influence tourism. Surprisingly, some tourism scholars have travelled little outside of their home country and, when doing so, have tended to do so in a highly controlled manner that limits their level of engagement. Confirmation bias is certainly something that rears its head in tourism research, especially when much special interest research is conducted by fans of the specialist interest, or when it seems much social impacts research is conducted by people who hold a dislike for tourism.

Table 2.1: Potential for person bias

The subject	The researcher
Acquiescence bias – yes saying *Social desirability bias* – providing a socially 'correct' answer even if it does not reflect the respondent's views *Habituation* – providing the same answers to questions worded in the same ways, something that is especially risky with Likert scaled questions *Sponsor bias* where the sponsor is seeking a certain outcome	*Confirmation bias* or self-fulfilling prophecy research where you set out to 'prove' something and them prove it *Culture bias* where research is filtered through a cultural lens *Question order bias* where one question inadvertently directs the person to a certain response in following questions *Use of leading questions* and / or word bias *Halo effect* – seeing something positive based on a single attribute, which is especially problematic in qualitative research

Source: Sarniak, 2015.

Rules

Research is driven by a series of the accepted rules, traditions, methods and accepted protocols as will be discussed in the next chapter. Academic researchers stress that research must be grounded within some discipline-based concept, which in turn drives how research is undertaken. A great deal of conservatism drives research, as few people are willing to challenge accepted methods or try alternative methods to examine the phenomenon. As a result, a level of inertia sets in where research is replicated using the same method which, not surprisingly, produces the same results. Much concern is expressed about 'scientific' research and the need to follow quantitative methodologies in order to demonstrate the legitimacy of tourism as a field of study. The focus severely limits the ability to explore an activity that is essentially experiential and non-quantitative in nature.

The end result is a stagnation of ideas. Rather than exploring ideas and looking for integration, many tourism concepts and ideas are treated in isolation. Mazanec (2009: 313) is particularly scathing, noting how we conduct research is "delaying or even hampering scientific progress and that is why I address them as myths – beliefs that lose substance when scrutinized thoroughly." Deery, Jago, and Fredline (2012: 65) comment that research on social impacts of tourism "appears to be in a state of 'arrested development,' [where] there is a sense that the advances in understanding the impacts of tourists on host communities is incremental at best, or potentially circular."

Ends

The purpose of the research is a third factor. The pursuit of knowledge is never interest-free. Questions relating to whom the research is being conducted for, whether it is commercial or conceptual, and whether it is deep or shallow need to be considered. In addition, the ulterior motive behind the research needs to be considered, especially when one considers funding or various research assessment exercises. The need to produce research that satisfies certain standards may compromise its creativity.

Two other related features emerge. The first is the commodification of research, while the second is the medium used. Most published academic research follows a standard formula involving introduction, literature review, method, results and discussion. This format works well for empirical research that builds on existing knowledge, but often constrains conceptual research or papers that try to develop new ideas. Likewise, academic publishing may unwittingly constrain research. The reliance on the traditional paper-based model, even when journals go online and use a pdf format, constrains research creativity (Alker, Donaldson and Mussell, 2016). Different journals serve different purposes and, therefore, seek different types of manuscripts (Cheng, Li, Petrick and O'Leary, 2011). It is important to understand the purpose of the journal before assessing the research.

Increasingly too, we are being asked to publish certain types of papers in certain types of journals to satisfy the neo-liberal agenda of accountability through metrification of tourism outputs in accordance with state-organised research assessment exercises. Fennell (2019:424) argues that this practice leads to gamesmanship whereby the pursuit of standards, audits, and administrative objectives becomes more important than the pursuit of knowledge for the sake of knowledge. The result is that the ends (productivity) count more than the means (good ideas and novel research). We produce a lot but what real knowledge does it add to our understanding of tourism? It is for this reason that so many of the references cited in this book appear old – academics during the critical stages of the development of tourism as a field of study were interested in ideas, more so than pumping out papers.

Position

As Tribe (2006) mentions, position relates to both geographic as well as intellectual positions. He notes that the seemingly innocuous concept of geographical location influences the thinking of tourism in a number of subtle ways. He illustrates its impacts on ethnocentricity, departmentalism and academic tribalism, where ethnocentrism involves the promotion of the interests of one's own ethnic and cultural group, and a corresponding demotion of those of other groups. Departmentalism describes the organizational effects of universities on research, where tourism programs are imbedded in different schools or departments which influence which type of staff who are hired, the perspective of research undertaken and the interpretation of tourism. Tribe (2010) writes about tribalism and recognizes that researchers are located within academic communities, each with its own behaviour patterns, rules, norms and hierarchies operating within disciplinary communities. Leaders within a community may exert undue influence on the community.

Ideology

Finally, ideology refers to the common sense set of beliefs (often implicit) permeating society which guides thought and action. Like it or not, most of the concepts in tourism have been developed either from an American and European ideological perspective, where the individual rules supreme, or from a European social responsibility perspective. Little research has been developed from a collectivist Asian perspective. Likewise, Western tourism theories and concepts are grounded strongly in the Judeo-Christian philosophy and not in other philosophies.

Tourism as an inter-discipline

The issue is complicated even further by the lack of single disciplinary focus that has driven our conceptual development of tourism. Instead, we prefer to call tourism an inter-discipline. Jafari and Ritchie's (1981) work is still relevant today although the 'wheel' described in his paper has been expanded significantly by Goeldner and Ritchie (2012) and Cheng et al. (2011). The model adopted by Cheng et al. (2011), shown in Figure 2.2, illustrates how different disciplines have been added to the study of tourism over the years. Jafari argues that tourism as a field of study is a true multi-discipline. Unlike anthropology or psychology that have a single disciplinary base, the study of tourism has been influenced heavily by a range of disciplines. Each discipline brings its conceptual baggage to tourism, and this informs and guides research. This diversity explains the richness of tourism research, but it also explains the great difficulty in coming up with a central, unified model of tourism, or even a single agreed upon definition. Instead, each discipline defines it in such a way as to serve its own unique needs.

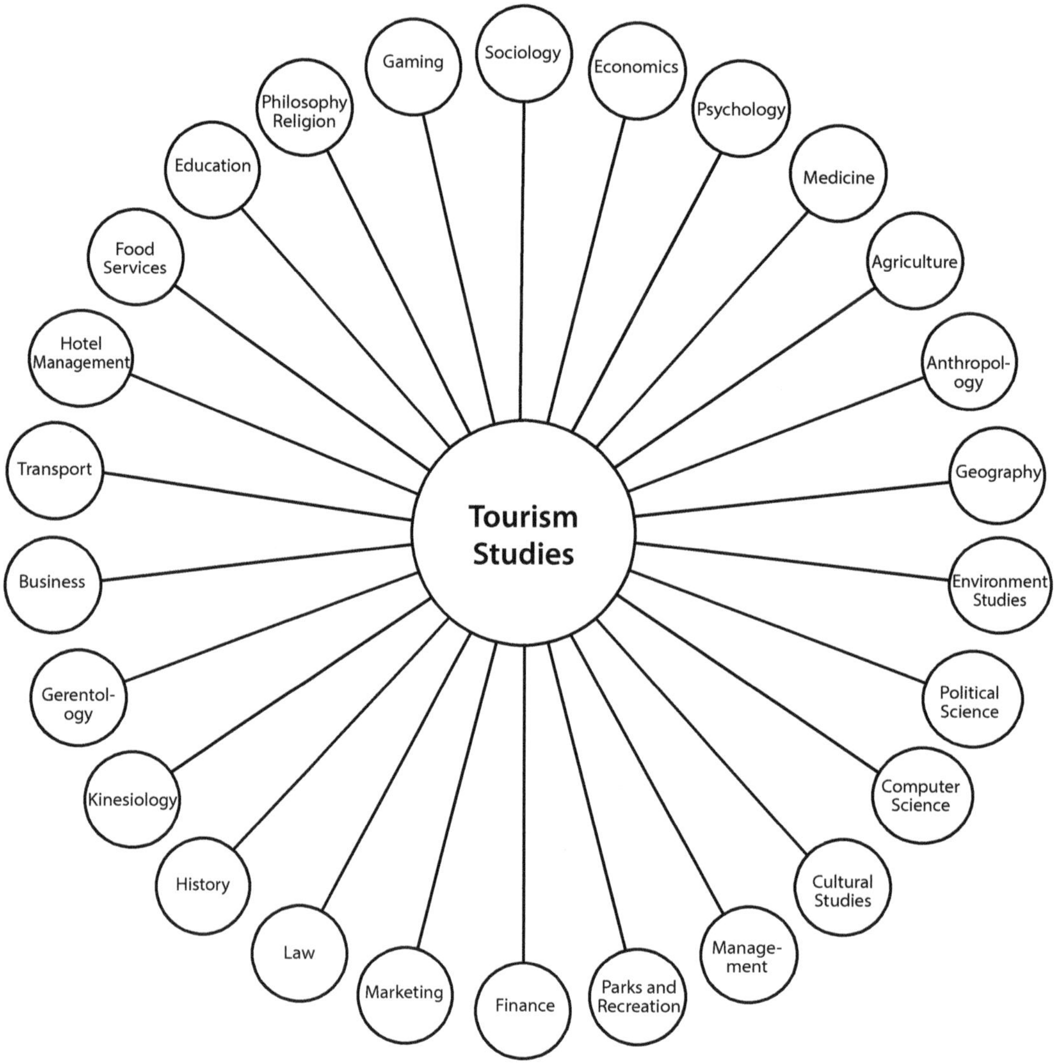

Figure 2.2: The evolving framework of Tourism Studies. Adapted from Cheng et al., 2011.

Indeed, the range of disciplines and fields of study that inform tourism is immense. Crouch and Perdue (2014) suggest early research was based on five core disciplines of economics, sociology, psychology, geography and anthropology, but was also informed by other fields of study such as archaeology, religion, language, history, and political science, as well as professional fields including law, architecture, management, 'recreology', and communications. A 1991 special edition of *Annals of Tourism Research* published papers on the state of tourism research from 10 different disciplines, including: anthropology, ecology, economics, geography, history, leisure and recreation, marketing management, political science, psychology, and sociology.

Cheng et al. (2011) analysed 59 journals and found papers covered 29 disciplines. To make matters even more confusing, Crouch and Perdue (2015) analyzed journals more thoroughly using the Australian four digit research field system and identified that our research fits into 50 broad disciplines and fields of study as shown below:

- Agricultural and veterinary sciences
- Anthropology
- Applied economics
- Applied mathematics
- Architecture
- Atmospheric sciences
- Banking, finance, and investment
- Biological science
- Business and management
- Civil engineering
- Cognitive sciences
- Commercial services
- Communication and media studies
- Computation theory and mathematics
- Cultural studies
- Earth sciences
- Ecology
- Econometrics
- Economics
- Engineering
- Environmental engineering
- Environmental science and management
- Environmental sciences
- Fisheries science
- Forestry sciences
- Historical studies
- History and philosophy of specific fields
- Human geography
- Human movement and sports science
- Information and computing sciences
- Information systems
- Law
- Library and information studies
- Marketing
- Mathematical sciences
- Multidisciplinary
- Numerical and computational mathematics
- Other economics
- Other studies in human society
- Philosophy
- Physical geography and environmental geoscience
- Policy and administration
- Political science
- Psychology
- Public health and health services
- Sociology
- Specialist studies in education
- Statistics
- Transportation and freight services
- Urban and regional planning

Why has tourism become such a multi-discipline? The answer can be traced to its origins as a field of enquiry, the background of academics studying and teaching in tourism, and the nature of tourism itself. In its simplest form, Dann (1997) asserts there is no single truth underlying tourism enquiry and as such, the field must be explored from various perspectives. Taillon (2014) adds no agreed-upon definition of tourism exists due primarily to its multidisciplinary nature. He along with Ateljevic (2014) comment that the field has grown historically in a fragmented and disjointed manner driven by different disciplinary perspectives, institutional and political agendas and journals with different foci. Indeed, Kunwar (2018) asserts that no single discipline can accommodate, treat or understand tourism. Instead it must be studied only if disciplinary boundaries are crossed and if multi-disciplinary perspectives are sought and formed. Laws and Scott (2015) point out the tourism has to be considered as a system of interlinked parts with the body of knowledge described as a mosaic, rather than disjointed fragmented parts.

Historically tourism studies has struggled to find a single disciplinary home and instead began as a way that academics from different disciplines could study to gain deeper insights into their own disciplines. As a result, tourism was conceptualized in different ways to such an extent that each discipline has developed its own set of definitions to describe and therefore examine this phenomenon (Tribe, 1997). Xiao and Smith (2006) illustrate that academics from well-established disciplines such as sociology, geography and anthropology were the first to explore tourism in the late 1970s, while the field broadened to include management, economics and environmental issues in the 1980s. Research interest further broadened in the 1990s as leading journals ran special issues on arts, pilgrimage, heritage and a range of other topics. Xiao and Smith (2006) also illustrate how the themes involved in tourism scholarship have ebbed and flowed over the years. Psychology, anthropology, geography, planning and sociology dominated in the earlier stages of the development of the field but seem to have declined recently. In their place, marketing, economics, education and other topics have risen in prominence.

One challenge that has beset the development of the field is how to balance the duality of tourism as a social science and a business management stream (Ritchie, Sheehan and Timur, 2008). Most programs, regardless of which department or faculty they are based in, include both elements in their curriculum.

The backgrounds of academics add to the multi-disciplinarity of the subject area. Ateljevic (2014), for example, points out that 70% of tourism programs in the UK are located in business schools, while almost 90% of individual academics have a background in social science disciplines of sociology, economics, geography, anthropology and the like. Xiao, Li and Lin (2011) and Weiler, Moyle and McLennan (2012) analyzed doctoral dissertations and found that social sciences have informed the greatest number of dissertations, even during the present when programs are housed in business schools.

Perhaps, though, the single greatest reason why tourism is a multidiscipline rests in the fact that the range of topics we explore is so broad that they cannot be encapsulated within any single discipline. Some 30 years ago, Gunn (1991) speaking at a Canadian Association of Geographers conference talked about the need for multidisciplinary tourism education, for he felt that the component parts of tourism could only be understood if several different disciplines were applied. He stated the element of travel had to be considered from the perspective of transportation, political science and geography while visitor activities encompassed consumer behaviour, psychology, sociology, history and marketing. Economic impact could only be studied from a finance, economics, management, entrepreneurship or business sciences perspective, while the impacts of tourism needed both business and environmental fields of study.

Liszewski (2010) concurs by arguing that tourism is a system of elements comprising of the person, the process of tourism, the destination entrance of places, activities, and the impacts or consequences of tourism. Different disciplines are best suited to examine the processes, mechanisms, rules and regularities of each of the component parts.

■ A cautionary approach

On the one hand the multidisciplinary nature of tourism explains its richness as a field of inquiry. But, on the other hand, the very multidisciplinarity also means that we must take a cautionary approach when considering tourism. Each discipline or field of study defines it in its own terms, conflates ideas and adopts a self-serving approach to its study. As a result, different disciplines can use the same term but be talking about quite different phenomena. What an economist considers as tourism or a tourist may be quite different from the views of an anthropologist, a sociologist or marketer. This issue will be discussed in much more detail in the next chapter. But, it has led to much confusion and silofication of tourism, leading Pritchard (2012) to comment that the field remains beset by many ontological, epistemological and methodological shortcomings.

Myth creation and tourism

What it means is 'tourism' can mean almost anything to anyone. It is not surprising that Benckendorff and Zehrer (2013), Echtner and Jamal (1997) and Laws and Scott (2015) say tourism is a fragmented area of study, with Gibson (1998) adding it is typified by multiple communities of discourse with historically little cross fertilization between communities. The result is weak theory development (Mazanec, 2008; Tribe, 2006). Franklin and Crang (2001) add that as we explore smaller and smaller units of analysis, each author tries to create his or her own myths about the importance of the research area. Each discipline has staked a claim to certain areas of enquiry, conceptualized them, examined them and, as Tribe (2006) argues, constrains research from within that discipline's knowledge force field. The end result is that tourism studies lacks a common language under which objects are defined and explored.

As a result, many myths have been formed around what tourism is and how it works. McGee (1985) states myths are demonstrably false beliefs that are widely held, long-standing and never subjected to deep inspection, while Franz (2006) identifies them as wrong ideas widely held by authority figures. Kompier (2006) is much more succinct, classifying them as believed truths. McKercher and Prideaux (2014) argue that what constitutes a demonstrably false belief is a matter of interpretation, for some myths are simply factually incorrect, while many others have some essence of truth, providing they are seen from within a proper context. Myths are created when simple truths are taken out of context, repeated over time and narrowed to such a simplistic formula as to become nothing more than jargon. As Heydenryck (1993) states, the original truth becomes so diluted that only a kernel of truth remains. Barthes (1984) adds that myths distort facts, so much so that the symbolic or metaphorical value of the myth often exceeds its absolute truth (Alexander and Archer, 2000). They are common in all areas of academic study, for more than any other reason they serve to legitimize a field of study (Hansen and Williams, 1999).

A number of causes of myths have been identified by various authors (Kompier, 2006; Lance, 2011; Mazanec, 2009; Mercer, 2006; Vandenberg, 2006). They include:

- The origins of a discipline, where the work by many early scholars remains dominant, even though their ideas have been confronted by contradictory evidence;
- Inertia;
- Entrenchment by the peer review process and doctoral student training;
- Results that are too good not to be true;
- Poor research;
- Failure to read and place research in the proper context;
- Self-fulfilling prophecy research, where research sets out to 'prove' something and does;
- Fallacy of misplaced rationalism (Schaeffer, 2008), where a great deal of time and effort is devoted to pursuing elaborate, rational explanations for phenomena, without first demonstrating their existence;
- False attribution of cause and effect without considering other causal agents;
- Stakeholders.

McKercher and Prideaux (2014) identified over 40 myths that fell into six broad categories, as shown in Table 2.2.

Table 2.2: Myths about tourism

Type of myth	Examples
Self interest	Tourism is the world's largest industry Tourism is an industry Tourism is a discipline
Negative foundation	Tourism destroys tourism Tourism perpetuates colonialism / tourism is neo-colonialism Tourism generates some positive economic benefits, but they are offset by massive social, cultural and environmental impacts Tourism leads to conflict Tourism commodifies culture
Reactive positive stakeholder	Tourism is a path to economic independence Tourism revitalizes culture Tourism protects the environment Tourism provides a range of positive social benefits People damage the environment, not tourism
Convergent	Size, importance and differences of ecotourism/ecotourists and other forms of alternative tourism/tourists Employment Small is preferable A traveller is better than a tourist
Too good not to be true	Special interest tourism as the new mainstream Tourism and peace Economic and employment benefits of tourism Tourists seek authentic experiences Ecotourism and ecotourism operators are more ethical than others Tourism is a clean sector Sustainability and the triple bottom line can be operationalized Quality of tourism jobs Multiplier effect of tourism
Inherited	Relationship between satisfaction, loyalty, intent to return and actual repeat visitation rates Carrying capacity works in tourism Willingness to pay Tourism lacks theory

Source: McKercher and Prideaux, 2014.

The self-interest myths about the size of tourism, its existence as an industry and arguments about it as a discipline serve the valuable role of legitimizing it as a field of study, especially in its early days when tourism research was held in low regard and seen as a frivolous or inconsequential issue (Nash, 1979). As a result, promulgation of these myths helped create a shared sense of belonging among academics who were interested in this topic (Dann, Nash & Pearce, 1988; Tribe, 1997). Negative foundation myths formed the basis of the social science inquiry into tourism as a response to pro-development oriented attitudes in the 1970s, that had little regard for the adverse impacts tourism could have on the environment and host communities. Doxey's (1975) Irridex is probably the best known, though least read of these types of myths. The positive reactive stakeholder myths, especially about the new type of tourist engaging in a new type of ethical behaviour, were a response to the negative origin myths, while convergent myths sought balance between the two ideologies. The too good not to be true myths emerged at a time when tourism studies exploded globally in the 1990s and a whole series of new academics discovered sustainable development and special interest tourism. Lastly inherited myths were those that were imported from other disciplines and applied to tourism, often poorly.

Of course, there is often a kernel of truth behind these myths. No doubt, some forms of tourism have created a range of adverse social and environmental impacts, while others have enhanced local environments and benefitted local communities. But, we cannot treat these myths as absolutes. Instead, some may have an element of truth, some may be untrue and others may be partially true. The key issue is one of context.

Defining tourism – Why definitions are important

Before addressing the impossible task of defining tourism, a brief discussion of the need for and the benefits of definitions is important. It is necessary to come up with some agreed-upon terminology before you can begin to identify, discuss and analyze objects, concepts or things. The English language is very subtle. It is very easy to use the same term to discuss completely different concepts. We will look at this more thoroughly when we talk about the idea of sustainability, where two competing ideological approaches both use the phrase sustainable development, but for completely different ends.

A definition, according to the English Oxford dictionary is "an exact statement or description of the nature, scope, or meaning of something." Lisle (2017) indicates it is a statement or series of statements that explain a word's meaning that acts as a linguistic token that represents something. Gupta (2015) suggest definitions vary in function and can provide, for example, precision in meaning, accuracy, clarity or fecundity. Ultimately, they describe the essence of the thing, enabling us to use a common vocabulary when discussing or reading about an issue (Whitfield, 2012).

Are tourists different than travellers? Much of what we read about tourism is tinged by value judgements suggesting one group of tourists is more legitimate than another. Tourism is an activity that occurs rarely in our daily life, is driven by personal motives and provides a range of personal benefits. We want to feel special when we travel and moreover, feel we are having a unique experience. Yet, we are often confronted by large crowds of other people having the same experience, as shown here in the Forbidden City in Beijing, China. How do many people react? They tend to make value judgements about others that somehow, they are an inferior type of tourist who is having a lesser experience. Indeed, the term 'tourist' often carries with it a negative connotation as Leiper (2004) noted. How do people distance themselves from being labelled tourists, while simultaneously dismissing or delegitimising others? They label themselves as travellers, and others as tourists. In doing so, 'traveller' infers greater standing and simultaneously distances oneself from everyone else. What is the real difference between a tourist and a traveller? None, we are all tourists. The only difference is semantic, where you tend to label yourself as being superior and label others as being inferior. From my perspective, I am a traveller, you are a tourist! But from your perspective, you are the traveller and I am the tourist!

As such, a good definition helps us assess the situation better and make better decisions, whereby imprecise or weak definitions make it difficult to even agree on what is being discussed and how to address or resolve a problem (Andrees, 2014; Veerasamy, 2013). Owens (2015) highlights how variations in definitions may actually change what we are talking about or who we are referring to, and therefore limit the generalizability across different situations.

Veerasamy (2013) states definitions have to satisfy two criteria. First, they must be precise and not convoluted and have as few elements as possible to capture the essence of the item. Second, they must also be falsifiable in that the definition not only captures what is embodied by the term but also what is not embodied or what is excluded from the term.

Ideally, then a good definition of tourism or the tourist will provide the types of benefits identified in Table 2.3. A good definition will provide a common understanding of the subject and then allow the use of common language to describe what we mean by tourism or the tourist, so that different groups using the term can discuss the same topic. We can then set boundaries that can define what is or is not included in tourism, and to distinguish tourism or the tourist from things that are not considered tourism or the tourist. In doing so we can describe and classify the phenomenon, count it if needed, analyze it and examine it according to the definition, and also define or determine what it is that we wish to examine when we look at this broad concept called tourism. Andrees (2014) writing about international labour, also indicates clear definitions are necessary to measure change, for only if we can quantify a problem are we able to understand it and see how it evolves over time.

Table 2.3: Benefits of definitions in tourism

Components of a good definition
☐ precise
☐ uses clear terms
☐ falsifiable
Benefits to tourism / tourist of a good definition
☐ have a common understanding of the concept
☐ use common language so different groups are talking about the same thing
☐ set agreed upon boundaries to describe what is and is not included in tourism or who is/is not a tourist
☐ describe and classify
☐ count
☐ analyse it according to the definition
☐ determine how to examine
☐ determine what to examine

The impossible task of developing a precise definition of tourism

On the surface, tourism appears to be a simple concept. The Oxford English Dictionary defines the term as the 'theory and practice of touring, travelling for pleasure.' This simple concept, however, is exceedingly difficult to define in any meaningful manner, for hundreds of definitions have been proposed over the years and none has captured the essence of the subject clearly and precisely. Two big challenges emerge. The first is that the boundaries of what constitutes tourism and non-tourism are imprecise and fuzzy, while the criteria to distinguish between tourists and non-tourists are even more blurred. The second is that tourism embodies both business considerations, where absolute and precise terms are required, and social science considerations, where a loose idea of the phenomenon enables better research than any absolute definition (Tribe, 2006).

■ The core concept – a journey away and back by someone

Let's go back to the origins. The term *tour* traces its roots to the Greek term 'tornos' and the Latin 'tornare' which describes a lathe or a circle. Leiper (1979) reminds us then that the essential feature of tourism is a round trip that ultimately means returning to the point of departure. The suffix *ism* refers to the action, process of typical behaviour, while the suffix *ist* refers to the individual doing the action (Leiper, 2004). Combining *tour* and *ism* suggests tourism is the action or movement around a circle, while *tour* and *ist* is the person doing the action.

■ Heuristic vs technical definitions

Beyond that simple concept though, definitions begin to run into trouble, for no single definition can capture the full essence of tourism and still satisfy all criteria of a good definition. The definitional challenge is further exacerbated by the use of circular arguments to define tourism. As Morley (1990) discussed, tourism has often been defined through the use of such terms as tourist and the supply of tourist facilities. These terms, in turn have relied on tourism as a means of helping to define them. Ultimately, Leiper (2004) feels two main types of definitions exist: technical and heuristic. Each is quite different from the other and yet, both are essential to capture the essence of what tourism is.

Technical definitions

One of the great challenges facing governments, tourist organizations and tourism academics is to define the size and scope of the tourism industry. As a consequence, a number of technical and statistical definitions have been devised that have tried to use absolute boundaries to define the term. The first such definition was developed by the League of Nations in 1937 and described a tourist as a person who visits a country other than that in which he habitually lives for a period of at least 24 hours (Leiper, 1979).

In 1963, the United Nations Conference on Travel and Tourism developed definitions of international visitors and tourists that, with some modifications, were adopted by the World Tourism Organization. An 'international visitor' is any person who travelled to a country other than that in which he had his usual residence, the main purpose of whose visit is other than the exercise of an activity remunerated from within the country visited and who is staying for a period of one year or less (WTO as cited in Morley, 1990). This definition distinguished between international tourists and international excursionists. An international tourist was defined as a person staying at least one night, but not more than a year in a country, whereas an international excursionist was someone who did not stay overnight in a country.

Other technical definitions of both domestic and international tourism have applied arbitrary and artificial criteria of a minimum distance travelled (varying between 30 and 160 km), a minimum length of stay requirement and finally, trip purpose. Some definitions specifically exclude business travel, while others embrace it.

The oft cited UNWTO definition of a tourist for the satellite accounting purposes as "any person travelling to a country other than that of his/her usual residence, who stays there at least 24 hours (or at least one night) and whose purpose of trip corresponds to leisure (pleasure, holidays, health, studies, religion and sports) or business, family, mission, meetings) and is other than the exercise of an activity remunerated from within the country visited" is a classic example of a technical definition.

These definitions serve a useful purpose in helping governments and tourism bodies come to grips with the size of the tourism sector. As well, they serve to establish common criteria by which global tourism can be measured. From a geographical perspective, such definitions recognize the spatial aspects of tourism and the spatial interaction that arises from the tourist's movement from one place to another and back home again (Pearce, 1987).

But, while the use of an arbitrary definition serves statistical purposes, it has limited utility for other uses. As Morley (1990) suggests, the real world is rarely statistically convenient. While it may be useful to try to distinguish between locals and visitors, in reality it may be almost impossible to make that distinction. The use of an arbitrary distance or minimum stay criterion may mean that two people doing exactly the same activity in the same space and time may be defined differently.

To overcome this problem some researchers propose rejecting such criteria and are beginning to discuss tourism motivated trips regardless of the distance travelled or time spent away from home (Jansen-Verbeke and Dietvorst, 1987).

Heuristic approaches

Heuristic definitions strive to help learning. They tend to be broader in scope and serve three main purposes: to help focus thinking; remove ambiguities; and aid communication. Leiper (2004) feels that heuristic definitions conform more to popular notions of tourism and do not have to conform to technical definitions. Additionally, people should feel unconstrained through the use of these definitions. The first such definition was developed as early as 1942 by Hunziker and Kraph (as cited in Burkart and Medlick, 1974). They defined tourism as "the sum of the phenomena and relationships arising from the travel and stay of non-residents in so far as they do not lead to permanent residence and are not connected with any earning activity."

Similarly, Jafari (1987) suggested that tourism was the study of man away from his usual habitat, of the industry which responds to his needs, and of the impacts that both he and the industry have on the host's socio-cultural, economic and physical environments. Pearce recognized the link between tourism and recreation and suggested that "tourism may be thought of as the relationships and phenomena arising out of the journeys and temporary stays of people travelling for leisure or recreational purposes (Pearce, 1987:1)."

These definitions appear to have greater legitimacy in the study of tourism as a social phenomenon for they attempt to recognize its component parts. But, as Leiper (1979) among others has argued, while being superficially attractive, they tend to be too broad, too vague and lack a clear spatial focus.

In response to some of these criticisms, a systems approach to defining tourism has emerged. Mill and Morrison (1985) adopt a marketing approach and define four segments of tourism as the market, travel, the destination and marketing. Mathieson and Wall (1982) discuss a supply and demand approach to tourism and identify a number of internal and external factors. McIntosh and Goeldner's (2006) system approach focuses on the economic or industry side of the concept.

While these conceptual approaches have helped focus the argument more clearly, they still have their weaknesses. Again, they tend to focus on the discipline of the authors. In the case of Mill and Morrison, it is marketing; Mathieson and Wall adopt a geographer's perspective focusing on the spatial dimension; Pearce integrates tourism with the leisure paradigm; and McIntosh and Goeldner tend to take an industry perspective.

The need to consider both technical and heuristic definitions

Both technical and heuristic definitions have their own relative strengths and weaknesses Indeed, as Table 2.4 indicates, the very strength of one represents the inherent weakness of the other. In fact, you cannot consider defining tourism without addressing both heuristic and technical components.

Table 2.4: Comparing technical and heuristic definitions of tourism

	Technical	Heuristic
Strength	Sets clear boundaries Can count tourists Can document tourist businesses Can calculate statistics and economic impacts Essential for census and statistical organisations Can calculate income, leakages, employment, etc. Allows business disciplines to study tourism	Inclusive Focus on the tourist Consider motivation Considers experience Considers impact Allows non-business disciplines to study tourism
Weakness	Exclusive Does not consider motivation Does not consider the tourist Does not consider experience Does not consider impact	Fuzzy boundaries Cannot count tourists Cannot conduct any economic or statistical analysis

Technical definitions tend to be simple, explicit and set clear boundaries to differentiate what is and what is not a tourist or tourism. They tend to have set time limitations and also have a specific distance criterion of either a minimum distance for domestic tourism or crossing international boundary. Many of these definitions also have fairly narrow prescriptions on who is a tourist. They are essential if one wishes to document the number of tourists, associated tourism businesses and the economic impacts, both positive and negative from tourism. They are also falsifiable, in that it is easy to identify what is not tourism and who is not a tourist.

Their biggest weakness though is that the individual is reduced to nothing more than a number. Moreover, the arbitrary nature of the definitions means some aspects of tourism and some tourists are either excluded or included who may not be counted. For example, someone who is away for only 23 hours may really be a tourist, but if the criterion is set at 24 hours, the person would not be counted. The same applies if an arbitrary distance threshold is not met, or if arbitrary time thresholds (such as 12 months) are exceeded. They also do not consider the experiences people have, why they travel and what impacts their actions may have on host communities or the environment.

Heuristic definitions differ from technical definitions in a number of important ways. First, no physical distance needs to be specified. Second, there is no minimum or maximum time duration. Third, the boundaries of who are tourists can be much more flexible, while fourth, the boundary of what constitutes tourist attractions can also be flexible. These types of definitions are essential when considering the social and environmental impacts of tourism, motivations, and the experience generated. They are precise within their domain of trying to understand impacts, motives and the industry and also use clear terms, although they are often convoluted. But, they are not falsifiable, for one cannot define who is not a tourist or what is not tourism unless some boundaries are embedded. But setting boundaries defeats the purpose of a holistic or heuristic definition.

Fundamentals of tourism

The preceding review of some of the approaches used to define tourism reflects the challenges faced by people from multi-disciplinary backgrounds who share an interest in tourism. Perhaps, it can be argued that it is impossible to develop a precise definition of tourism. Perhaps as well, it is not necessary to do so and instead focus discussions of what tourism is on common themes rather than specific terminology. If fact, the title for this section is a misnomer, for we will not be defining tourism. To do so may take an entire book in itself.

Instead, Netto (2009) identifies fundamentals of tourism that summarise the key elements that encompass both technical and heuristic considerations (Table 2.5). Some of these elements are self-explanatory, while others require brief elaboration. The subject is the tourist, for without tourists there would be no tourism. This principle is fundamental in all of the concepts that will be discussed throughout the book. Moreover, the tourist has some clear characteristics that distinguish him or herself from non-tourists. The ideas of displacement and return relate to the core Latin concept where tourism involves a journey away and back. Motive is critical, for as Netto (2009:57) states "there will always be a reason to travel be it explicit or not." Hospitality represents the in-destination and in-transit zone sectors that serve the interests of the tourist. Tourism is comprised of a range of experiences that can be active or passive. Tourists will need to communicate with other tourists, workers in destination regions and others in order to have hospitality and experiences. Finally, any kind of tourism uses some kind of technology, with transportation and communications technology being the most important.

Table 2.5: Fundamentals of tourism

☐ Subject	☐ Principle of hospitality
☐ Displacement	☐ Principle of experience
☐ Principle of return	☐ Principle of communication
☐ Principle of motivation	☐ Principle of technology

Source: Netto, 2009.

Pleasure tourism growing out of leisure as an extreme form of leisure

Let's confuse matters even more by adding an additional dimension that distinguishes pleasure travel from leisure. Pleasure tourism evolved from leisure and recreation and is intrinsically linked to the leisure paradigm, as outlined in Figure 2.3 (Murphy, 1985). Tourism and recreation share many of the same facilities and compete for the same consumer dollar and level of government support (Jansen-Verbeke and Dietvorst, 1987; Mathieson and Wall, 1982; Mill, 2008). Both also produce common social and psychological outcomes (Fedler, 1987), with similar motivations for participation (Iso-Ahola, 1983). And, as Smith and Godbey (1991) suggest, both fields share the same heavy ideological baggage, be it a search for authenticity, a spiritual search for self-fulfilment, or an existential quest for meaning in an industrialized society.

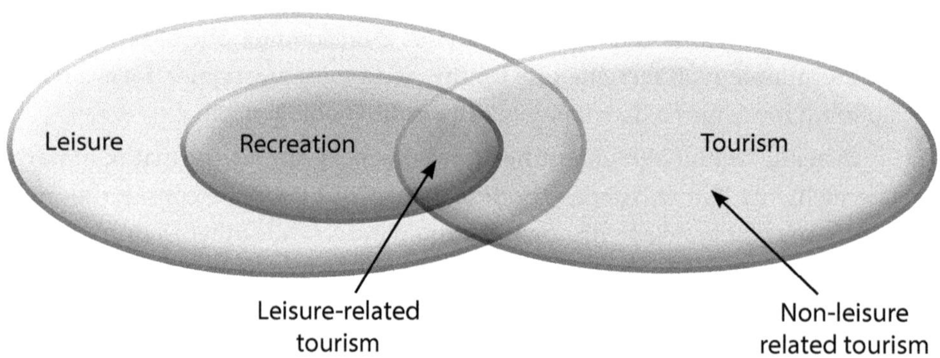

Figure 2.3: Tourism as an extreme form of leisure. *Source*: Murphy, 1985.

Indeed, pleasure tourism has been called an extreme form of leisure while business travel and some forms of unethical tourism, such as child sex tourism, fall outside what would normally be considered as part of the leisure paradigm. Jansen-Verbeke & Dietvorst (1987), state that leisure and tourism are strongly interwoven. Colton (1987) discusses the symbolic interactionist's view of the link between leisure recreation and tourism. Leiper (1979) suggests that tourism offers a special medium for recreation. Mathieson and Wall (1982) and Pearce (1987) discuss the linkages between leisure, recreation and tourism. They assert that tourism constitutes one end of the leisure spectrum, representing an extreme form of recreation.

Proponents of the link between leisure and tourism, and of the leisure spectrum approach, argue their case on the following basis. Leisure theory suggests that leisure is essentially free or unobligated time. Recreation is a voluntary activity that is done during leisure time, principally for enjoyment, intrinsic and extrinsic benefits. Because tourism generally occurs over large amounts of free time (days or longer) and usually involves some type of recreational activity, it therefore is seen to lie at one end of the leisure spectrum.

Tourism is an extreme form of leisure. Tourism is part of the leisure paradigm, for both tourists and leisure participants share the same facilities and derive similar benefits from participation. Moreover, each activity vies for a share of the consumer dollar. Indeed, many so-called tourist attractions could not survive without a core local leisure market providing a majority of their visitors. Beechworth, a largely intact historic gold mining town in North-eastern Victoria, Australia, relies heavily on the local day trip market from a nearby large country town to support its many shops, restaurants and historic attractions. These visitors both supplement and complement the overnight domestic tourism market from elsewhere in Australia. The difference between leisure and tourism is often subtle and depends on which definition of tourism one uses. It is much easier to distinguish between the two if one adopts a statistical or technical definition, for the parameters are absolute (time and distance in particular). It is much harder to make that distinction if a heuristic definition is adopted.

It has been argued that modern mass tourism could only develop in a society that had leisure time. Historically, leisure was an activity that was strictly the purview of the affluent. The Industrial Revolution changed all that, leading to the growth of the evolution of leisure time and a recognition by 19th century society that recreation, or non-work time, played a vital role in enhancing productivity of staff (Godbey, 1985; Krause, 1984). Therefore, the key difference between tourism and leisure is that tourism requires MORE than leisure: more time, money, effort, commitment, risk, etc. As a result, tourists' expectations are more and the product must also be more, as shown in Table 2.6.

Tourism requires more! What can satisfy the local leisure market may not have strong enough appeal to satisfy a tourist, for tourism involves more time, effort, commitment, risk, possible cost and a range of other factors. As such, tourists expect something more: bigger, better, rarer, a better experience, etc. The product, therefore, must also offer more if it is to appeal to the tourist market. Many local destination areas make the mistake of assuming that something that satisfies local residents' needs will also appeal to tourists. These two examples, one of a small seasonal theme park on eastern Ontario, Canada and the other of Hong Kong Disneyland highlight the difference. The local theme park attracts children from nearby communities but has very little drawing power outside its immediate hinterland. Hong Kong Disneyland, on the other hand, is a multi-billion-dollar development that has broad drawing power from within the Asia Pacific region. It offers an experience that can only be had at a handful of similar theme parks throughout the world, and as such appeals to an international audience.

This issue is critical for many destination management organisations for they often think that what they have to offer will satisfy tourists' needs, because it will satisfy the needs of local residents. But, they forget the 'more' element. Few tourists will travel to see a movie, but they will travel to see a film festival, because it offers more, and can meet higher expectations created by greater involvement.

Table 2.6: Tourism requires MORE

Tourism requires MORE than leisure	Tourism expectations are MORE than leisure	Tourist experiences must be MORE than leisure
☐ Time ☐ Effort ☐ Expense ☐ Commitment ☐ Pre-planning ☐ Skill ☐ Risk	☐ More excitement ☐ More novelty ☐ More intense (either more active or more passive)	☐ Bigger, better ☐ Different ☐ Unique ☐ Not available at home

Conclusion

So, what is tourism? This chapter explored a range of issues but explicitly did not try to define the term other than to appreciate some broad features – that it involves a round trip taken for some reason that creates some type of impact, however you want to define it. Who is a tourist? Again, we have taken the decision not to try to define the term other than to appreciate that a tourist is the person who takes the trip.

Is this a satisfying outcome? No. As stated at the outset of this chapter, it would be nice to come up with absolute terms. But that is impossible given the number of disciplines that explore tourism, the force field lens under which tourism is investigated and the impossibility of rectifying the inherent dissonance between heuristic and technical definitions.

None the less, this chapter does set the stage for examining a range of issues explored throughout the rest of the book, for each of the different theories, concepts and models developed adopt unique disciplinary foci. Flexibility in defining the term, is thus required to make sense of why we think about tourism the way we think about it.

Discussion points / Exercises

1. Tribe (2006) discusses the force field of tourism and how it influences research. This force field is evident even in the journals that publish tourism research. Identify any five journals and go online to review their mission statements, aims, scope, and/or objectives. Develop a chart showing how each uses the force field of tourism to create its own version of tourism knowledge.
2. Select any 10 of the discipline areas and fields of study identified by Crouch and Perdue (2014) and explain how tourism research can be informed by these areas.
3. Many myths have been created about tourism. Identify as many as you can and explain the type of myth from Table 2.2 and why the myth has emerged.
4. Develop your own definition of tourism. Analyse whether it reflects a technical or heuristic viewpoint and try to identify the limitations of the definition from the perspective of the viewpoint not considered.
5. Tourism is MORE. As a result, tourists' expectations are GREATER, and the type of service or facilities they demand must be BETTER than they can find at home. Analyse your own destination's offerings by identifying attractions and then analysing them by identifying how tourists' expectations are greater than local residents, and how they must demand better offerings than found at home.

Points to consider for revision

- How the knowledge force field gap influences what we think about tourism and why we think about it the way we do.
- Multiple disciplines and fields of study have each made their own contribution to tourism, but also may have defined tourism in different ways to suit their needs.
- As a result, many myths of tourism have emerged that do not stand up to rigorous scrutiny.
- Tourism is extremely difficult to define for it has to consider both technical and heuristic elements, which in many ways are mutually exclusive.
- Pleasure tourism traces its roots to leisure, but it requires more, meaning the product on offer must provide a better, bigger or different experience than a local leisure or recreational product, in order to appeal to tourists.

■ Suggested additional readings

Tribe, J. (2006) The Truth about Tourism. *Annals of Tourism Research,* 33(2): 360 – 381

McKercher, B. and Prideaux, B. (2014) Academic Myths of Tourism. *Annals of Tourism Research.* 46: 16-28.

References

Alexander, D. & Archer, S. (2000) On the myth of "Anglo-Saxon" financial accounting. *The International Journal of Accounting,* 35(4), 539-557.

Alker, J., Donaldson, C. & Mussell, J. (2016) The future of academic journals. *Journal of Victorian Culture,* 21(1), 92–94. https://doi.org/10.1080/13555502.2015.1127281 <downloaded Dec 18, 2018>.

Andrees, B. (2014) Why definitions matter. *International Labour Organisation.* https://www.ilo.org/global/about-the-ilo/newsroom/news/WCMS_234854/lang--en/index.htm <downloaded Jan 30, 2019>.

Ateljević, I. (2014) Mapping a history and development of Tourism Studies field. *Turizam,* 62(1), 75-101.

Barthes, R. (1984) *Mythologies* translated by Annette Lavers, New York: Hill and Wang. Excerpt from http://xroads.virginia.edu/~DRBR/myth.html <downloaded June 24, 2013>.

Benckendorff, P. & Zehrer (2013) A network analysis of tourism research. *Annals of Tourism Research.* 43:121 – 143.

Burkart A. & Medlick, S. (1974) *Tourism: Past Present and Future.* Heinemann.

Cheng, C., Li, Z., Petrick, J. & O'Leary, J. (2011) An examination of tourism journal development. *Tourism Management,* 32, 53-61.

Coles, T., Hall, C. M. & Duval, D. (2006) Tourism and post-disciplinary enquiry, *Current Issues in Tourism,* 9(4-5), 293-319.

Colton, C. (1987) Leisure, recreation and tourism: a symbolic interactionism view, *Annals of Tourism Research,* 14, 345-360.

Crouch, G. & Perdue, R. (2015) The disciplinary foundations of tourism research: 1980 – 2010. *Journal of Travel Research,* 54 (5), 563-577.

Dann, G. (1997) Paradigms in tourism research. *Annals of Tourism Research,* 24, 472-474.

Dann, G., Nash, D., and Pearce, P. (1988) Methodology in tourism research, *Annals of Tourism Research,* 15, 1-28.

Deery, M., Jago, L. & Fredline, L. (2012) Rethinking social impacts of tourism research: A new research agenda. *Tourism Management,* 33(1), 64-73.

Doxey, G. V. (1975) A causation theory of visitor-resident irritants: methodology and research inferences. In *Travel and Tourism Research Association Sixth Annual Conference Proceedings* (195-198), San Diego, September.

Echtner, C. & Jamal, T. (1997) The disciplinary dilemma of tourism studies. *Annals of Tourism Research,* 24(4), 868-883.

Fedler, A. J. (1987) Introduction: Are leisure, recreation and tourism interrelated, *Annals of Tourism Research*, 14(3), 311-314.

Fennell, D. (2019) The ethics of excellence in tourism research. *Journal of Travel Research*, 52(4), 417-425.

Franklin, A. & Crang, M. (2001) The trouble with tourism and travel theory? *Tourist Studies*, 1(1), 5-22.

Frantz, J. (2006) Academic myths. http://frantzmd.info/Social%20Commentary/Academic_Myths.htm <downloaded Sept 24, 2012>.

Gibson, H. (1998) Sport tourism: A critical analysis of research. *Sport Management Review*, 1, 45-76.

Godbey, G (1985) *Leisure in Your Life: An Exploration 2nd Edition*. Philadelphia: Saunders College Publishing.

Goeldner, C. & Ritchie, B. (2012) *Tourism: Principles, Practices and Philosophies*. Hoboken: John Wiley and Sons.

Gunn, C. (1991) The need for a multidiscplinary tourism education. In Bratton, R.D., Go, F. & Ritchie, J.R.B. (eds), *New Horizons in Tourism and Hospitality Education, Training and Research*, Calgary, Canada: World Tourism Education and Research Centre, pp. 27-33.

Gupta, A. (2015) Definitions. *Stanford Encyclopedia of Philosophy*. https://plato.stanford.edu/entries/definitions/ <downloaded Jan 30, 2019>

Hall, C. M., Williams, A. M. & Lew, A. A. (2014) Tourism: Conceptualisations, disciplinarity, institutions and issues. In Lew, A., Hall, C. M. & Williams, A. M. (eds) *The Wiley Blackwell Companion to Tourism*, Chichester: John Wiley.

Hansen, L. & M. Williams (1999) The myths of Europe: Legitimacy,community and the 'crisis' of the EU. *Journal of Common Market Studies*, 37(2), 233-249.

Heydenryck, K. (1993). A systems model for political communication: A case study of academic mythmaking. *Critical Arts: South-North Cultural and Media Studies*, 6(2), 27-48.

Iso-Ahola, S. (1983) Towards a social psychology of recreational travel, *Leisure Studies*, 2(1), 45-56, DOI: 10.1080/02614368300390041.

Jafari, J. (1987) Tourism models: the sociocultural aspects. *Tourism Management*, 8(2), 151-159.

Jafari, J. & Ritchie, B. (1981) Toward a framework for tourism education: Problems and prospects, *Annals of Tourism Research*, 8(1), 13-34.

Jansen-Verbeke, M. & Dietvorst, A. (1987) Leisure, recreation, tourism: a geographic view on integration, *Annals of Tourism Research*, 14(3), 361-37.

Kompier, M. (2006) The "Hawthorne effect" is a myth, but what keeps the story going? *Scandinavian Journal of Work, Environment and Health*, 32(5), 402-412

Krause, R. (1984) *Leisure and Recreation in Modern Society, 3rd Ed*. Scott Forseman and Co. London, pp 19-49.

Kunwar, R. (2018) Tourism education, curriculum spaces, knowledge production, and disciplinary pluralism. *The Gaze Journal of Tourism and Hospitality*, 9, 83-155.

Lance, C. (2011) More statistical and methodological myths and urban legends. *Organizational Research Methods*, 14(2), 279-286

Laws, E. & Scott, N. (2015) Tourism research: Building from other disciplines. *Tourism Recreation Research*, 49(1), 48-58.

Leiper, N. (1979) The framework of tourism: towards a definition of tourism, tourist, and the tourist industry, *Annals of Tourism Research*, 6(4), 390-407.

Leiper, N. (1999). A conceptual analysis of tourism-supported employment which reduces the incidence of exaggerated, misleading statistics about jobs. *Tourism Management*, 20(5), 605-613.

Leiper, N. (2004). *Tourism Management (3rd ed).* French's Forest: Pearson Education Australia.

Lew, A. (2012) Tourism is NOT the World's Largest Industry – so Stop Saying it is. *TGJournal's Tourism Place*. http://tourismplace.blogspot.hk/2008/04/tourism-is-not-worlds-largest-industry.html <downloaded Dec 7, 2012>.

Lisle, D. (2017) Logic: The importance of definitions. https://biblicalscienceinstitute.com/logic/logic-the-importance-of-definitions/ <downloaded Jan. 30, 2019>.

Liszewski, S. (2010) Tourism studies: Situated within multiple disciplines or a single independent discipline? *Tourism*, 20(2), 37-44.

Mathieson A. & Wall, G. (1982) *Tourism: Economic, Physical and Social Impacts*, Longman, London.

Mazanec, J. A. (2009) Unravelling myths in tourism research, *Tourism Recreation Research*, 34(3), 319-323.

McGee, R. (1985) Lies we live by: Some academic myths and their functions. *Teaching Sociology*, 12(4), 477-490.

McIntosh R., & Goeldner, C.R. (2006) *Tourism: Principles, Practices, Philosophies*, Wiley and Sons, New York.

McKercher, B. & Prideaux, B. (2014) Academic myths of tourism. *Annals of Tourism Research*, 46, 16-28.

McKercher, B. & Tung, V. (2015) Publishing in tourism and hospitality journals: Is the past a prelude to the future? *Tourism Management*, 50, 306-315.

Mercer, D. (2006) The trouble with paradigms. A historiographical study on the development of ideas in the discipline of castle studies. *Archaeological Dialogues*, 13(1), 93-109.

Mill, R. (2008) The inter-relationships between leisure, recreation, tourism, and hospitality, in Brotherton, B. and Wood, R. (eds) *The SAGE Handbook of Hospitality Management*, London: Sage.

Mill R. C. & Morrison A. M. (1985) *The Tourism System: An Introductory Text*, Prentice Hall, NJ.

Morley, C. (1990) What is tourism? *The Journal of Tourism Studies*, 1(1), 3-8.

Murphy, P. (1985) *Tourism: A Community Approach*, Routledge, London.

Nash, D. (1979) Tourism as an anthropological subject. *Current Anthropology*, 22(5), 461-481.

Netto, A. (2009) What is tourism? Definitions, theoretical phases and principles. In Tribe, J (ed.), *Philosophical Issues in Tourism.* Bristol: Channel View Publications.

Owens, J. (2015) Why definitions matter: Stalking victimization in the United States. *Journal of Interpersonal Violence*, 31(12), 2196-2226.

Pearce, D. (1987) *Tourism Today: A Geographic Analysis*, Longman, Harlow, Essex.

Pritchard, A. (2012) Critical tourism research, In Ateljevic, I., Morgan, N, & A. Pritchard (eds.), *The Critical Turn in Tourism Studies; Creating an Academy of Hope.* Oxon: Routledge, pp 11-13.

Ritchie, J. Sheehan, L. & Timur, S. (2008) Tourism Sciences or Tourism Studies? Implications for the design and content of tourism programming. *Teoros*, 27(1) http://teoros.revues.org/1621 <downloaded Dec 2, 2019>.

Sarniak, R. (2015) 9 types of research bias and how to avoid them. www.quirks.com/articles/9-types-of-research-bias-and-how-to-avoid-them <downloaded Feb. 1, 2019>.

Scheaffer, R. (2008) The fallacy of misplaced rationalism. *Skeptical Inquirer*, 32(4), 23-24, 47.

Smith, S. & Godbey, G. (1991) Leisure, recreation and tourism. *Annals of Tourism Research*, 18(1), 85-100.

Smith, S., Xiao, H., Nunkoo, R. & Tukamushaba, E. (2013) Theory in Hospitality, Tourism and Leisure Studies, *Journal of Hospitality Marketing and Management*, 22, 875-894.

Taillon, J. (2014) Understanding tourism as an academic community, study or discipline? *Journal of Tourism and Hospitality*, 3(3), 5. https://www.longdom.org/open-access/understanding-tourism-as-an-academic-community-2167-0269.1000131.pdf <downloaded May 28, 2019>.

Tribe, J. (1997) The indiscipline of tourism, *Annals of Tourism Research*, 24(3), 638-657.

Tribe, J. (2006) The truth about tourism, *Annals of Tourism Research*, 33(2), 360-381.

Tribe, J. (2010) Tribes, territories and networks in the tourism community. *Annals of Tourism Research*, 37(1), 7-33.

Vandenberg, R. (2006) Introduction: Statistical and methodological myths and urban legends: Where, pray tell, did they get this idea? *Organizational Research Methods*, 9(2), 194-201.

Veerasamy, V. (2013) The importance of good definitions (or: how to think clearly). www.referralcandy.com/blog/importance-of-good-definitions/ <downloaded Jan. 30, 2019>.

Weiler, B., Moyle, B. & McLennan, C. (2012) Disciplines that influence tourism doctoral research: The United States, Canada, Australia and New Zealand. *Annals of Tourism Research*, 39(3), 1425-1445.

Whitfield, G. (2012) The importance of proper definition, https://piadvice.wordpress.com/2012/06/13/the-importance-of-proper-definition/ <downloaded Jan 30, 2019>.

Xiao, H. & Smith, S. (2006). The making of tourism research. *Annals of Tourism Research*, 33, 490–507.

Xiao, H., Li, M. & Lin, E. (2011) Diffusion patterns and knowledge networks: an inductive analysis of intellectual connections in multidisciplinary Tourism Studies, *Journal of Travel & Tourism Marketing*, 28(4), 405-422, DOI: 10.1080/10548408.2011.571574.

3 The Structure of Tourism

By the end of this chapter, the reader will be able to:

- Describe the key elements of systems models
- Compare and contrast different open and closed ended models of tourism
- Demonstrate how each of us operates within our own unique tourism system
- Analyse the concept of a destination as a key component of any model
- Discuss how space change can lead to adverse impacts of tourism.

Introduction

This chapter discusses the component parts of tourism and various models that have been developed to illustrate how the components interact. As Goeldner and Ritchie (2006) note, though, the complex phenomenon of tourism is extremely difficult to describe succinctly, meaning models often become exceedingly complex in their effort to capture the components, processes and outcomes of tourism. We often talk about the 5 As of tourism – attractions, accommodation, activities, access and amenities – as if they capture the entire scope of this topic. While such a way of thinking of tourism's components is important, it cannot fully capture the nuances of what tourism is and how it works.

Tourism as a system

Leiper (1979) and Mill and Morrison (1985) were the first to argue that tourism functions as a system, where the parts contribute to the whole, but whose whole is far greater than the sum of its parts. A system is defined as a set of elements standing in interrelation among themselves and their environments (Bertalanffy 1972: 417). As Lohman and Netto (2008: 3) indicate, to be complete, a system must possess the following components: an environment – the system's location; units – the system's parts; relationships – how the parts interact; attributes – the quality of the units in the system itself; inputs, outputs and feedback; and a model.

Leiper (2004: 22), reviewing the biological origins of systems theory, noted it begins with the assumption that anything complex can be conceptualised as a series of interrelated systems that can be arranged in a hierarchy so that each system has its subsystems and superior systems. He then proceeds to state that elements are the basic building blocks of systems, and do not need to be further dissected to understand what the system is and how it works. Essentially, two types of systems were proposed in the early days of tourism scholarship. On the one hand, closed systems argued that there was little interaction with the environment, while open systems suggested tourism could have predictable outcomes.

While Serra (2016) notes that systems thinking recognises the interdependence among the elements of a system, all too often a reductionist approach is taken, whereby the system is broken down into its component parts and each is examined individually. The traditional Newtonian-Cartesian approach serves a purpose of identifying constituent elements of any tourism system, but does not explain how the system functions as a holistic whole.

It is for this reason that traditional systems approaches seemed flawed, for as Leiper (2004) noted, many models had become nothing more than a set of elements that came into play when people went on trips. Today, tourism is now recognised as an adaptive, open complex system guided by the rules of complexity theory (Baggio, 2008; Faulkner and Russel, 1997; Zahra and Ryan, 2007). The reason is that complex systems do not have predictable reactions, cannot be decomposed, have nonlinear interactions, high sensitivity to initial conditions, are dynamic, adaptable to the environment, produce emergent structures and behaviours, and can become chaotic (Jere Jakulin, 2017:209). We will discuss this issue in greater detail in Chapter 9, where understanding complex systems helps us to understand the evolution of destinations.

■ Gunn's Functioning Tourism System

Clare Gunn, one of the pioneers of tourism developed a Functioning Tourism System model in the early 1980s and then modified it slightly over the years. His model (Figure 3.1) highlights the fact that every part of tourism is related to every other part (Gunn and Var, 2002: 34). He feels the two main drivers of tourism consist of a demand and supply side and that in order to satisfy consumer demand the destination must perceive a range of developments and services to cater to their needs.

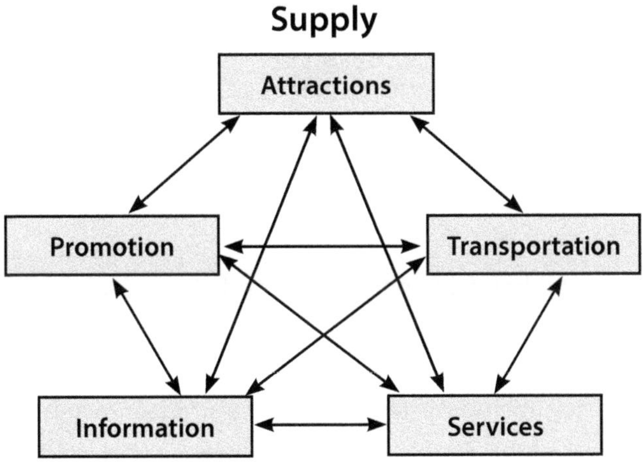

Figure 3.1: Gunn's Functioning Tourism System. *Source*: Gunn and Var, 2002.

The demand side of tourism involves four major factors. The tourist must be motivated to travel, have the financial ability to pay for goods and services, and have the time and physical ability to travel. These considerations, directly or indirectly, also result in the identification of ideal geographic or demographic market segments.

Supply side components are somewhat more straightforward and typically represent the five As of tourism. Destinations must have suitable attractions. Attractions are the most powerful component of the tourism system, for they provide the pull to satisfy the tourist's push. Everything else acts as a facilitator to enable consumption of attractions and is positioned in a subordinate place in the model and categorised under the broad term of services.

None the less, travel service businesses represent the greatest economic contributor to tourism. They include accommodations, food service, transport, the travel trade and other ancillary businesses. Information acts as the catalyst that links the consumer to the attraction and/or the destination. Information includes standard promotional material, but also, Gunn emphasises it also involves interpretation, environmental messages and the like.

While not formally included in the model, Gunn also recognises a range of externalities that influence the tourism system. These externalities range from government policy, through community attitudes, natural and cultural resources, leadership and the role that competition plays in shaping systems.

Leiper's models

Neil Leiper's various incarnations of his systems model were developed between the late 1970s and the early 1990s and suggested a tourism system involved both geographic and social elements (Serra, 2016). His first model (Leiper, 1979), shown in Figure 3.2, adopted a strictly geographic focus to describe how tourism works. He does acknowledge the contribution of Gunn (1972) to the development of the model. Three elements were defined. They include:

- The **tourist generating region**, which represents the permanent residential base of tourists, the source of tourism demand and the place where the trip begins and ends. It includes the basic geographic features as well as the social push elements that drive people to travel. As a result, major marketing functions are conducted there.
- The **tourist destination areas** which are the locales where tourists stay temporarily. They include physical space, and also encompass those attributes that appeal to tourists and may pull them there. Destination areas are also the location of many parts of the tourist business, including accommodation establishments, services, entertainment and recreational facilities
- The **transit region** representing the space tourists must pass through or over to get to the destination and return home again. Transit areas can also include stopover points which might be used for convenience or because of the existence of attractions (Leiper, 1979: 396, 397).

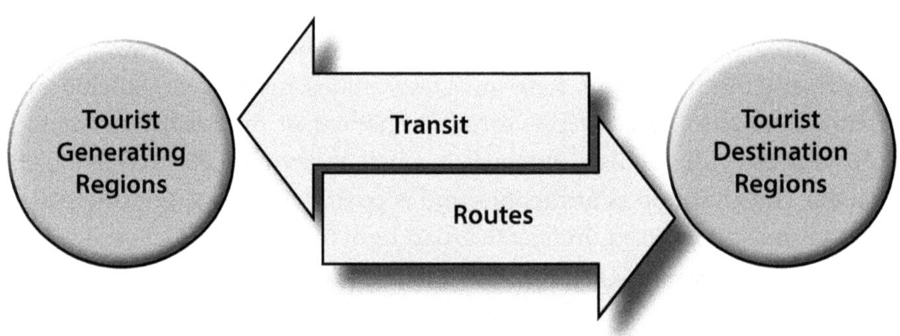

Figure 3.2: Leiper's Whole Tourism System. *Source*: Leiper, 1979.

This systems model incorporates five elements: the tourist, the generating region, transit routes, destination regions and a tourist industry. They are arranged in spatial and functional connections (Leiper, 1979: 404), with each potentially playing a role of impacting on other elements of the system. These elements operate within a broader socio-political and geographic environment. Physically, tourism involves a journey away from home, a stay at a destination and a return, through a transit region. Economically it also involves an exchange of money for a good and service; and socially, hopefully, involves an exchange of understanding between host and resident.

While the elements of the system did not change, the structure of the model did, as he later found it to be somewhat too geographically focussed. Instead, in his doctoral thesis, he developed the Systematic Model for Interdisciplinary Tourism Studies (Leiper 1992), as shown in Figure 3.3. This model explicitly incorporated the social elements of the tourists themselves and the industry that serves their needs. As he stated (Leiper 1992: 333) the model:

> indicates a structural method for interdisciplinary studies. Each side of the pentagon represents one element in a whole tourism system. The pentagon's environments are disciplines which may be useful in studying some facet of tourism. The method begins with links between elements. For example, the question "why do people travel?" begins with the travel generating region and its links with tourists (its residents who make trips), transit routes, destinations, and the industry. Lines inside the pentagon represent the process: studying any one element involves considering its links with the other four.

In this manner, he proceeds to state that no research about tourism can afford to focus exclusively on one element in whole tourism systems but must recognize, implicitly or explicitly, connections between elements and between a system and its environments (Leiper 1992: 334).

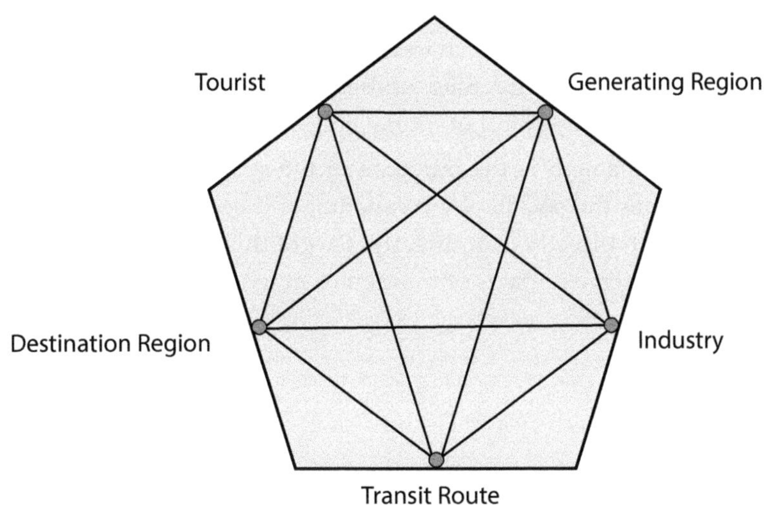

Figure 3.3: Systematic model for interdisciplinary tourism studies. *Source*: Leiper, 1992.

Importantly, Leiper's models show how all five components are linked individually and collectively. He suggests that every tourism system must have at least one of these elements. The model is particularly useful for it shows how tourism can be analysed from different disciplinary perspectives. For example, an economic perspective follows the outflow of income from the generating region through transit regions and destination regions and also through the use of the tourism industry, while a geographic perspective enables one to analyse how tourists negotiate the geography (social, political and economic) between home, transit routes and destination areas.

Mathieson and Wall's models

Mathieson and Wall (1982) proposed an open tourism system that consisted of three elements (Figure 3.4). The first is a basic dynamic element that involves the decision to travel and the range of social, economic and institutional factors that affect these decisions. It consists of tourism demand and the various forms of tourism available. They note demand consists of three major components: actual or effective demand which equates with participation; potential demand which includes those people motivated to travel but who do not because of temporal or financial reasons and; deferred demand which consists of those people who could travel if motivated but do not because of lack of knowledge, opportunity or facilities (Matheison and Wall 1982: 16).

The second is the static or destination element, where the stay in the destination interacts with its economic, environmental and social systems. Considerations here include the carrying capacity of the destination, the geopolitical structure, level of development and also the social organisation of tourism. These features are balanced off against the type of tourist attracted to the destination, which include the socio-economic characteristics of the tourist, his or her level of use, length of stay and activity. Each of these is important for they influence the magnitude, frequency and kind of interaction.

The third element is the outcome of the system or a consequential element, which includes the social, environmental and economic systems tourism comes in contact with directly or indirectly. Part of this element also includes efforts to control the adverse impacts of tourism or magnify the positive benefits.

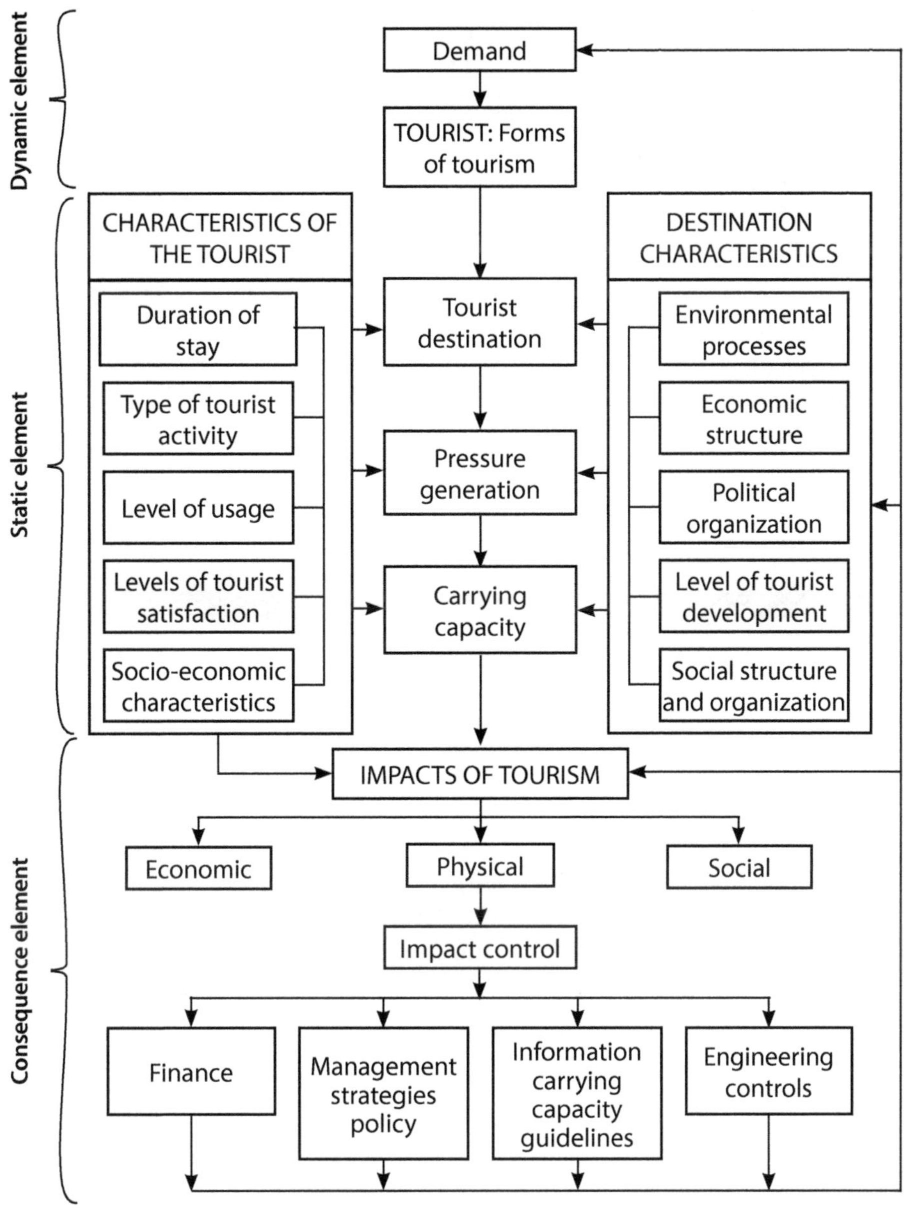

Figure 3.4: Mathieson and Wall's open tourism system model. *Source*: Mathieson and Wall, 1982.

■ Mill and Morrison's marketing model

Robert Mill and Alastair Morrison (1985) proposed a more marketing oriented model to explain how tourism works, as shown in Figure 3.5. Flognfeldt (2005) notes the content of the model does not differ much from the Leiper models, but adds marketing as a specified tool of convincing people to travel to a certain destination. As a system of a spider web, analysis could start at any box in the model.

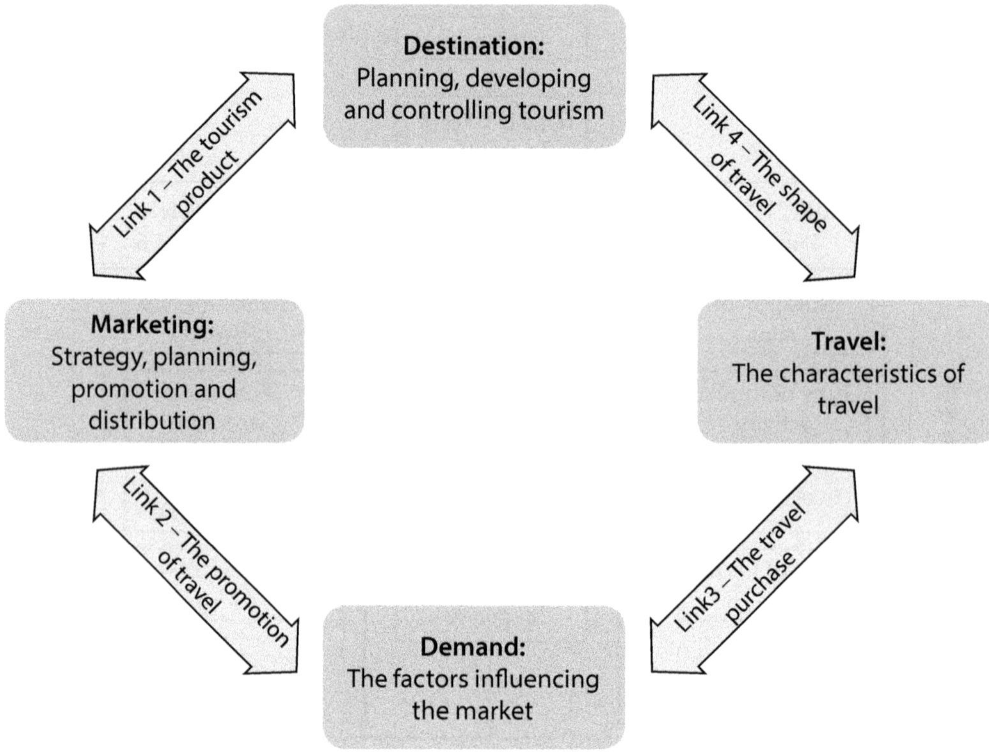

Figure 3.5: Mill and Morrison marketing system model. *Source*: Mill and Morrison, 1985.

Mill and Morrison realized what a difficult task it was to model tourism, partly because there is no universally agreed upon definition, partly because of the diversity of tourism and also because there are so many players involved, some of whose primary function is tourism, some with tourism as a secondary function and also some who would likely say they were not in tourism. They also highlight the risk of labelling tourism as an industry.

Their model involves four inter-linked component parts in a closed system:

- *Market or Demand* – A consumer behaviour approach to market demand emphasizing both the external and internal influences on travel including the alternatives to travel, the market inputs of tourism suppliers, and the process by which a buying decision is reached;
- *Marketing* – An examination of the process by which the destination area and individual suppliers market their products and services to potential customers with an emphasis on the effective use of distribution channels;
- *Destination* – An identification of the procedures that the destination area should follow to research, plan, regulate, develop, and service tourism activity;
- *Travel* – A description and analysis of major travel segments, travel flows, and modes of transportation used.

Again, like Leiper, they state that without the individual, no tourism can occur. Leisure travel occurs to satisfy internal needs and wants. Tourists do not take vacations just to relax and have fun, to experience another culture or to educate their children. They take vacations in the hope and belief that the vacations will satisfy, either wholly or partially, various needs or wants. As a result, adopting a marketing approach to describe tourism is appropriate. As they illustrate, realizing that motivation to satisfy needs and wants is at the heart of travel, the difference between seeing a destination as a collection of palm trees and hotel rooms and seeing it as a means of satisfying needs and wants of tourists. Understanding the market, or more specifically the needs and wants of the market also provide insights into why specific destinations are chosen.

Travel represents the important link between the individual and the destination. The mode of travel taken, the distance travelled and a variety of other factors are influenced by social, demographic, motivational, trip purpose, time and accessibility factors. Chapter 5 looks at tourist flows and a range of intervening factors that influence the movement of individuals from their home to a destination or between destinations.

The destination contains a mix of attractions, facilities and activities that cater to the needs of the tourist. Attractions are what draw visitors to a destination while facilities and services cater to the needs of the tourist. Attractions are discussed in Chapter 4.

Marketing is a function by which destinations or attractions communicate to the market and by which the market pursues information about a destination. Thus, marketing has both a supply function as well as a demand function. Stated another way, marketing reflects a two-way communication process between the parties.

The main benefit of the model is its appreciation that marketing in tourism is substantially different from marketing in other areas.

- First, tourism is a service, where intangible experiences are being sold. More than that, though, the intangible experience being sold is a dream or illusion.
- Second, tourism does not have an inventory like other businesses. The tourism product cannot be stored and sold at a later date. What is not sold today, can never again be sold.
- Third, tourism supply is relatively fixed. A hotel cannot magically add 300 extra rooms for the peak season and then it, just as magically, make them disappear during off-peak times.
- Fourth, the diverse nature of tourism means that the experience being sold is an amalgam of several products and services.

- Fifth, distance and the reliance on travel intermediaries further complicate the marketing and communication process. For the most part, tourism is still reliant on the on and offline travel trade. Yet, the trade does not buy products and instead, acts as middle men who on-sell existing products to the consumer.
- Finally, tourism demand is unlike that for most other businesses. It is highly elastic, seasonal in nature, subject to the external factors and is highly volatile.

Thurot

Thurot (1980) too notes tourism is a multifaceted activity that involves both the demand for tourism services by the travelling public and the supply of tourism facilities to satisfy this demand. His model shown in Figure 3.6 recognises that that virtually every place in the world can be both a generating region, a main or secondary destination and a transit zone in its own right. Different elements of the commercial tourism sector could perform unique functions depending on the role the place plays.

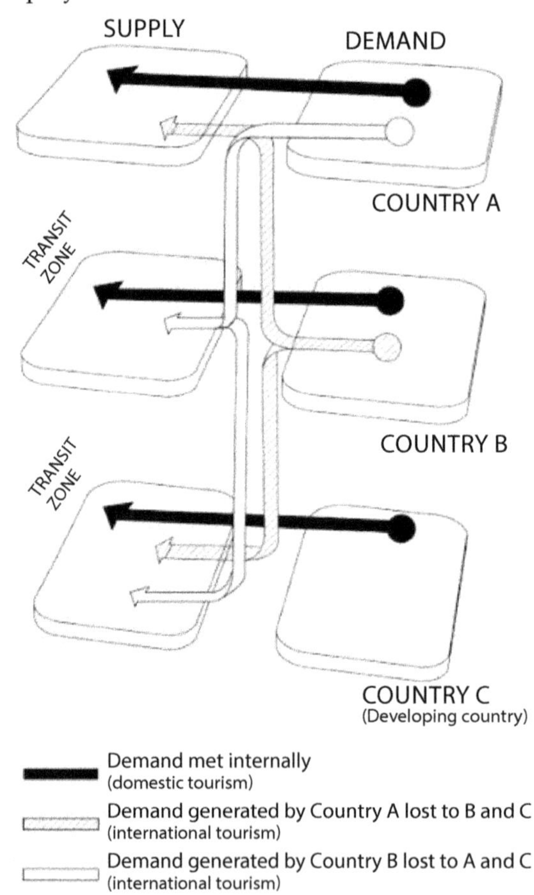

Figure 3.6: Thurot's model of the relationship between supply and demand. *Source*: Pearce, 1989.

Hall

The treatment of tourism as a closed system is the greatest weakness of the above models, for they fail to appreciate both inputs and inevitable outputs or impacts of tourism on host communities and the local and global environment. These models may imply impacts and make no mention of them. Hall's (1995) model (Figure 3.7) recognises impacts are a function of both supply and demand.

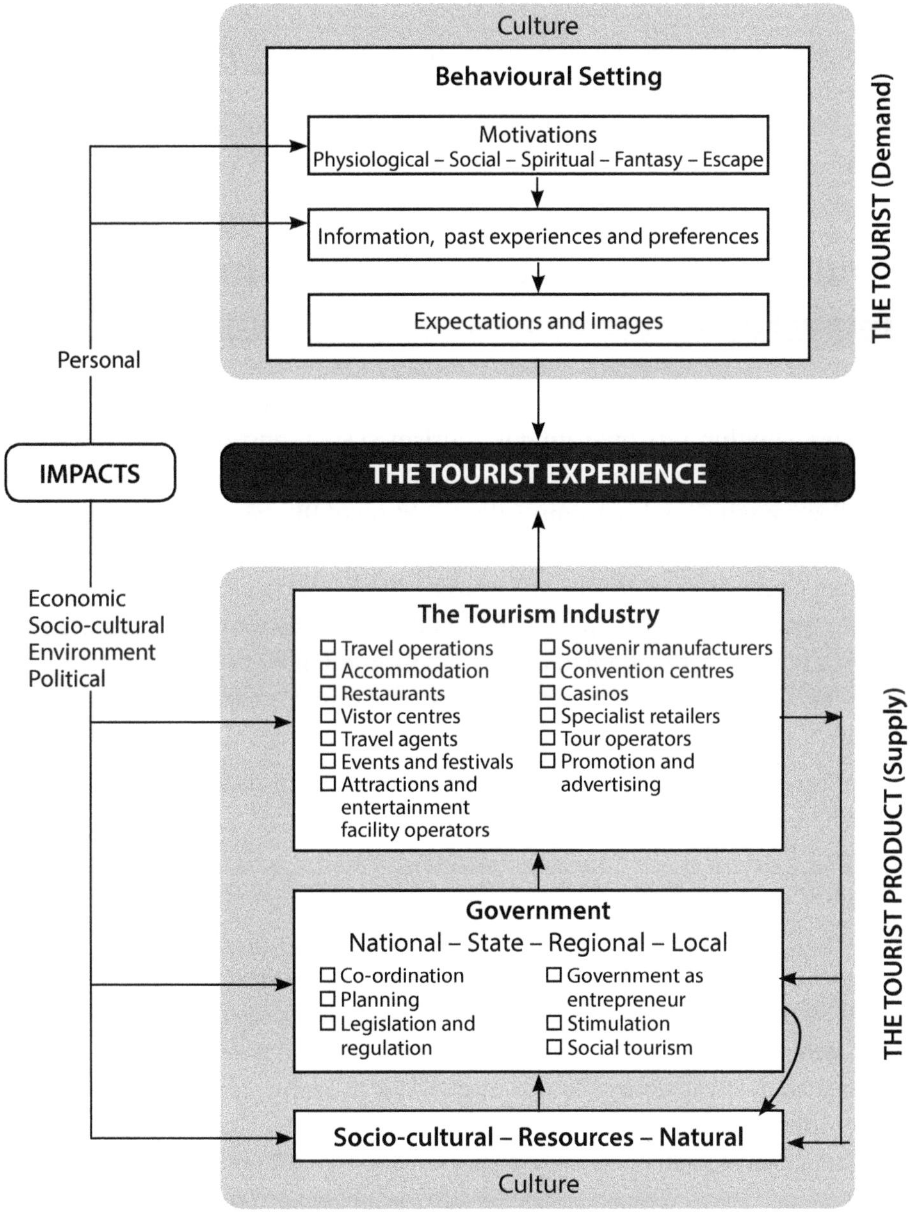

Figure 3.7: Hall's tourism experience model. *Source*: Hall, 1995.

He basically uses a combination of Pearce's evolving travel career model, Maslow's (1943) motivations and Cohen's (1972, 1979) sociological perspective works in defining demand factors. Supply factors relate to the range of tourism resources available, government's role and the tourism industry. He argues the capacity of any nation or region to deliver a tourism product will depend on the composition and nature of its tourism resources.

Combined, these two elements of supply and demand provide the basis for the tourism experience and the likely outcomes or impacts of tourism. The tourist brings his or her own cultural bias, motivations, knowledge and expectations to a destination which seeks to provide a variety of products, services and experiences to satisfy the needs of the tourist. How well these two match and how well tourism is managed will influence the range of positive and negative impacts felt.

Each tourist functions within his or her own unique system

A key feature of the models that most people do not appreciate is that each of us functions within our own unique tourism system and this overlaps with other people's tourism systems (Leiper, 1990). We tend to talk about a tourism system as a collective whole, with a simplistic repetition of the models suggesting a generic system exists that is comprised of the various component parts. But, since the system begins and ends with the tourist, then each person will operate within his or her own tourism system. We cannot talk about one tourism system. Instead, we must consider somewhere in the neighbourhood of eight billion separate tourism systems when all domestic and international travel is considered.

Each system will overlap at many points along the way, but each is discrete. Overlap can occur when we share the same flight, stay at the same hotel or visit the same attractions.

The existence of individual tourism systems was demonstrated in some research produced in Hong Kong analysing what people did in this destination. A departing visitor survey asked respondents to indicate what broad array of activities they pursued. They were then analysed individually to see the degree of overlap. Table 3.1 shows the results. It was surprising to note that there were virtually no common movement patterns other than with business travel and dedicated festival goers. For example, 216 different combinations of activities were identified among the 221 people who went to beaches, and 299 different combinations of activities were noted among the 315 people who visited historic sites.

Table 3.1: Multiple tourism systems.

Activity/Attraction	People participating in the activity	Different activity sets involving the item	% of different combinations
Beaches	221	216	97.7
Business	585	127	21.7
Churches, temples and other religious sites	298	285	95.6
Country parks and natural areas	270	259	97.7
Cross border tourism	323	288	89.1
Ecotourism	37	37	100.0
Festivals	133	42	30.7
Historical attractions	315	299	94.9
Museums & art galleries	246	235	95.5
Shopping	1107	850	76.8
Sightseeing	865	727	84.0
Theme parks	383	346	89.4
All	1304	1002	76.8

Each tourist operates within his or her own discrete tourism system, although it may not seem so when crowding is such an issue at popular sites, like here in Pompeii, Italy. In reality, each independent tourist, or small group of independent tourists behaves uniquely. They follow their own tourism system either at a macro level (from their home community, the use of travel intermediaries, transport providers, and in destination accommodation, services and facilities), or at a micro level as noted in their movements through a destination. Leiper (1990) reminds us that each system overlaps at multiple points, creating an impression of a single tourism system.

Concept of a destination

Chapters 4 and 5 discuss the key components of attraction and access, respectively, while the latter half of the book is dedicated to the tourist. It is worth noting that tourism occurs at a destination level. Yet, little work has been conducted conceptualising the destination. The following section examines the destination.

The geographic scope of what is considered a 'destination' is as variable as tourism itself. Countries, states, regions and cities have all been labelled as destinations. America, California, Napa Valley, Sonoma! Such imprecision is understandable because the mandate of marketing bodies varies by the geographic scale. But it does little to help us understand how and where tourism is consumed.

In fact, according to the UNWTO, the local destination is the key to tourism for it is the focal point of much of the development and also is the locus where many of the impacts of tourism are felt directly. It defines a local destination as:

> a physical space with or without administrative and/or analytical boundaries in which a visitor can spend an overnight. It is the cluster (co-location) of products and services, and of activities and experiences along the tourism value chain and a basic unit of analysis of tourism. A destination incorporates various stakeholders and can network to form larger destinations. It is also intangible with its image and identity which may influence its market competitiveness. (UNWTO, nd)

The issue of networking is especially relevant, for destinations are not uniform. Instead, tourism facilities tend to be clustered spatially into tourist nodes or precincts unevenly throughout a destination (Tunbridge and Ashworth, 1996; Dredge, 1999; Pearce, 2001). Each node or precinct usually has some type of capstone facility or facilities that forms the core of the area. They may be hotels, attractions, shopping outlets, transport hubs or meeting and convention facilities (Dredge, 1999). The number of tourist nodes depends on both the size and geomorphology of the destination (Pearce, 1998).

Destinations consist of three types of space as shown in Figure 3.8 (McKercher, Wang and Park, 2015). 'Tourism space' is designated for the primary use of tourists. Locals are also welcome in tourism place where they can behave as tourists and release subterranean values (Matza and Sykes, 1961). 'Non-tourism space' represents areas of a destination where tourists are not supposed to visit. Residential and industrial areas along with retail and recreational precincts designed mostly for local residents are considered as non-tourism space. Tourists may transit through non-tourism space but are not welcomed there. 'Shared space' is hybrid place where both tourists and locals are welcome, providing tourists adhere to certain norms of social behaviour. How much of each place exists depends on the volume of tourists, the size of the local population and the spatial configuration of tourist nodes. Each type of place is clearly signposted semiotically by a variety of markers that signal its identity and legitimacy of use by certain groups.

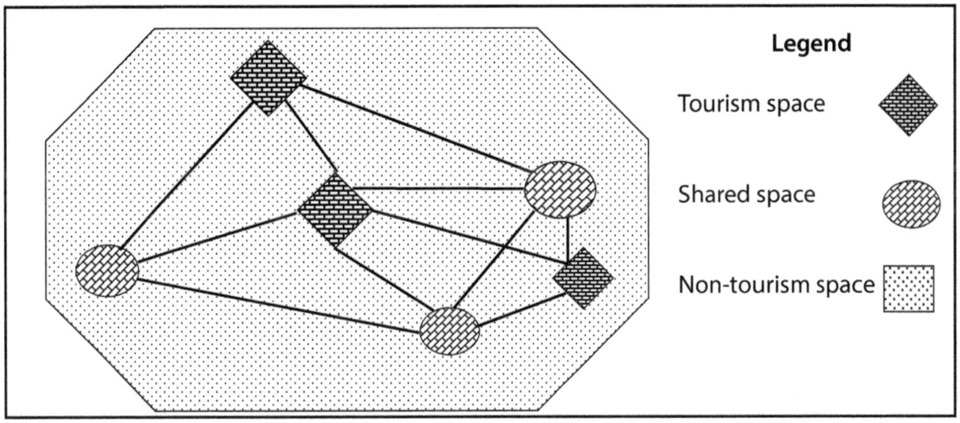

Figure 3.8: Conceptualization of the destination as consisting of three types of space.
Source: McKercher, Wang and Park, 2015.

Tourism space is something that is either constructed or signified for and by tourists as a medium through which the tourist experience is negotiated (Crouch, Aronsson and Wahlstrom, 2001). Wall (1996), suggests tourism space can exist as points, lines and areas. He asserts points represents individual, isolated attractions, lines represent the linear routes that join sets of attractions, and areas represent concentrations of attractions that create nodes.

But tourism space also involves more than simply a geographic or psychological space. It also involves both mental and social space. Mental space is signified semiotically by a series of markers that create an appealing destination image (Timothy and Boyd, 2006), that says it is something worth visiting (MacCannell, 2001). Social space is signified by the behaviour of fellow tourists. We would all like to believe we do things out of free will. In reality we look for signs or signals that guide our behaviour unconsciously (Donaire and Gali, 2008), to such an extent that Andreu, Kozak, Avci and Cifter (2005) call this pattern the 'herding effect'. Regardless of the cause, we tend to go where other tourists go and avoid places where they do not go.

Shared place is hybrid space suitable for both tourists and locals. Hayllar, Griffin and Edwards (2008) described the idea of shared place as being part of the everyday fabric of the destination such as transport hubs, shopping and dining districts, where both tourists and locals strive to have the same goals met. Snepenger, Murphy, O'Connell and Gregg (2003), for example have written about how tourists and locals use and share the same shopping space. Much of the downtown core of urban destinations, as well as trendy shopping and dining districts reflect shared place.

Examples of semantically themed tourism space, shared space and non-tourism space can be found in all destinations. These images from Macau, SAR China, illustrate the concept. The first image is of the canals in the Venetian Casino. It is designed for both tourists and locals, but tourists dominate. Locals who visit get to play at being tourists, while the attraction and surrounding shops are built with tourists' needs and wants in mind. The second image is of shared space, at the A Ma Temple, one of the historic temples in Macau. Here tourists are welcome providing they behave in accordance with the norms of local residents. The third image is of an historic church located in a residential enclave in one of the outer areas of Macau. It is signposted as non-tourism space that serves the needs of local residents. Tourists are not welcomed and are not supposed to visit.

Most parts of a destination are comprised of non-tourism space, though. It represents a void that is either intimidating to tourists or has no meaning to them. They may transit through such space, but should have no reason to stop there (McKercher, Wang and Park, 2015).

Most of the adverse impacts of tourism are caused when space changes. Non-tourism space can be invaded by tourists and transformed into shared or exclusive tourism space, to the detriment of the livelihood of local residents. Shared space can be so overwhelmed by tourists that it becomes tourist space. Most of the time, these three types of place are in equilibrium, especially when tourism becomes an accepted part of the social fabric of the community. Each coexists and each respects its boundaries. It is only when systems are thrown out of equilibrium that adverse impacts are felt, until the system again reaches a steady state in accordance with complexity theory (Russell and Faulkner, 1999).

Conclusion

This chapter reviewed some of the models used to classify tourism as either a closed or open ended system. Essentially, all the models start with the tourist and recognize a transport, destination and activity component. Only open ended systems, though, explicitly model impacts of tourism and the role that feedback loops play in tourism supply and demand. Importantly, Neil Leiper recognized that each of us functions within our own tourism system. This understanding makes it difficult to model systems at any more precise a level than the macro models shown above.

The chapter finished with a brief introduction of the destination and the recognition that it is not homogeneous. Instead, it consists of different types of space, with only a minority of space devoted exclusively to tourists. Most of the destination is not relevant to tourists and therefore, tourists have no reason to visit.

■ Discussion points / Exercises

1. Systems consist of at least five elements. Identify and discuss these elements within a destination of your choice.
2. Interview a number of tourists in your home community and try to identify the individual tourism system each operates in and where these systems overlap.
3. Destinations consist of three types of space. Analyse your destination and map out tourism, shared and non-tourism space. Show the transit links that join different types of tourism and shared space. Analyse the findings to determine the differences between the tourist's perception of the destination and that of local residents.

Points to consider for revision

- Leiper argues tourism functions largely as a closed system that contains at least five elements of: the tourist, the generating region, the transit region, the destination area and the businesses that serve tourists in each region.
- Mill and Morrison also argue tourism functions as a closed marketing-oriented system that considers the market, marketing, the destination and the act of travel.
- Thurot models tourism in such a way that a place could simultaneously be a destination, source market or transit region.
- Hall feels tourism is an open system comprised of both supply and demand considerations that result in a tourist's experience and that, in the end, produce a series of impacts, positive and negative.

Additional recommended reading

Hall, C. M. and Page, S. (2010) The contribution of Neil Leiper to tourism studies, *Current Issues in Tourism*, 13(4), 299-309. DOI: 10.1080/13683500.2010.482652.

McKercher, B., Wang, D. and Park, E. (2015) Social impacts as a function of place change. *Annals of Tourism Research*, 50, 52:56.

References

Andreu, L., Kozak, M., Avci, N. & Cifter, N. (2005) Market segmentation by motivations to travel. *Journal of Travel and Tourism Marketing*, 19(1), 1-14.

Baggio, R. (2008) Symptoms of complexity in a tourism system. *Tourism Analysis* 13(1), 1-20.

Bertalanffy (1972) The history and status of General Systems Theory. *The Academy of Management Journal*, 15(4), 407-426.

Cohen, E. (1972) Toward a sociology of international tourism. *Social Research*, 39(1), 164.

Cohen, E. (1979) A phenomenology of tourist experiences, *Sociology*, 13, 170-201.

Crouch, D., Aronsson, L. & Wahlstrom, L. (2001) Tourist encounters. *Tourist Studies*, 1(3), 253-270.

Donaire, J. & Gali, N. (2008) Modeling tourist itineraries in heritage cities. Routes around the Old District of Girona. *PASOS*, 6(3), 435-449.

Dredge, D. (1999) Destination place and planning. *Annals of Tourism Research*, 26(4), 772-779.

Faulkner, B. and Russell, R. (1997) Chaos and complexity in sourism: In search of a new perspective, *Pacific Tourism Review*, 1, 93-102.

Flognfeldt, T. (2005) The tourist route system – models of travelling patterns. *Belgian Journal of Geography*, 1(2), 35-58. https://journals.openedition.org/belgeo/12406 <downloaded Aug 15, 2019>.

Goeldner C. R. & Ritchie J. R. B .(2006) *Tourism Principles, Practices and Philosophies 9th* Ed, John Wiley and Sons.

Gunn, C. (1972) *Vacationscape: Designing tourist regions.* Austin: University of Texas.

Gunn, C. & Var, T. (2002) *Tourism Planning: Basic Concepts and Cases.* Toronto: Routledge.

Hall M (1995) *Introduction to Tourism* 2nd Ed., Melbourne: Longman Australia.

Hayllar, B., Griffin, T. & Edwards, D. (2008) Urban tourism precincts: Engaging with the field. In Hayllar, B., Griffin, T. & Edwards, D. (Eds.), *City Spaces – Tourist Places: Urban tourism precincts* (pp. 1-19). Sydney: Routledge.

Jere Jakulin, T. (2017) Systems approach to Tourism: A methodology for defining complex tourism system. *Organizacija*, 50(3), 208-215.

Leiper, N. (1979) The framework of tourism: towards a definition of tourism, tourist, and the tourist industry. *Annals of Tourism Research*, 6(4), 390-407.

Leiper, N. (1990) *Tourism Systems; An Interdisciplinary perspective.* Department of Management Systems (Occasional Paper #1). Palmerston North: Massey University

Leiper, N. (1992) *Whole Tourism Systems: Interdisciplinary Perspectives on structures, functions, environmental issues and management.* PhD thesis Faculty of Business, Massey University, 367 pp.

Leiper, N. (2004) *Tourism Management (3rd ed).* French's Forest: Pearson Education Australia.

Lohman, G. & Netto, A. (2008) *Tourism Theory: Concepts, models and systems.* Wallingford: CABI.

MacCannell, D. (2001) Tourist Agency. *Tourist Studies*, 1(1), 22-37.

Maslow, A. H. (1943) A theory of human motivation. *Psychological Review*, 50(4), 370-96.

Mathieson, A. & Wall, G. (1982) *Tourism: Economic, Physical and Social Impacts*, Longman, London.

Matza, D. & Sykes, S. (1961) Juvenile delinquency and subterranean values. *American Sociological Review*, 26, 712-719.

McKercher, B., Wang, D. & Park, E. (2015) Social impacts as a function of place change. *Annals of Tourism Research*, 50, 52-56.

Mill R. C. & Morrison A. M. (1985) *The Tourism System: An Introductory Text*, Prentice Hall.

Pearce, D. (1989) *Tourist Development 2nd Ed.* Harlow: Longman.

Pearce, D. (1998) Tourist districts in Paris: structure and functions. *Tourism Management*, 19(1), 49-65.

Pearce, D. (2001) An integrative framework for urban tourism research. *Annals of Tourism Research*, 28(4), 926-946.

Russell, R. & Faulkner, B. (1999) Movers and shakers: chaos makers in tourism development, *Tourism Management*, 20(4), 411-423.

Serra, F. (2016) Systems. In Jafari, J. & Xiao, H. (Eds) *Encyclopedia of Tourism.* Springer

Snepenger, D., Murphy, L., O'Connell, R. & Gregg, E. (2003) Tourists' and residents' use of a shopping space, *Annals of Tourism Research*, 30(3), 567-580.

3

Thurot, J. M. (1980)*Capacité de Change et Production Touristique,* Etudes et Mëmories no. 43. Centre des Hautes Etudes Touristiques: Aux-en-Provence.

Timothy, D. & Boyd, S. (2006) Heritage tourism in the 21st Century: Valued traditions and new perspectives, *Journal of Heritage Tourism,* 1(1), 1-16.

Tunbridge, J. E. & Ashworth, G. J. (1996) *Dissonant Heritage: The management of the past as a resource in conflict.* Wiley and Sons, Chichester

UNWTO (n.d.) *Definitions Committee on Tourism and Competitiveness (CTC).* http://cf.cdn.unwto.org/sites/all/files/pdf/unwto_tourism_definitions.pdf <downloaded Feb. 3, 2019>.

Wall, G. (1996) Perspectives on tourism in selected Balinese villages. *Annals of Tourism Research,* 23(1), 123-137.

Zahra, A. & C. Ryan (2007) From chaos to cohesion—Complexity in tourism structures: An analysis of New Zealand's regional tourism organizations, *Tourism Management,* 28(3), 854-862.

Part 2 - How Tourism Works

The first part of the book set the stage, by describing what tourism is and how the tourism sector is organised. The opportunity exists now to focus on understanding how tourism works as both a social and commercial activity. It is important to appreciate the key factors that drive tourism and influence the intersection between supply and demand.

Chapter 4 discusses the role of attractions or demand generators, without which the pull of the destination may not exist. Traditionally, attractions have been thought of in fairly narrow supply side terms. Recently, though, the UNWTO and others have started to think of them from a demand perspective, meaning that almost anything a tourist does in a destination could be considered as an attraction. Yet, the epistemological discussion of attractions, activities and products has been narrow traditionally. This chapter introduces the idea that attractions can be considered as existing on one of a number of tiers in a taxonomic structure from the very broad to the very specific.

Chapter 5 considers the key element of access, for without good access, the ability of a destination to compete in the global marketplace is limited. The chapter begins with a discussion of time and how varying perspectives of the use of time influence tourist behaviour. It then proceeds to examine distance decay, the first law of geography, and its related concept of market access. The chapter concludes with a review of various itinerary models.

Chapter 6 examines some of the political issues that affect international tourism flows. It must be appreciated that nowhere in international law does anyone have the right to enter a country. Instead, it is a privilege granted by both the departing and receiving nation or territory. This means, of course, that access to international travel is a fundamentally political act that is also fundamentally unequal. Bilateral air service agreements are introduced next, for without understanding them, one cannot understand international tourism. The chapter concludes with a brief discussion of the role that the World Trade Organisation's General Agreement on Trade in Services (GATS) has on tourism.

Chapter 7 focuses on the political economics of tourism and examines different approaches to forecasting. This chapter explores how political economy and economic organisation affect both people's desire to travel and the ability of the economy to supply suitable experiences.

4 Attractions

By the end of this chapter the reader will be able to:

- Conceptualise an attraction
- Critique the tourism attraction system and identify and apply markers
- Understand the hierarchy of attractions and the same attraction can play different roles for different market segments
- Analyse a taxonomy of attractions
- Apply the tourism attractions taxonomy in real world situations.

Introduction

Tourism is driven by attractions and access. Attractions explain why people come to a destination, while access influences the volume of visitors. The two are interrelated for strong attractions create demand to enhance access, while weak access effectively limits demand and thus the size and scale of attractions. The opposite is also true. Weak attractions create little necessity to improve access, while strong access creates the opportunity to enhance the breadth of attraction offerings.

Tourism attractions play a fundamental role in the successful development of any destination. They serve as demand generators that give the customer a reason to visit. Moreover, consumption of attractions is often considered to be the central theme of the tourism experience being offered. Goeldner and Ritchie (2012) argue that attractions drive people to travel and represent the main motivations for travel. More importantly, they note that without attractions there would be little need for all other tourism services and goods. Thirty years ago, Gee, Makens and

Choy (1989) stated that both the quality and quantity of attractions affect travel decisions. Mill and Morrison (2012) illustrate that attractions at a primary destination have to have sufficient breadth and depth of appeal to encourage visitors to stay in a single destination for many days. Holloway and Humphries (2016: 16) also note destinations must have adequate attractions to appeal to a large number of tourists and add further, the more attractions a destination has the easier it becomes to market.

Their features should be of a quality experiential character, unique, exciting, and offer one of a kind experiences that appeal to the target market (EPGC, 1995). Bull (1991) feels that attractions play one of three roles in tourism. They can be:

- an intrinsic part of the trip, in which the demand for the attraction is established before the trip commences;
- a major motivator for a trip or for selecting a destination;
- an optional, discretionary activity engaged in at a destination.

However, like most things in tourism, attractions are quite misunderstood, with Leask (2010) noting that even arriving at a universally accepted definition is a challenge and is a reflection of the vast array of issues influencing management. Issues such as classification, inclusive or exclusive definitions, size, aesthetic appeal, management regime and the target market all come into play. It is for this reason that some prefer the use of the term 'visitor attraction' instead of 'tourist attraction' and why in this book, we are using the term 'product.'

A bigger issue, though, is that the term has been misused, with the concept of what constitutes varying levels of attractions and activities conflated and therefore confused. Holloway and Humphries (2016) observe even the term 'destination' is used synonymously with attraction. This chapter sifts through many of the issues relating to the conceptualisation of attractions and how they work. It begins by trying to define what an attraction is and how the tourism attractions' system works. Then it examines how the term has been abused, so that its use encompasses far more than the original meaning. This discussion then leads to the need for some type of structured, taxonomic system to classify the various concepts of an attraction that encompasses its breadth. Finally, we look at the role attractions in the broadest sense play in drawing visitors to a destination.

Conceptualising attractions

The idea of what constitutes an attraction was discussed in our literature some 30 years ago or more but has received relatively little attention since. Scholars agree in general that an attraction consists of a single entity bounded by time or space that is visited by tourists. There have been many attempts to categorise attractions over the years. Some have focussed on defining groups and sub groups according to the traditional triad of natural, cultural and built attractions, with, for example, built attractions being divided into purpose-built and non purpose-built (Holloway, 2002). Goeldner and Ritchie (2012) identified five broad classes of attractions, including cultural, natural, events, recreation and entertainment, while Walker and Walker (2011) note a continuum of attractions from natural to built, entertainment-oriented. Douglas Pearce (1989) identified the source of attractions as including purpose-built facilities, features that are part of the regional fabric, places that have been transformed from their original function to serve tourism use or features of existing facilities that supplement or complement their original function.

If no tourists visit, then the place cannot be called a tourist attraction. Phantom demand affects many places, such as the largely abandoned National Cultural Centre in the island nation of Tonga. Phantom demand is demand that appears to exist based on either flawed research or overly-optimistic expectations of the drawing power of the attraction. In reality, little or no demand is generated. Phantom demand affects many small and medium sized cultural, heritage and eco attractions.

Three features are worth considering here. First, if no tourists visit, then it could not be called a tourist attraction (Leiper, 1990). This point seems moot, but lies at the heart of the issue of phantom demand affecting many so-called attractions, where they were built for tourists but tourists show a disdain for them. Second, the narrow geographic or temporal focus means that a region or an extended environment is not considered as a single, holistic attraction (Pearce, 1991), even though attractions may have a linear component (Denstadli and Steen Jacobsen, 2011; Hardy, 2003). By extension, a shopping district would not be considered an attraction per se, but instead, each shop might be considered as an attraction in its own right. Third, the issue of how do activities pursued fit into the equation has to be asked. Technically, activities represent the consumption of a set of like attractions, such as shopping and sightseeing and, therefore, cannot be considered an 'attraction' in themselves. But destination management organizations promote activities as attractions, suggesting that attractions may be conceptualised as existing at different levels. More on this later.

The conceptualisation of attractions began with a supply side perspective, but has only recently shifted to considering them from the demand side. Phillip Pearce (1991: 46), for example, offered a supply based operational definition of "a named site with a specific human or natural feature which is the focus of visitor and management attention." By his own definition, good scenery is not a tourist attraction per se, but a scenic lookout which is named, managed and well used is.

Leask (2008) more or less adopts this approach in her substantial volume of work on visitor attractions. Her model, shown in Figure 4.1 identifies the main attractor of visits, such as the resource in question, interpretation, events and augmented services provided as the core of the model. From here, she looks at some of the structural issues, whether it is natural or built and whether admission is free or has a charge attached to it. The ownership structure follows, with two broad types of public or private ownership. But as Page (2019) notes, four different ownership styles may be evident, including private, public, charity and voluntary ownership. Lastly, she considers the primary market or market mix. Some visitor attractions cater mostly to the local market, while others cater to a regional, national or international market.

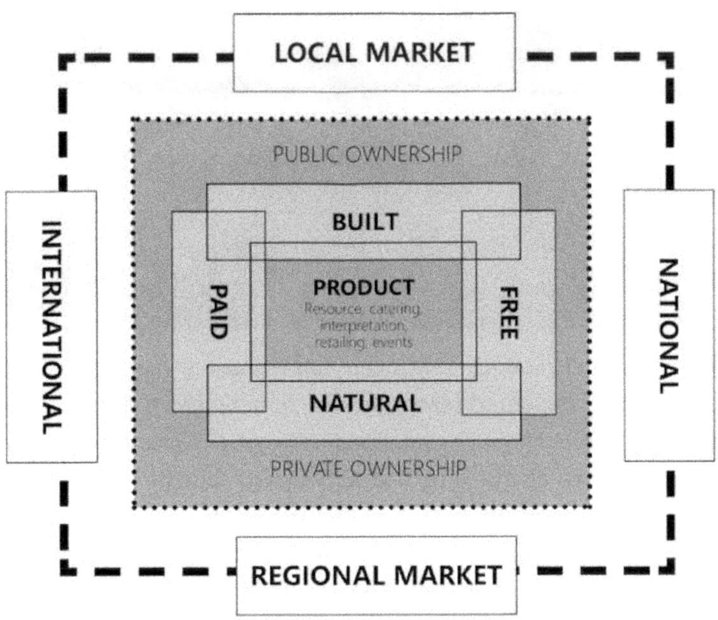

Figure 4.1: Model of visitor attractions. *Source*: from Leask, 2008.

Within the cultural tourism space, for example, Timothy (1997) identifies four different communities of cultural tourist, based on the level of connectivity to the place in question. World heritage attractions may draw large numbers of tourists that invoke feelings of awe, but they probably do not invoke feelings of deep personal attachment. National cultural assets symbolize a society's shared recollections that represent much about how it expresses its identity through valorization of durable national icons that reflect national pride. Domestic tourists may feel a sense of pilgrimage by visiting these places and paying homage to national symbols that represent their shared identity. Local and personal sites engender progressively stronger feelings of connectivity and likely facilitate different depths of experience by the visitor.

Lew (1987) also supported the supply side argument, but takes a more liberal view when he suggests attractions consist of all those elements of a 'non home' place that draw discretionary travellers away from their homes. Importantly, he disagreed with Pearce that they needed to be signed, signalled and managed for tourists, and instead suggests that the suite of potential attractions is much broader.

Traditionally, tourist attractions have been defined using a supply side perspective. They are either named or unnamed places, events or activities that draw tourists to a destination. Recently, the UNWTO has adopted more of a demand-side perspective, noting attractions are defined as much by the activities of tourists as by suppliers. This reconceptualization has fundamentally broadened the scope of what can be considered as an attraction to include, for example, people for the visiting friends and relatives' market (as above), places of business for the business tourism market, schools and universities for the educational tourist and many more.

Recently, though, the focus has shifted to considering attractions from the perspective of the consumer. Koutoulas (2015) reminds us that tourism products exist to satisfy the needs of tourists and for no other reason. He, therefore, suggests we need to look at them from the demand side. The UNWTO (2008a, b) also adopts a demand side approach when it states attractions are defined as much by the activities pursued by visitors as by the supplier, meaning that almost anything could be considered as an 'attraction' if it is used by tourists.

This change in perspective is important for it clarifies a number of ontological issues. First, it expands the idea of what constitutes an attraction beyond the traditional triad of location specific built, natural and cultural features. Instead, it recognises that any feature that attracts tourists may be considered as an attraction. Thus friends could be an attraction for the visiting friends and relative market; a business could be an attraction for a business tourist; and the like. Second, at a more generic level, it expands the idea of an attraction to include non-location-specific experiences or activities pursued by tourists. In this manner, sightseeing, exploring, shopping, dining, relaxing on a beach and many other activities are identified as attractions at some level. Even though consumption still occurs at specific site (a scenic spot, a specific shop, beach or restaurant, etc.), the collective set of activities that constitute the experience suggests that attractions may be categorised in a hierarchical or taxonomic manner from the specific (a shop) to the more general (shopping). Third, the definition acknowledges that attractions do not have to be commercial or managed, in recognition that much of what tourists consume is part of the public domain, available free of charge (McKercher, 1993). Finally, the definition eliminates scale and spatial considerations as defining variables. Attractions can be as small as someone's back yard or as large as a theme park or ski resort.

A hierarchy of attractions is seen to exist at most destinations. The most important attractions play a critical role in shaping the image of a destination and in influencing visitors to come to a region (Mill and Morrison, 2012). Many years ago, Bull (1991) argued that there is often a degree of obligation involved in visiting attractions; the more dominant the attraction, the greater the sense of obligation to visit. This case still applies today, for in many cases, the desire to visit a specific attraction may define the entire trip. On the other hand, the purchase decision becomes increasingly discretionary for lower order attractions, until visits to the lowest order ones are typified by low involvement purchases or by little effort required on behalf of the visitor. Lesser attractions complete the tourist experience by providing a range of activities for visitors while at the destination, but do little to create an image of an area or to pull visitors to it. In some instances, they simply represent a productive use of free time and nothing more (Kantanen and Tikkanen, 2006).

Traditionally, they have been conceptualised simplistically as Primary, Secondary or Tertiary attractions (Leiper 1990). Primary attractions are what draw people to the region. One of the goals of tourism planning, therefore is to try to create primary attractions that will provide it with that unique draw card. Secondary attractions generally have regional appeal. People will travel shorter distances to see them, or if they are already in the region will make a specific trip to visit the attraction. Tertiary attractions, on the other hand, have limited appeal. While primary attractions may drive development of a region, secondary and tertiary attractions are important players in completing the attraction mix and in encouraging people to stay in the region longer.

Of course, like most things in tourism this type of classification is overly simplistic, for it assumes the market is homogeneous and will therefore rate all attractions the same way. It ignores the fact that tourists consume attractions to satisfy their needs and that needs vary depending on the segment or even individual involved. Mill and Morrison (2012) further recognize that attractions appeal to different groups of tourists in a different manner. Certain attractions can be regarded as primary attractions for one segment of the travel market, but as lower order attractions for another group of travellers. These groups will view the area's attractions differently.

Moreover, needs vary. In some cases, needs are singular and specific, meaning that the likely range of attractions that can satisfy the needs is also small. On the other, people often travel to have many broad and unspecified needs met, such as sightseeing, spending time with family or escape. In such situations, it may be difficult to ascribe visitations to any specific single attraction or attractions. Instead, the perceived existence of more generic attraction classes or even attraction families that can satisfy needs may be the level of specificity that people want.

Tourist attractions systems

The preceding discussion suggests that attractions function as stand-alone entities. However, renowned tourism scholar Neil Leiper (1990) challenged this idea and argued that a tourist attraction does not exist in isolation, and instead functions as a component of a tourist attractions system. The system includes three elements: the tourist; the attraction or nucleus; and some type of marker or catalytic element that connects the tourist to the attraction or the attraction to the tourist. His model is shown in Figure 4.2. This contribution has not garnered the attention it deserves, but it is none the less a very important conceptual contribution to the field. Each of the components is discussed briefly.

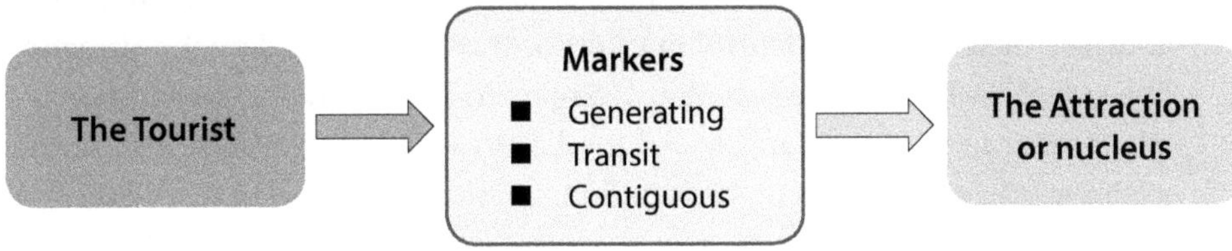

Figure 4.2: The Tourist Attractions System. *Source*: Leiper, 1990.

The tourist

Like everything else in tourism, the tourist lies at the heart of the tourist attractions system, for without tourists, places would not be thought of as tourist attractions. Leiper suggests the essence of tourist behaviour involves a search for satisfying leisure experiences that can only be met by searching for suitable attractions or experiences. This quest involves looking for various signs, symbols and other markers that signify the place as something worth visiting (Leiper, 1990:371).

The nucleus

The nucleus or specific attraction is "any feature or characteristic of a place that a traveler contemplates visiting or actually visits" (Leiper, 1990: 372). Because tourists' needs and wants are individualistic, what constitutes a nucleus is also equally individualistic. Nuclei can have varying geographic scales, consist of a temporal dimension and exist along a hierarchy of importance to the individual. Leiper suggests the geographic scale can range from a single entity to touristic precincts where a range of attractions and activities are bundled in close proximity. The temporal dimension relates to short duration events that attract tourists, such as festivals, trade shows and the like. The hierarchical consideration relates to the importance tourists place on any given nucleus. A primary nucleus is the key feature that draws tourists to an area, while a secondary nucleus is known pre-visit but does not influence the visit decision, and a tertiary nucleus is unknown pre-visit but is visited none the less.

Markers

It is the marker, though, that is the most important feature of the tourist attractions system for it represents the catalytic link between the tourist and the attraction that triggers motivation (Clark, 2009). Markers are defined as "items of information about a phenomenon that is a potential nuclear element in a tourist attraction (Leiper, 1990: 377)." The contribution of earlier work by MacCannell (1976) is acknowledged. Culler (1988: 5) defines markers as "any kind of information or representation that constitutes a sight as a sight: by giving information about it, representing it, making it recognizable." MacCannell (2001) adds that markers signal the sight as something worth seeing.

What constitutes a marker is quite broad as Wong and McKercher (2013) summarise. Most of us can identify markers as those developed by the attraction itself, the travel trade and travel media. These include advertising, promotional materials, press releases, web and social media presence, and others. But, markers can also include other forms of information, including common knowledge that is not attributable to any single source; movies, novels and magazines (Mercille, 2005); intangible heritage associated with destinations (Bessiere, 1998); information provided by key informants, including staff at tourist information centers (Wong and McKercher, 2013); souvenirs that remind the individual and others of the sight (Culler, 1988); word-of-mouth from other tourists; lifestyle television programs; and increasingly the web and social networking sites (Tussyadiah and Fesenmaier, 2009). Interestingly, more seems to have been written about non-commercial markers than about commercial ones, as attested by the citations above.

Markers are arguably the most important component of the tourist attraction system, for they signal a place as being worthy of visiting. They can take many forms including those produced by the commercial tourism sector as well as a wide array of non-commercial markers. Two powerful markers are shown here at the Temple of Horus in Edfu, Egypt, Souvenirs provide a meaningful memory for tourist by marking a place as having been visited, but also act as a marker for potential future visitors. Crowds of people are often the most powerful on-site marker. Tourists often visit places only once and do not wish to miss highlight attractions. Crowds signal a place as something worth seeing.

In fact, markers may be more important than the attraction itself (MacCannell, 2001) for without them, the tourist would not become aware of the attraction. Markers, therefore, stimulate the tourists' desire to look for experiences that match their needs and wants. Clark (2009) reminds us that markers often contain information or present an image about what might be experienced at the sights targeted. As a result, a key goal of markers is to act as image enablers (Jenkins, 1999). Effective attractions systems work when markers stimulate visitation, while the system may become defective when no markers are present, when they send conflicting messages or do not encourage visitation.

Leiper suggests markers can be found in three locations, in generating markets, in transit and in the destination (or as he labels them, contiguous markers). It is simpler, though, to consider them as existing at two levels, those found prior to arrival and those available once in the destination. Generating markers seem to be most important for large scale attractions that can afford to have a presence in source markets (Litvin and Mouri, 2009). They also play a key role in behaviour for tourists decide what they want to do prior to departure, and if their time is limited are unlikely to modify their behaviour much (Shoval and Raveh, 2004). Richards (2002), for example, found almost half of all cultural tourists surveyed indicated their decision to visit cultural sites was made prior to departure.

Lower order attractions have to rely on in-destination markers for their owners/managers do not possess the resources to engage in awareness building through generating markers (McKercher and Lau, 2007). Their inability to reach source markets prior to departure means that they are disadvantaged, as people who are unaware of the activity are unlikely to participate unless they happen to stumble across it on their way to other attractions.

The role of markers is variable. Tourists visit a destinations' most popular attractions, irrespective of whether they are receptive to generating or in-destination markers. Those who are unaware of major attractions prior to arrival, become aware of them and, more importantly, visit. By contrast, the conversion rate for planned visits to middle or lower order attractions declines, irrespective of pre-departure awareness levels. Inertia appears to play a critical role in determining the behaviour of tourists. The most popular attractions will be visited because of their popularity, while lesser attractions will generate fewer visits by the simple virtue of their lower popularity. Thus, at a crude level, pre-departure awareness levels may provide the best indicator of likely visitation rates among tourists.

A taxonomy of attractions

The concept of an attraction has moved a long way from its humble origins as a specific entity bounded by space or time into a brand category that includes a vast array of like entities categorised under broad umbrella concepts. In fact, the term 'attraction' is used so loosely now that its meaning can include anything that tourists may consume from a specific product to an entire category of products or activities that encompasses but extends well beyond the specific. As an example, an historic site is classified as an attraction in the traditional sense of the word. But, by the same token, heritage, consisting of many historic sites, archeological sites and historic landmarks scattered around a destination, and even the more generic term 'culture' which includes not only historic sites, but other forms of tangible and intangible heritage, have also gained legitimacy as 'attractions'. Yet, historic sites are constituent elements of heritage and heritage is a constituent element of culture. As one moves from the specific the general, then, the set of attractions, activities and entities that can be included grows exponentially. Conversely, as one moves from the general to the specific, what constitutes an attraction becomes more and more clearly defined.

Two challenges emerge. First and foremost, consumption still occurs at the specific attraction, event or entity level. Thus, while a destination may sell 'culture' or 'heritage' as part of its attraction mix, in reality, the tourist will not consume 'heritage' but instead will consume specific sites under the subset of heritage. But having said this, the second consideration is: do we know what motivated the tourist to visit the destination in the first place? In some cases, it may be the specific attraction, while in many other cases, it will be the more generic attraction sets of, say, 'heritage' or 'culture.' In the latter case, the specific attraction may be relatively unimportant in the visit decision, providing enough opportunities exist to consume different heritage or cultural places.

So, the long and short of it is that attractions may be seen to exist in a hierarchical form from the specific site at the bottom to sets of like attractions, to sets of similar attractions, to sets of thematically related attractions, and so on. The set of thematically similar attractions and attractions sets increases significantly as one moves up the classification system, and likewise decreases significantly until reaching a single item as one moves down the system. The classification of each tier of the hierarchy may be conceptually valid.

It is worth repeating that tourism remains a fragmented and disjointed activity (Benckendorff and Zehrer, 2013; Echtner and Jamal, 1997; Laws and Scott 2015) typified by multiple communities of discourse with historically little cross fertilization between communities (Gibson, 1998). Nowhere is this issue more true than with attractions, where the need to develop a classification system has been long recognised (Mehmetoglu, 2004), but not acted upon. Instead, a series of independent studies examining, different product types has been undertaken, such as Young et al.'s (2007) classification of VFR tourism with overlap, gaps and no internal cohesion (McKercher, 2016a)

In response, McKercher (2016a) proposed a taxonomy of tourism products in order to add some structure to the understanding and classification of attractions. His system is described below. Taxonomic systems are empirically based and classify items using observable and measureable characteristics. They are formed based on the idea of a natural classification, where members of each tier of a taxonomy are on the average more similar to each other than they are to members of other taxa at corresponding levels (Sokal, 1986). Reasonably homogenous groups are formed with reasonably sharp boundaries. It is for this reasons that the use of a common vocabulary is important (Hansman and Hunt, 2004; Hedden, 2010). Essentially, then a taxonomy is a hierarchical classification system to indicate natural relationships between items at different levels (Fenneman, 2013). Taxa are arranged vertically to reflect increasing specificity, with the top tier encompassing a broad comprehensive category that captures the essence of all groups and sub groups that fall under it. Lower level tiers are disaggregated into one or more narrower cohorts, until one reaches the bottom where individual entities are found (anon, 2016).

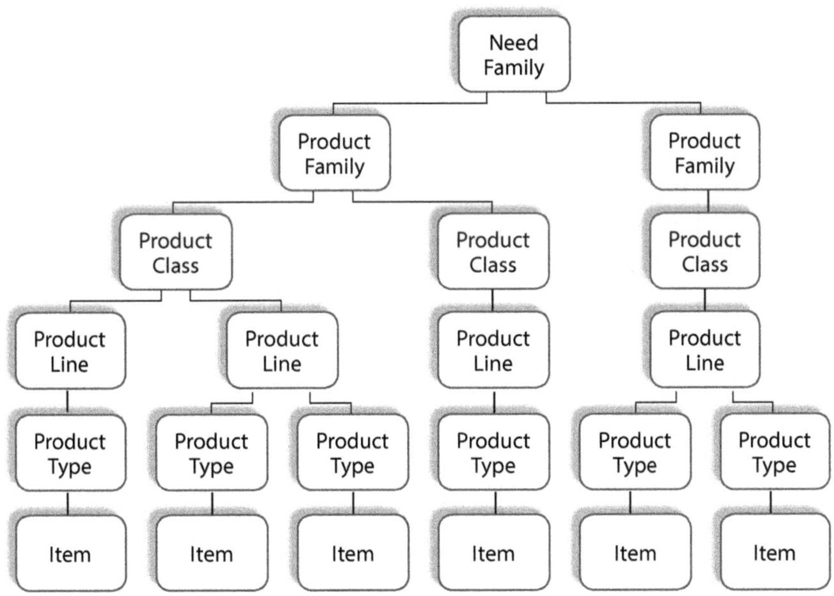

Figure 4.3: Product taxonomy. *Source*: McKercher, 2016a, from Kotler and Keller, 2012.

A modified version of Kotler and Keller's (2012) marketing-oriented product taxonomic classification system was applied. It is shown in Figure 4.3, and is based on the ability of a product to satisfy a consumer's need. At the top is a broad Need Family (for example the need to quench one's thirst). As one moves down through the taxonomy one gets increasingly specific. For example, the

Product family may include, dairy, sodas or juices. All subsequent products then fall under these families. Using sodas as an example, the Product Class might be flavoured sodas (colas, non-colas, etc.), while the Product Line may be sugared and sugar free. The Product Type could be the various brands, while the Item may be different sizes available. Thus, one can go from the broad family to the specific item by reading down the taxonomy or see the relationship between specific items and the need family by reading up.

Figure 4.4: Partial taxonomy of tourism products. *Source*: McKercher, 2016a.

A similar structure has been proposed for tourism attractions that again link broad Need Families at the top with specific items or attractions at the bottom, or conversely link specific items at the attraction at the bottom to more generic needs at top. The idea is that the more specific a need is, the fewer attractions exist that can satisfy that need. For example, if your need was to have a historic theme park experience, then your options would be limited in most destinations. But, if you need was more general, then a vast array of built attraction Product Families could exist to satisfy that need.

Five broad attraction Need Families are identified. They include Pleasure, Personal Quest, Human Endeavour, Nature and Business. The first three layers of the taxonomy, encompassing the Need Family, Product Family and Product Class are shown in Figure 4.4. Pleasure captures most of what we could consider as encompassing leisure tourism. Personal quest reflects more meaningful touristic experiences people seek. Human endeavour captures the full range of cultural and heritage activities, the creative arts as well as dark tourism. Nature looks at passive and active outdoor recreational opportunities, while business is self-explanatory.

Where does this distillery in Lake Schliersee, Germany fit into the attractions taxonomy? The term 'attraction' is one of the most misused in tourism, for it has been used to define anything from a single entity to a broad family of products. A taxonomy of tourism products has been developed to attempt to add some structure to our categorisation. It is based on Kotlerand Keller's (2012) product taxonomy and tries to order attractions and/or products into family trees. Thus, at the bottom or most specific level of the family tree, this distillery could be considered as an Item (specific distillery). But it also can fit into a Product Type (whisky), Product Line (distilleries), Product Class (drink), Product Family (food and drink) or Need Family (pleasure). The nature of the product becomes more clearly defined as one moves from the Need Family down through the taxonomy to the Item and increasingly inclusive as one moves from the Item up to the Need Family.

Figure 4.5 gives a more detailed description of the Pleasure Need Family by drilling down to more specific layers. The categories shown here begin with the Product Family, and then show examples of the Product Class, Product Line and Product Type. Individual Items are too numerous to mention here but represent the specific temporal or physical space consumed.

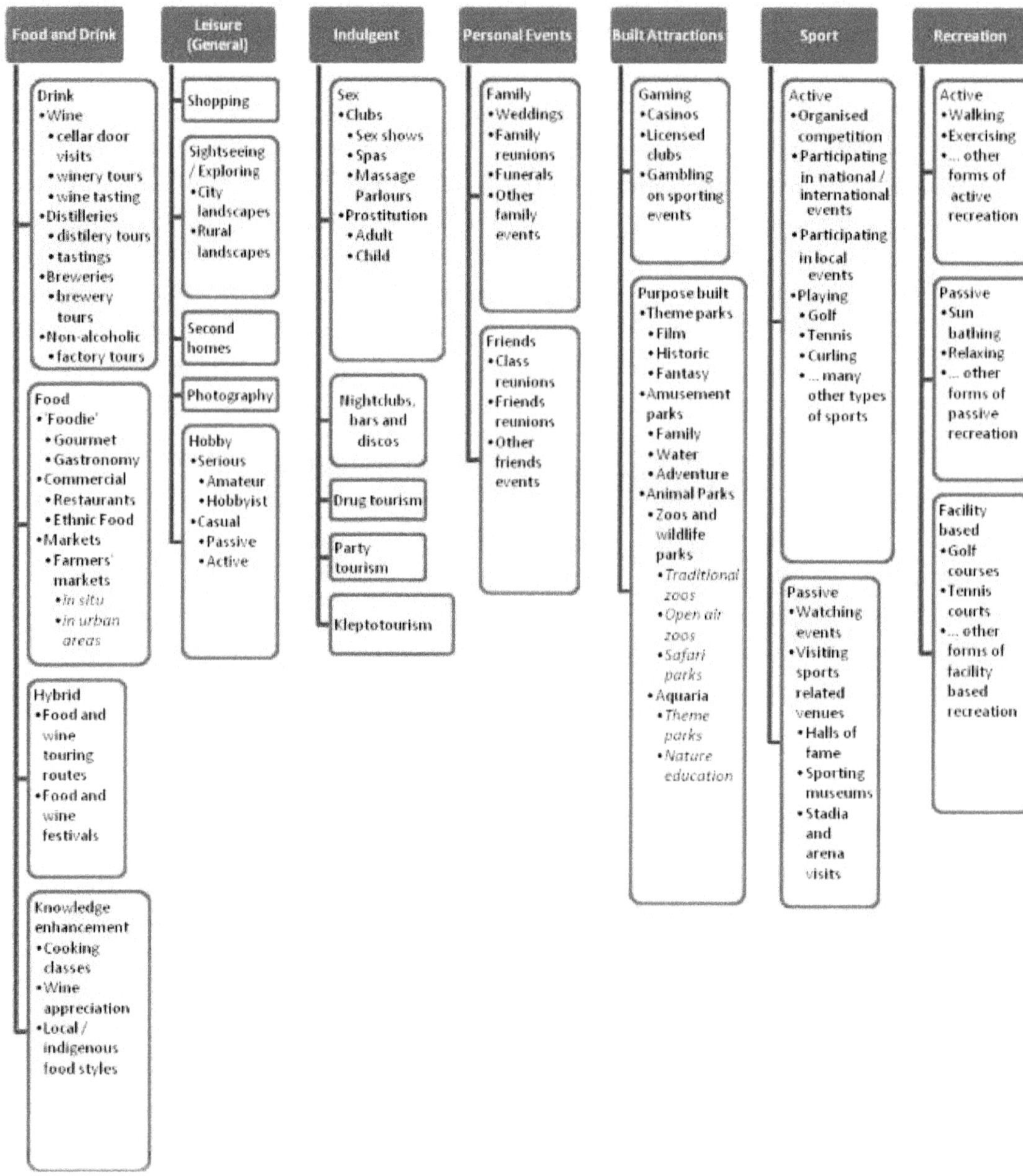

Figure 4.5: Detailed 'pleasure' need family taxonomy. Source: McKercher, 2016a.

Why is such a system important? To begin, it adds some clarity to what we normally call attractions by recognising they can be conceptualised at different levels of specificity. More importantly, it recognises that consumers' needs vary as do their motives for visiting places. A 'foodie' may be interested in sampling the culinary delights of a region, without reference to any specific place. As such his or her needs sit at the top of the taxonomy, and further mean that a wide array of attraction Classes, Lines and Types may satisfy their needs. By contrast, someone who seeks only Michelin starred restaurants has much more specific needs that can be satisfied by fewer choices. At an extreme someone may travel to a destination because of a specific attraction, be it a restaurant or a culinary festival. In this case, the need is precise and can only be satisfied at the item level.

Ranking attractions by needs and proximate cause

The taxonomy also raises the fairly simple but profoundly important question of do attractions actually attract tourists? Do people come to a destination because of a specific attraction or because of a collection of like attractions (i.e. shopping or dining), where the individual attraction plays little or no role in the visit decision? This question is important for destinations strive to create attractions under the assumption of "if you build it they will come".

In reality, tourists' needs are quite diverse. What constitutes an attraction to them may also be quite diverse. Two features are critical when considering the role of attractions in the visitation decision: proximate cause and the relationship between needs and level of the attractions' hierarchy.

In law, proximate cause (*sine qua non* or 'but not for') considers whether an event would have occurred without this action happening. It is defined as the nearest cause of an event (CILA, 2012) and can be considered as the active, direct, and efficient cause of something that sets in motion an unbroken chain of events (BD, 2016). In insurance terms, *sine qua non* means that if you had not driven through a red light, the accident would not have occurred. Thus, your actions were the proximate cause of the accident and you should be held liable for it. It does not matter if the other driver could have stopped but did not.

Why is it important in tourism? *Sine qua non* distinguishes the difference between what people do in a destination and why they come. Destination management organisations are much more interested in why people come and less so than what they do. The two may be quite different, for we do a lot of things that may have little to do with the reasons why we visited. Here is an example. One of the authors was invited to a wedding of a former student in Thailand. While there, he attended the wedding, visited royal palaces, the ancient capital, did some sightseeing, shopping and dining and generally was very active during the stay. In this case, though, *sine qua non,* or the 'but not for', the wedding he would not have visited, and therefore, would not have done all the others things that made it an enjoyable trip.

Sine qua non is never tested in visitor surveys, other than asking generic questions about trip purpose (usually limited to business, pleasure, visiting friends and perhaps a special event) and maybe some generic motivational questions. But it should be, for it is important to know why people come. Visiting popular attractions represents what we do, but does not necessarily act as a proxy for why we come. Proximate cause, therefore, distinguishes the difference between what features of the destination draw people from what they did. The 'attraction' or 'attractions' that is/are the cause of why people come then becomes the primary attraction or attractions, while everything else becomes a lower order attraction that complements the trip. This concept aligns closely with the UNWTO's conceptualisation of an attraction as anything that tourists do in, or draw them to, a destination. Considering proximate cause makes, for example, friends and family the primary attraction even if people are incredibly active in a destination, a specific business that someone visits during the source of a business trip, a specific meeting or conference, or something more general.

This observation also raises the importance of matching people's needs to the specific tier of the aforementioned attraction's hierarchy, for needs do not have to be specific. Tourists needs vary from the generic to the specific and usually include a variety of both. Indeed, the possible range of needs is as equally vast at one level and as it is highly individualistic at another. Stebbins and Hartel (2006), writing about serious leisure suggest the trip may be motivated to have a single, specialist need met. The same applies to most forms of business travel. Others though suggest most travel decisions are influenced by a basket of needs (Herbert, 1996; Peters and Weiermair, 2000) that may comprise of a mix of specific and generalized needs (Mannell and Iso-Ahola, 1987; Dunn Ross and Iso-Ahola, 1991) or simply generalized needs (Crompton, 1979).

It is likely that specific needs can only be satisfied by the type of specific product found at the base of the taxonomy, whereas more generalist needs can be satisfied by attraction sets found at the higher order of the taxonomy. Where a combination exists, the balance between specific and generalist attractions comes into play. Alternately, when needs are general or unspecified, almost any destination and any activity within a destination can satisfy them potentially (Caccomo and Solonandrasana, 2002; Botti, Peypoch and Solonandrasana, 2008). Indeed, Crompton (1979) determined that particular attributes of destinations are relatively unimportant if people traveled to have broadly based social psychological needs of regression, escape and relationship development met.

Let's offer another example to explain this concept. Say you 'need' a holiday and just want to relax by escaping your home and exploring a destination, but have no deeper desire to travel. Your needs are very generic and therefore, almost any destination that fits your financial and time budgets as well as interests can satisfy this need. You may be very active while travelling, but it does not matter what you do as long as the desire for an escape is met. Hence, broad Need or Product Family sets of attractions can satisfy your need. But, if you are a film buff and want to visit a certain film festival before the films are released commercially, your needs are specific and can only be satisfied by either a product type (film festival) or maybe even a single product item (Cannes).

A model linking the specificity of need with the tier of the tourism attractions taxonomy to better understand the role that individual attractions play in the overall travel decision has been proposed (McKercher, 2016b). Ultimately, all touristic consumption occurs at the Item level of the product taxonomy. You cannot go 'shopping' unless you visit specific shops or pursue heritage without visiting heritage sites. But, thinking of proximate cause, was it the opportunity to shop at any number of stores that attracted the person or was it the chance to shop at a specific store that drew her or him to the destination? The key point becomes one of determining what role, if any, the individual attraction played in driving visitation. In some cases, likely when needs are specific, the individual attraction is key. Were it not for that attraction, the person would not be visiting.

Equally, the importance of individual attractions diminishes as specificity of need diminishes. If you travelled for escape or to spend quality time with your family members, then specific attractions may have played little or no role in the trip decision. Instead, the decision would have been made based on whether a bundle of like attractions or even a group of attraction sets (pretty scenery, lots to see and do) would have influenced the decision.

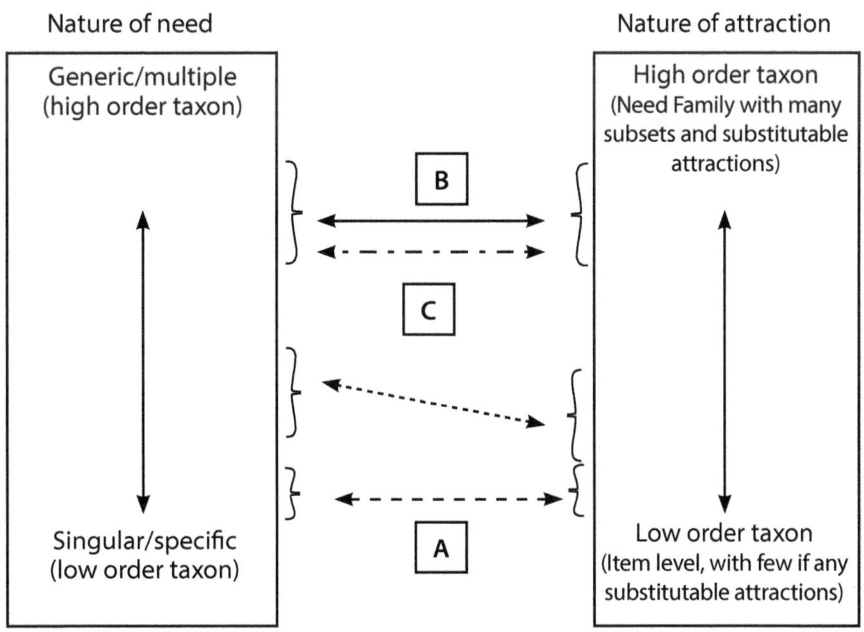

Figure 4.6: Matching a hierarchy of needs with the product taxonomy. *Source*: McKercher, 2016b.

Figure 4.6 shows this relationship. A specific need (example *A*) can only be met by a specific attraction. Case *C* shows how a mix of specific and generalist needs can be met by both attractions at a low order of the taxon as well as more generalist high order attraction sets. 'Bleisure' is a common example, where the person comes for specific business reasons and then extends the trip for less well defined pleasure reasons. Here, the belief that the destination can offer sufficient range of highly substitutable shopping and dining opportunity sets, in aggregate, will satisfy needs. Lastly, example *B* shows the common situation where one travels for more generalized needs of escape, exploration, spending time with family and friends and the like. Here, the specific attributes of the destination are less important than the mix of activities it offers.

This model shifts the conceptualisation of what constitutes primary, secondary or tertiary attractions from the traditional focus on the specific attraction itself to the tier of attraction that draws tourists. If needs are specific, then the Item level becomes the primary attraction. But, if needs are more generic, then the nature of the primary attractions shifts higher up the taxonomy.

Conclusion

Surprisingly little research has been conducted on attractions even though they are the focal point of tourism. This chapter reviewed the conceptualisation of attractions and an attractions' hierarchy before proposing a novel system to consider a continuum of attractions that could be organised in a taxonomic structure. This taxonomy also helps us understand better the role that single attractions play in tourism and explains better the link between motive and attraction. The idea will be developed further in Chapter 10 when motives are discussed.

One of the ironic features of tourism is that the vast tourism infrastructure exists to service people's desire to consume attractions, most of which are available free of charge or at a very low price. It costs nothing to walk down a street and experience a city's vibrancy, sights, sounds and smells. It costs nothing to enjoy views, sit on a beach, wander into a shop or spend time with friends and family. Entry into many publicly funded attractions is highly subsidised so one only pays a fraction of the actual cost of visitation. Indeed, the vast majority of what the average tourist spends is for amenities and services that facilitate the consumption of attractions and not for the attractions themselves. Few people in the tourism industry actually work in tourism attractions. Most of them work for businesses that facilitate, or make possible, the opportunity to visit attractions. Hotels, airlines, tour operators, and restaurants are not attractions. Yet, without these products tourists could not visit attractions.

Discussion points / Exercises

1. Select a major attraction in your area and document all types of markers that signify the attraction as being worthy of a visit. Make sure you include both commercial and non-commercial markers. Be as creative as possible.
2. Select a tourism Product Family of your choice and fill in the taxonomic structure by identifying the different Classes, Lines and Types as well as examples of specific products under each Type. Analyse the table to identify logical linkages upward and downward through the taxonomy
3. What is the proximate cause for pleasure tourists to visit your region? In doing so, identify where attractions would fit in the taxonomy to match their needs.

Points to consider for revision

- Like most things in tourism, it is extremely difficult to define an attraction precisely. Historically, efforts have focussed on a supply side approach by categorising or grouping like attractions and developing suitable models. More recently, a demand side approach has determined that almost anything a tourist does in a destination can be considered as an attraction, which broadens the concept substantially.
- Attractions do not operate in isolation. Instead, they are part of an attractions system that includes the tourist, the nucleus and markers. In many ways, markers may be the most important element of the system for they act as the catalyst that links the consumer to the product.
- The term 'attraction' or product has been used very loosely in tourism. In reality, attractions/products can be organised in a taxonomic structure that groups them according to their inter-relationships. At the top of the taxonomy is the broad Need Family, and then as one progresses downward, the taxonomy becomes more specific until one reaches individual Items. Conversely, working upward, one can go from the specific to the general.
- Proximate cause is a key element for any destination management organisation. Understanding the reason why people come (*sine qua non* – but not for) is far more important than documenting what they did.
- In many cases individual attractions play little role in the decision to visit multi product destinations. Instead, attraction sets comprised of like attractions play a more important role.

Additional recommended reading

Leask A. (2014) Visitor attraction management: A critical review of research 2009 – 2014. *Tourism Management*, 57, 334-361.

Leiper, N. (1990) Tourist attractions systems *Annals of Tourism Research*, 17(3), 367-384

McKercher, B. (2016) Towards a taxonomy of tourism attractions. *Tourism Management*, 54, 196 -208.

McKercher. B. (2016) Do attractions attract tourists? a framework to assess the importance of attractions in driving demand. *International Journal of Tourism Research.* DOI: 10.1002/jtr.2091

References

Anon (2016) Definition for three types of theme parks. http://www.familythemeparkvacations.com/three-types-of-theme-parks.html <downloaded Dec 2, 2019>.

BD (2016) Business Dictionary.com, http://www.businessdictionary.com/definition/proximate-cause.html <downloaded

Benckendorff, P. & Zehrer, A. (2013) A network analysis of tourism research. *Annals of Tourism Research*, 43, 121-143.

Bessiere, J. (1998) Local development and heritage: Traditional food and cuisine as tourist attractions in rural areas. *Sociologia Ruralis*, 38(1), 21-34.

Botti, L., Peypoch, N. & Solonandrasana, B. (2008). Time and tourism attraction. *Tourism Management*, 29(3), 594-596.

Bull, A. (1991) *The Economics of Travel and Tourism*, Pitman, Melbourne

Caccomo, J. & Solonandrasana, B. (2002) Réflexions autour du concept d'attraction touristique: analyse et taxonomie, *Revue de Recherche en Tourisme UQAM*, 3, 68-71

CILA (2012) *Introduction to the Insurance Industry*. Chartered Institute of Loss Adjusters. www.cila.co.uk/files/Certificate/Chapter%208.pdf <downloaded Mar 10, 2016.>

Clark, I. (2009). Naming sites: Names as management tools in indigenous tourism sites – An Australian case study. *Tourism management*, 30(1), 109-111.

Crompton, J. L. (1979) Motivations for pleasure vacations. *Annals of Tourism Research*, 6(4), 164-182.

Culler, J. (1988) The semiotics of tourism. In J. Culler, *Framing The Sign: Criticism and Its Institutions* (pp. 1-10). Norman: University of Oklahoma Press.

Denstadli, J. & Steen Jacobsen, J. (2011) The long and winding roads: Perceived quality of scenic tourism routes. *Tourism Management*, 32(4), 780-789.

Dunn-Ross, E. & S. Iso-Ahola (1991) Sightseeing tourists' motivation and satisfaction. *Annals of Tourism Research*, 18(2), 226-237.

Echtner, C. & Jamal, T. (1997) The disciplinary dilemma of Tourism Studies. *Annals of Tourism Research*, 24(4), 868-883.

EPGC (1995) *Packaging for and Selling to the United States Pleasure Travel Market*, Economic Planning Group of Canada, for Tourism Canada.

Fenneman J. (2013) What is taxonomy and where did it originate. *E-Flora BC* www.geog.ubc.ca/biodiversity/eflora/IntroductiontoPlantTaxonomy.html <downloaded Oct 6, 2013>.

Gee, C., Makens, J. & Choy, D. (1989) *The Travel Industry 6th E.*, Van Nostrand Reinhold, Melbourne.

Gibson, H. (1998) Sport Tourism: A critical analysis of research. *Sport Management Review*, 1, 45-76.

Goeldner, C. & Ritchie, B. (2012) *Tourism Principles, Practices and Philosophies*. Hoboken: John Wiley and Sons.

Hansman, S. & Hunt, R. (2004) A taxonomy of network and computer attacks. *Computers and Society*, 24(1), 31-43.

Hardy, A. (2003) An investigation into the key factors necessary for the development of iconic touring routes. *Journal of Vacation Marketing*, 9(4), 314-330

Hedden, H. (2010) *The Accidental Taxonomist* Medford: Information Today.

Herbert D. T. (1996) Artistic and literary places in France as tourist attractions. *Tourism Management*, 17(2), 77-85

Holloway, J (2002). *The Business of Tourism*, Prentice Hall, Upper Saddle River.

Holloway, J. & Humphries, C. (2016) *The Business of Tourism, 10th Ed.* Harlow: Pearson.

Jenkins, O. (1999) Understanding and measuring tourist destination images. *International Journal of Tourism Research*, 1, 1-15.

Kantanen, T. & Tikkanen, I. (2006). Advertising in low and high involvement cultural tourism attractions: Four cases. *Tourism and Hospitality Research*, 6(2), 99-110.

Kotler, P. & Keller, P. (2012) *Marketing Management 14th Ed.* Toronto: Prentice Hall.

Koutoulas, D. (2015). Understanding the Tourism Product. 10.13140/RG.2.1.2250.4806. Interim symposium of the Research Committee on International Tourism of the International Sociological Association 14-16 May, University of the Aegean, Mytilini, Greece

Laws, E. & Scott, N. (2015) Tourism research: Building from other disciplines. *Tourism Recreation Research*, 49(1), 48-58.

Leask A. (2014) Visitor attraction management: A critical review of research 2009 – 2014. *Tourism Management*, 57, 334-361.

Leiper, N. (1990) Tourist Attractions Systems. *Annals of Tourism Research*, 17(3), 367-384.

Lew, A. (1987) A framework of tourist attractions research. *Annals of Tourism Research*, 14, 553-575.

Litvin, S. & Mouri, N. (2009) A comparative study of the use of 'iconic' versus 'generic' advertising images for destination marketing. *Journal of Travel Research*, 48(2), 152-161.

MacCannell, D. (1976) *The Tourist: A New Theory of the Leisure Class*. New York: Schoken Books.

MacCannell, D. (2001) Tourist Agency. *Tourist Studies*, 1(1), 22-37.

Mannell, R. & Iso-Ahola, S. (1987) Psychological nature of leisure and tourism experience. *Annals of Tourism Research*, 14(3), 314-329.

McKercher, B. & Lau, G. (2007) Understanding the movements of tourists in a destination: Testing the importance of markers in the tourist attraction system. *Asian Journal of Tourism and Hospitality Research*, 1(1), 39-53.

McKercher, B. (1993) Some fundamental truths about tourism: understanding tourism's social and environmental impacts *Journal of Sustainable Tourism*, 1(1), 6-16.

McKercher, B. (2016a) Towards a taxonomy of tourism attractions. *Tourism Management*, 54, 196-208, http://www.sciencedirect.com/science/article/pii/S0261517715300431.

McKercher. B. (2016b) Do attractions attract tourists? A framework to assess the importance of attractions in driving demand. *International Journal of Tourism Research*, 19(1), 120-125.

Mehmetoglu, M. (2004) Tourist or traveller? A typological approach. *Tourism Review*, 59(3), 33-39.

Mercille, J. (2005) Media effects on image: The case of Tibet. *Annals of Tourism Research* 32(4), 1039-1055.

Mill, R. & Morrison, A. (2012) *The Tourism System 7th Ed.* Dubuque: Kendall Hunt.

Page, S. (2019) *Tourism Management 6th Ed* Oxon: Routledge.

Pearce, D. (1989) *Tourist Development 2nd Ed.* Harlow: Longman.

Pearce, P. (1991) Analysing tourist attractions *The Journal of Tourism Studies*, 2(1), 46-55.

Peters, M. & Weiermair, K. (2000) Tourist attractions and attracted tourists: How to satisfy today's 'fickle' tourist clientele? *Tourism Studies*, 11(1), 22-29.

Richards, G. (2002) Tourist attraction systems: Exploring cultural behaviour. *Annals of Tourism Research*, 29(4), 1048-1064.

Shoval, N. & Raveh, A. (2004) Categorization of tourist attractions and the modeling of tourist cities: Based on the co-plot method of multivariate analysis, *Tourism Management*, 25(6), 741-750.

Sokal, R. (1986) Phenetic taxonomy: Theory and methods. *Annual Review of Ecology and Systematics*, 17, 423-442.

Stebbins, R. & Hartel, J. (2006) The serious leisure perspective. www.seriousleisure.net/ <downloaded Feb 16, 2016>.

Timothy D. (1997) Tourism and the personal heritage experience, *Annals of Tourism Research*, 24(3), 751-754.

Tussyadiah, I. & Fesenmaier, D. (2009) Mediating tourist experiences: Access to places via shared videos. *Annals of Tourism Research*, 36(1), 24-40.

UNWTO (2008a) *International Recommendations for Tourism Statistics 2008 Draft Compilation Guide. Madrid: UNWTO.* http://unstats.un.org/unsd/tradeserv/egts/CG/IRTS%20compilation%20guide%207%20march%202011%20-%20final.pdf <downloaded Dec 2, 2019>.

UNWTO (2008b) *International Recommendations for Tourism Statistics 2008 Madrid: UNWTO.* http://unstats.un.org/unsd/tradeserv/IRTS%202008%20edited%20whitecover.pdf <downloaded Dec 2, 2019>.

Walker, J. & Walker, J. (2011) *Tourism Concepts and Practices.* New York: Prentice Hall.

Wong, C. & McKercher, B. (2013) Web markers for various tour products: The case of Hong Kong. *Tourism Management Perspectives*, 8, 126-130.

Young, C., Corsun, D. & Baloglu, S. (2007) A taxonomy of hosts visiting friends and relatives. *Annals of Tourism Research*, 34(2), 497-516.

5 Access and the Spatial Interactions of Tourists

By the end of this chapter, the reader will be able to:

- Distinguish between the resource and commodity values of time and analyse how they affect travel patterns
- Describe distance decay and analyse how it influences both absolute demand and behaviour of tourists
- Define the Effective Tourism Exclusion Zone and analyse its impact on tourism flows
- Evaluate segment transformation and the impact of distance on behaviour
- Define market access and assess how it influences tourist behaviour
- Explain different itinerary models

Introduction

While attractions drive tourism, access, or the ability of visitors to get to and travel within a destination, plays a key role in its success. If tourists cannot get to a destination, then they cannot experience the attractions therein. If they cannot move around the destination easily, their actions will be constrained. Moreover, if access is limited, the ability of the destination to provide viable products is also limited. This chapter examines the role that access plays in tourism.

Time

Before we discuss access, it is important to introduce the concept of time, for tourist flows can be seen as a function of the relationship between time and space (Leiper, 1990), with time, arguably, being the most critical factor affecting movements. As Pearce (1988) notes, time availability can directly constrain or expand the number and range of potential activities available and the depth at which individual activities can be experienced. Indeed, it is about the only absolute in tourism and is also the scarcest resource (McKercher and Lew, 2003). Usually the amount of time one has is fixed either by the necessity to return to work or by flight reservations and other itinerary constraints. It is rare for people to extend their holidays, although tourists can modify plans during their holidays if, for example, they encounter inclement weather (Becken and Wilson, 2013) or other factors cause plans to change. Time is even more constrained among business travelers.

Time cannot be saved or stored for future use. It can only be spent. However, the tourist has a great deal of discretion in how he or she chooses to spend time. From the perspective of tourist movements, time expenditure usually involves some trade-off between time spent in transit versus time spent in the destination or at desired attractions. How this trade-off is made depends both on how one values the act of travelling as well as the absolute amount of time available. Interestingly, the expenditure of time was discussed in detail 30 years ago but seems to have been forgotten more recently in this era of empirically driven approach to tourism research. Authors such as Chavas, Stoll and Sellar (1989), McKean, Johnson and Walsh (1995), and Walsh, Sanders and McKean (1990) were among the first to examine it, yet, as Hall (2005) discuses, it remains an underexplored, though vital area of tourism.

Essentially, travel time can be perceived in one of two ways. On the one hand, it can have a commodity or utility value, where the act of travelling has value in itself and where the journey to the destination is as or more important than being at the destination. People are willing to trade off time spent at the destination for travel time. On the other hand, it can be seen as a cost or a resource where its expenditure is calculated based on an opportunity/cost manner (Chavas et al., 1989). If the perceived cost outweighs the opportunity, travel time will be forsaken for more time spent at the goal. Here the trade off is seen as having a negative effect on the overall trip experience. Table 5.1 compares the two concepts.

Table 5.1: Different valuations of time

	Commodity	**Resource**
Value of time	Opportunity	Cost
Transit time	Has value in and of itself	Little value, needs to be minimized
Type of tourist	Process oriented	Goal oriented
Perceived amount of time available	Much	Little
Trade-off	Time spent getting to the destination means less time at the destination	Time spent at the destination at the cost of time spent travelling
Typical trip	Long haul Multiple destination Touring	Short haul Single destination

In a sense, therefore, one could argue that trips where time is seen as being a commodity are process driven where the journey is as important as the destination. People who are more likely to take this type of trip tend to have more time, engage in multi-destination trips often with no main destination and are motivated as much for exploration and discovery as they are for escape and spending time with friends and family (McKercher and Zoltan, 2014). Touring, sightseeing and exploring are key features of these trips. Most of these types of trips are independent in nature, rather than part of a package tour. Most long haul travel and much long duration short haul travel fall into this category, as does much adventure tourism, such a bicycle tours, hiking and canoeing trips. Here, the actual distance travelled each day is small, while the journey to get to the night's rest stop is the key goal of the trip. The same situation applies in destinations, where sightseeing and exploring, perhaps with no deeper goal in mind, are seen as valuable and rewarding ways to spend one's time.

Conversely, the type of trips where time is seen as a scare resource that must be spent efficiently can be described as goal driven where maximizing time in the destination or at a specific attraction is the primary objective. The person who values time accordingly tends to have a smaller time budget, a full itinerary and as such does not have the luxury of engaging in much touring. Instead, the choice is made to minimise travel time in order to maximize time elsewhere. Many families with young children engage in this type of trip, as do most short break, short haul tourists.

Of course, the time-space relationship is not absolute. Time and space can effectively be 'bought' depending on the mode of transit taken and the willingness of the individual to pay a price premium for greater distance. Time can be bought by either taking a direct route to the destination and thus avoiding sightseeing along the way or by purchasing more efficient transport links to connect the source to the destination area. Distance can also be bought for the same amount of time depending on the mode of transport taken.

The same situation also applies within a destination where effective trip planning, concentrating activities around primary attractions and the use of different transport modes influences time budget allocation (Cvelbar, Mayr and Vavpotic, 2018; Sirirak, and Pitakaso, 2018). Indeed, Botti, Peypoch and Solonandrasana (2008) suggest tourists make rational choices before deciding to go to a location about how much time they want to spend at each place, which attractions or activities represent primary and secondary objectives and whether they are interested in exploring other lesser known places.

The trend toward more short break trips and fewer long duration trips has significant implications for the overall sustainability of tourism and especially climate change related issues. As Gössling, Scott and Hall (2018) note, limited time budgets are making more and more people take short duration trips where they go to must-see attractions and destinations, leading to a greater concentration of visitors in time and space. The advent of low cost carriers, though, means that tourists can travel further in a much more carbon-intense manner than previously available.

Distance decay

Tobler's (1970: 236) First Law of Geography states that "everything is related to everything else, but near things are more related than distant things." It forms the basis for the distance decay concept. Distance decay states demand declines exponentially as distance increases, whether distance is measured by Euclidean distance, time, effort or cost. It has been found to apply in such a wide array of social and spatial situations that Waters (2017) says it complies with the three basic requirements of a 'law': universality, necessity and synthesis. Universality means it applies to all members of the class. Necessity indicates that the relationship is guided by some underlying principle, while the synthetic component joins at least two concepts, in this case consumer behaviour, physical and human geography.

The decaying effect of distance on travel demand was first noticed by Greer and Wall (1979) some 40 years ago. Since then, a number of studies have applied it in a variety of tourism (Hooper, 2015; Lee et al., 2011; McKercher and Lew, 2003; McKercher, Chan and Lam, 2008; Nyaupane and Graefe, 2008) and recreation contexts (Eagles et al., 2015; Hanink and White, 1999).

Distance decay (Figure 5.1) shows that demand for tourism peaks at some relatively close distance to the origin and then declines exponentially as distance increases. Most distance decay research shows demand drops exponentially immediately from the start. Tourism is unique though, in that demand rises for a short distance and peaks at some distance near to, but not immediately at the origin. This pattern recognises that people must travel a certain minimum distance to achieve a sense of escape or getting away from it all (Greer and Wall, 1979). The shape of the curve reflects the intersection of different supply and demand functions influenced by time availability.

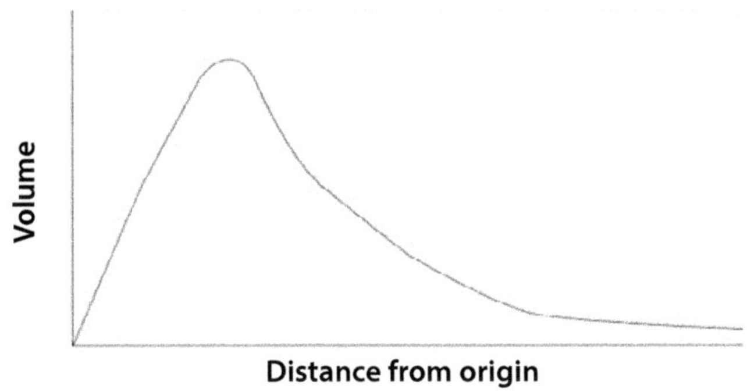

Figure 5.1: Classic distance decay curve

It is ubiquitous and enduring over time, as Lee, Denizci-Guillet, Law and Leung (2012) discovered when they looked at 10 years of outbound travel from Hong Kong. They found that, apart from minor variations, the overall pattern was broadly stable. Minor variations can be explained by changes in the general economic situation and/or the business cycle (Wong et al., 2016; Croes, Ridderstaat and Rivera, 2018), for when the economy is robust, people will travel further, while when it is fragile, distance travelled shrinks. Indeed, Butler (2008) says the bulk of tourists go to the same regions and do the same things they have always done, while Lorde, Li and Airey (2016) suggest the stability of the overall pattern is attributed to habit persistence.

Two features of distance decay must be appreciated. First and foremost, distance is not a deterministic variable. Nobody wakes up in the morning and says "I only want to travel *x* km while *x -1* km is not far enough and *x+1* km is too far." Instead, it represents a valid proxy variable that reflects the combined impact of a bundle of tangible and intangible attributes (Sirakaya and Woodside, 2005), and contextual influences (Decrop and Snelders, 2005), as well as physical geographic considerations. Factors include cost considerations, time availability, willingness to enter culturally distant environments, travel party composition, and a range of other factors.

The second feature of distance decay is that it can only be measured from the perspective of the source market and not the destination. Distance decay measures changes in demand from a source. It can be presented in absolute terms (actual number of tourists from the source who travel a certain distance threshold) or in relative terms (calculated as a share of tourists who travel a certain distance divided by the total volume of outbound tourists). The denominator remains constant, enabling valid like for like measures to be calculated.

Some researchers have tried to apply the concept by looking only at the share of arrivals to a destination, which is calculated by dividing the number of arrivals from the source market by all arrivals to that destination. Can you see the logical fallacy in this approach? Such a method cannot control for different population sizes of source markets, where the volume of arrivals may be much larger based solely on the fact that the source market generates a large volume of outbound tourism. Because the denominators are different, meaningful comparisons cannot be made. Instead, the use of arrival figures only can give a distorted impression of relative demand. Let's give a simple example of why this approach is so flawed. China with over 1.5 billion citizens generated about 135,000,000 outbound trips in 2016. Of these some 1.2 million people visited Australia, representing about 0.9% of total outbound travel. New Zealand, with a much smaller population, generated some 2.6 million outbound trips, of which 1.3 million were to Australia. Australia's share of New Zealand's outbound travel was, therefore, over 51%, meaning that proportionately, 60 times more New Zealanders went to Australia than did Chinese. Yet, if you look at arrival figures only, both generate about 15% to 16% of arrivals to the country. So, from the perspective of New Zealand outbound, Australia is a critical destination, while from the perspective of Chinese outbound, Australia is a relatively minor destination. In fact, 14 other countries and regions attract a greater share of Chinese outbound than Australia.

Three types of distance decay curve have been identified, as shown in Figure 5.2 (McKercher and Lew, 2003). The classic curve shows a peak near the source and a geometric decline. A second curve shows a 'plateauing effect' where demand is stable over some distance before declining. The cause of this type of curve is due to the limited accommodation or destination options which effectively forces people to travel further than they normally would prefer to go. In doing so, peak demand is extended until a threshold is met where demand drops exponentially. A third style has a secondary peak some great distance from the source, reflecting the unique drawing appeal of a distant attraction. This third style is caused by two factors. On the one hand the distant destination may have such great drawing power that its appeal overcomes the frictional effect of distance. Studies have shown, for example, that this pattern has been observed in the case of Australians' travel to the UK, Europeans' travel to Spain and travel from South America to the USA.

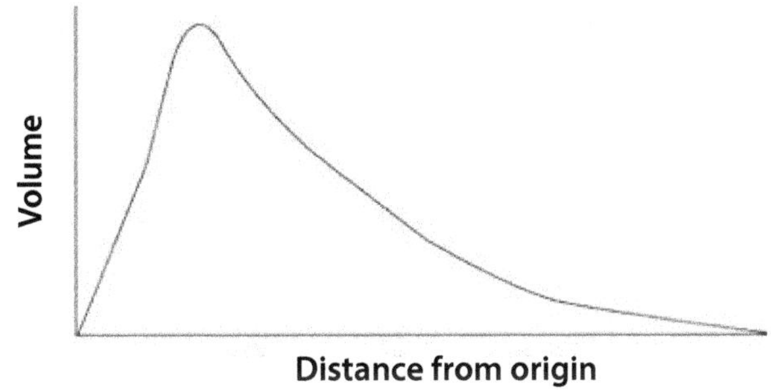

Figure 5.2a: Theoretical distance decay curve. *Source*: Bull, 1991.

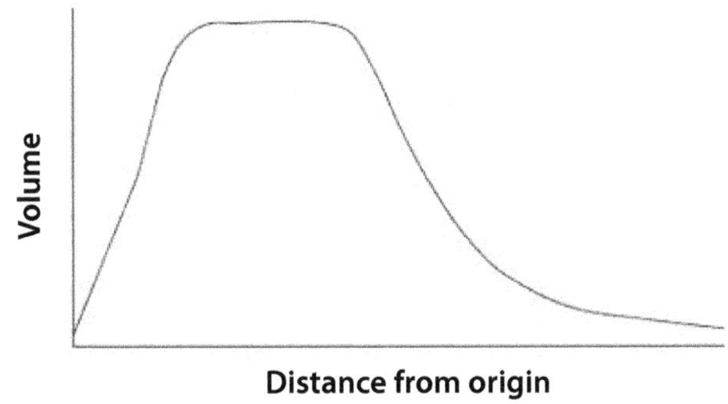

Figure 5.2b: Plateauing distance decay curve. *Source*: McKercher, 1998.

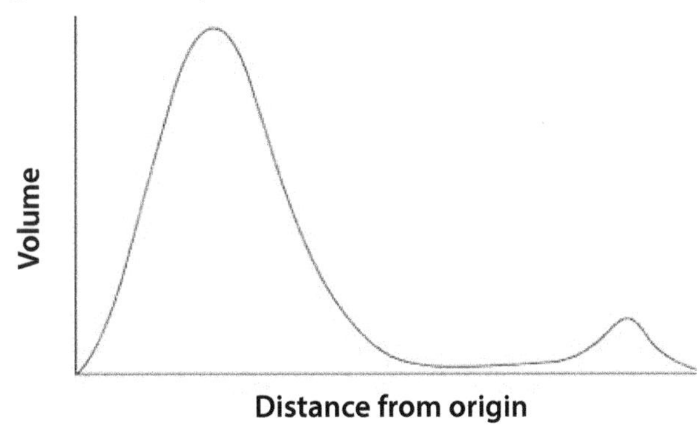

Figure 5.2c: Distance decay curve with secondary peak.

Effective Tourism Exclusion Zones

A second cause is the presence of an Effective Tourism Exclusion Zone (ETEZ) where little tourism activity occurs. McKercher and Lew (2003) noted that for every market, there is an effective area where little tourism demand exists. The causes of this effective tourism exclusion zone are varied and may include regions that are lacking in tourism development, various geographic features such as impassable mountain ranges, deserts or oceans, places at war where insecurity is rife, destination areas that prohibit residents of certain markets from visiting or simply destinations that offer products that are unattractive to the source market. Fotheringham (1981) and Eldridge and Jones (1991) label this concept 'space warping' where equivalent distances have spatially uneven effects on interaction.

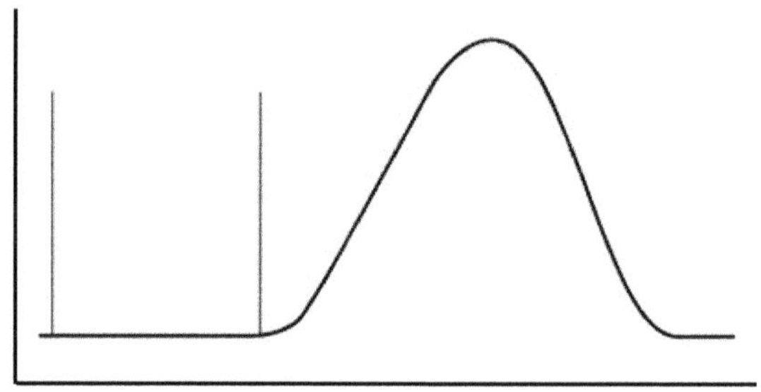

Figure 5.3a: Large ETEZ close to source market

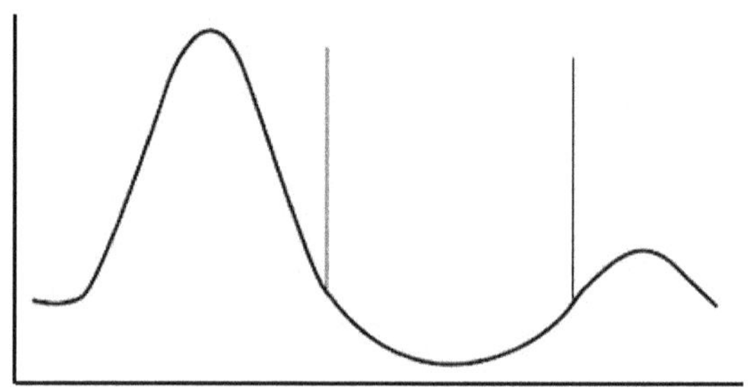

Figure 5.3b: ETEZ some distance from source market

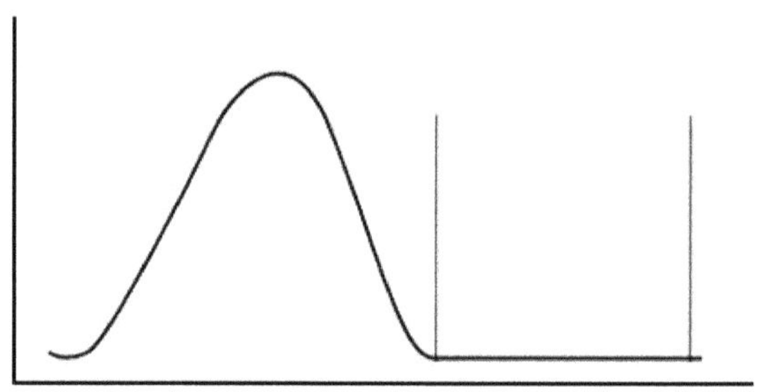

Figure 5.3c: Distance ETEZ has little impact on demand

The location of the ETEZ can have a dramatic effect on tourism movements, as shown in Figure 5.3. A small zone close to the source market will have little or no real impact on movements given that people have to travel a minimum distance anyways to have that sense of escape. A large ETEZ located close to the source market is typically found in isolated communities or, in the case of international tourism amongst island source markets (such as Japan, New Zealand and Australia), source markets with poor political relations amongst immediate neighbours or those whose neighbours are in a state of conflict. The effect is to elongate the area of low demand and to push the demand peak to more distant areas. A moderate ETEZ located some distance from the source market will tend to accentuate demand closer to the market and likely lead to a secondary peak while an ETEZ a great distance from the source market will likely have no impact on demand.

The inner edge of the standard ETEZ represents the limit of short haul travel where trips are typified by short break, short duration, repeat pleasure trips. The outer edge signifies the onset of long-haul travel which is typified more by longer duration, independent travel, multi-destination travel and more aspirational types of tourism.

The effect of distance decay on tourist flows

Figure 5.4 shows the results of the latest global study of tourism flows using 2016 UNWTO data (McKercher and Mak, 2019). The study analysed outbound travel from 92 source markets to over 4700 discrete origin-destination pairs. Travel to land neighbours dominates outbound travel, accounting for 53% of all departures. Almost 80% of all arrivals are to destinations with gateways cities less than 2000 km from a source market's gateway. Share of departures for travel greater than 5000 km was typically 3% or lower. In other words, more than half of all global travel is to immediate land neighbours and very little is to distant destinations.

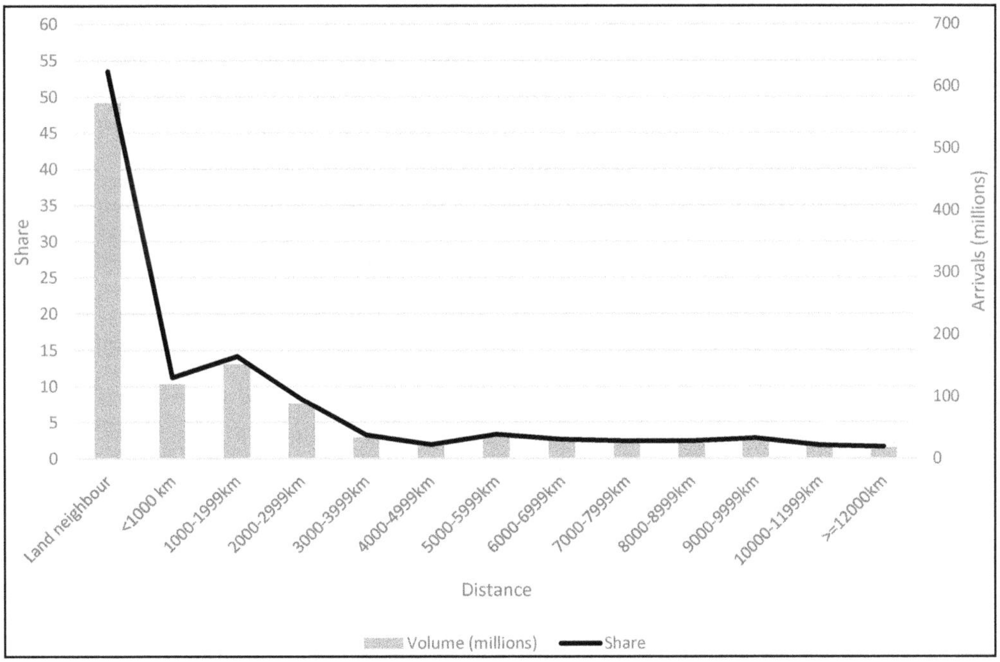

Figure 5.4: Arrival volume and departure share by distance 2016. *Source*: McKercher and Mak, 2019.

However, the rate of decay is not uniform as shown in Figure 5.5. Instead, it varies significantly by source market region. Most arrivals from markets in the Americas, Asia, Africa and former Soviet Republics are to neighbouring economies. By contrast the decay curve is much more elongated among European outbound markets. The impact of the Effective Tourism Exclusion Zone (ETEZ) is self-evident in the case of outbound travel from island nations.

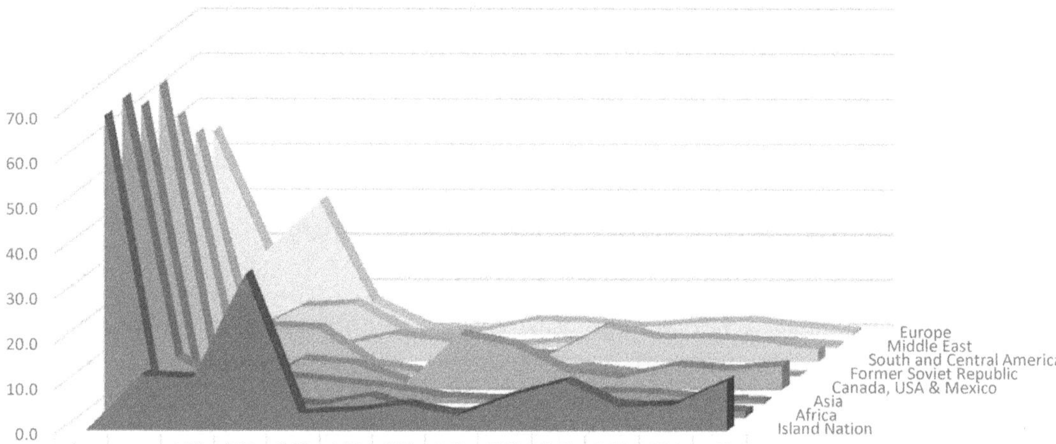

share border	< 1000km	1,000 to 1,999 km	2,000 to 2,999 km	3,000 to 3,999 km	4,000 to 4,999 km	5,000 to 5,999 km	6,000 to 6,999	7,000 to 7,000 km	8,000 to 8,999 km	9,000 to 9,999 km	10,000 to 12,000 km	>12,000 km	
	12.2	11.6	34.6	3.8	4.3	5.8	3.4	7.8	11.0	5.3	5.7	11.2	Island Nation
66.6	1.4	2.4	3.9	2.5	4.5	0.5	0.9	1.2	3.5	2.4	1.6	1.3	Africa
67.4	9.8	3.6	5.8	4.6	3.3	1.3	0.7	2.0	2.4	2.1	0.8	0.3	Asia
62.6	3.5	0.4	6.4	5.5	1.3	12.1	9.5	3.8	2.5	5.2	4.3	5.3	Canada, USA & Mexico
64.2	2.4	10.8	10.2	2.4	0.6	0.7	1.1	1.6	0.2	0.3	0.1	0.0	Former Soviet Republic
54.4	3.6	3.1	3.3	5.9	4.6	1.8	1.4	7.7	5.2	5.2	4.3	2.7	South and Central America
47.4	3.2	9.0	9.6	3.1	2.5	0.5	1.3	1.0	0.7	1.4	1.3	0.4	Middle East
44.8	18.3	29.3	6.9	1.2	0.4	2.8	2.5	1.4	2.4	2.7	1.5	0.4	Europe

Figure 5.5: Outbound share by region. *Source*: McKercher and Mak, 2019.

■ Segment transformation

In effect, distance acts as a filtering agent, benefiting some segments while acting as a constraint to travel by other segments. Anyone who can travel can travel short haul, but not everyone can or is willing to travel long haul. Distance, therefore, effectively filters out many types of tourists. Some segments are extremely sensitive to the decaying effect of distance, while others are much more robust. At its extreme, some segments may effectively disappear once a certain distance threshold is reached, while others may not emerge until a threshold is crossed. As shown in Figure 5.6, the solid line shows overall demand, while the dotted lines show the hypothetical demand curve for different segments. Segments A and B are particularly affected by distance and peak close to home before demand drops rapidly. Segments C and D are more decay resistant and peak progressively further from home before declining. Segment E is particularly distance resistant and emerges later than the others, peaking a great distance from home and only declining slowly. This decaying effect explains changes in market mix at different destinations, which in turn should explain different behavior.

Which segments are more sensitive and which are more robust? Market segments that are sensitive to distance include the young professional female or male segments (due to a combination of finances and limited vacation time), the least affluent members of a community (due to costs), families with young children (due to cost considerations, especially for transport, and an unwillingness to place children in a culturally unfamiliar environment), and older, less well educated individuals (again due to cost considerations and unwillingness to enter culturally strange environments). Those that are more robust and are, therefore, more willing to travel longer distances include more affluent, well-educated empty nesters who have the financial ability and accrued leave time to travel long distances and who also are more willing to enter strange environments. In addition, adventuresome backpackers also travel long haul, even if they do not have large financial reserves. In essence, then, the segments that were most likely to be affected by the impact of distance on demand were akin to individuals who could be found in the psychocentric end of Plog' (1974) spectrum while those who were more likely to do so share attributes common the Allocentrics or Venturers.

5

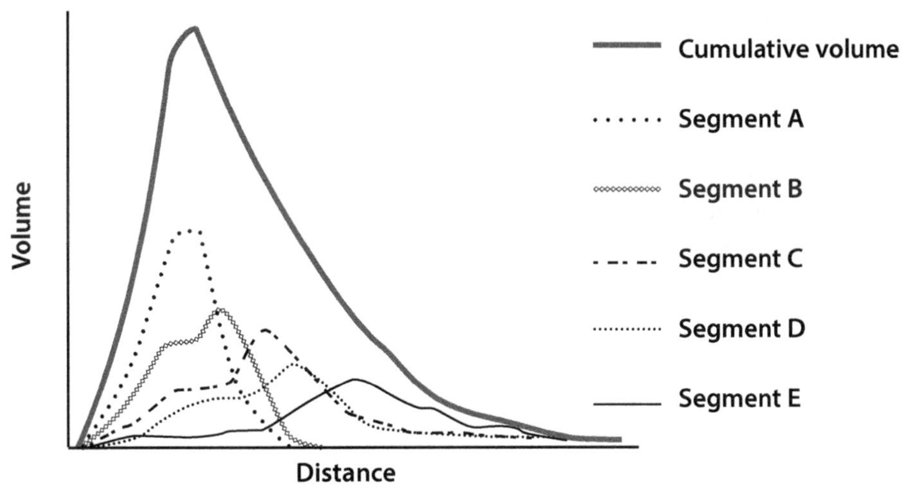

Figure 5.6: The filtering effect of distance on market segments. *Source*: McKercher, 2008.

■ Impact of distance by type of behaviour

Distance decay is an interesting geographic concept, with most studies stopping at the demand curve stage. But it has a much more profound impact on tourism than most people appreciate, for the filtering effect of distance influences who is likely to visit a destination, what their motives are and what their resultant behaviour is. This relationship is shown in Figure 5.7. Because anyone who can travel can travel short haul, the short haul market is quite democratic and consists of all segments. But, the long haul market is much more exclusive. Remember, as well, the inner edge of the ETEZ also demarcates the upper boundary of the short haul market, while the out boundary demarcates the start of the long haul market.

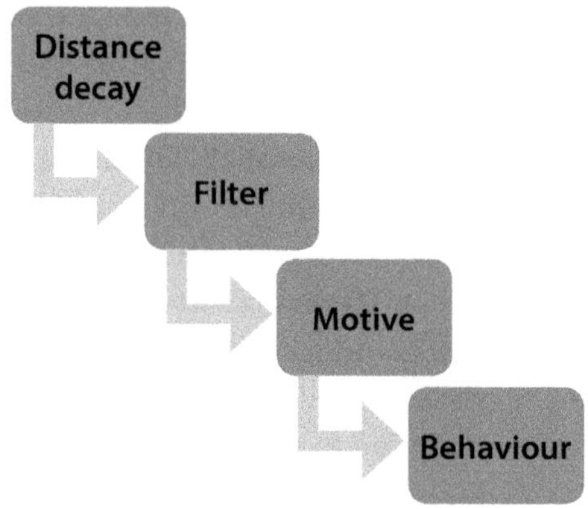

Figure 5.7: The implicit effect of distance on behaviour

Motives of different segments vary as well, and so if some segments are over or under represented by distance, then the average motive profile of the market will also vary by distance. Short haul travel is typical of the type of modal travel identified by Graburn (1983, 2018). It is the regular type of annual holiday people take where the primary motives are hedonism, escape, rest, relaxation and spending time with family. By contrast, Yeoman and Lederer (2005) write about the aspirational nature of much long haul travel and how it is viewed as a rare, often once-in-a-lifetime occurrence. In addition, Tideswell and Faulkner (1999) and Hwang and Fesenmaier (2003) observed that long haul tourists sought to have multiple trip purposes satisfied, while short haul tourists tended to take single destination, single purpose trips.

These differences in profile and motive translate into quite different behaviour patterns. Relatively speaking, long haul tourists also more likely to explore the cultural features of a destination, and in particular to visit historic sites, museums, outdoor markets and participate in culturally-oriented activities. By contrast tourists from nearby markets engage in activities that reflect the underlying motives of rest, relaxation and escape, including shopping, dining and visiting theme parks.

There is one other feature of the filtering effect of distance that must be appreciated and that is the markets targeted by destination management organizations. Table 5.2 identifies market segments targeted by the Hong Kong Tourism Board, by source market. Hong Kong is used here as an example, but most other destinations target similar segments, although the names vary somewhat. What is apparent is the change in target market by distance. Why? Destination marketing organizations target specific market segments. But equally specific segments are also predisposed to visit destinations. The challenge for the DMO is to identify those segments that are most likely to visit, even if they are not as large or attractive as others that are less likely to visit. It targets the younger and family segments among the short haul market because they are most likely to respond positively to calls to action. But, notice that it targets a much older segment among long haul tourists. It is not that long haul markets do not have attractive younger segments. It is, though, that these segments are far more likely to travel short haul. By contrast, achievers, affluent seniors and backpackers are able to travel long haul, and so they are targeted.

Table 5.2: Hong Kong's target markets

	China	Indonesia	Malaysia	Philippines	Singapore	Thailand	Japan	Korea	Australia	USA	Canada	UK	France	Germany
Young (16 – 30) - backpackers	X												X	X
Young professionals (20s to late 30s)		X	X	X	X	X	X	X	X					
Family (any age travelling with children)	X	X	X	X	X	X								
Mid-career (31 – 45)	X							X	X	X	X	X	X	X
Achiever (46+)		X			X				X	X	X	X		X

Source: HKTB Market Summaries, 2018.

Market access

Market access is a more difficult concept to grasp. It has received almost no attention from tourism academics since Pearce (1989) first introduced the idea, yet is of vital importance. It is related to distance decay, but instead of looking at the absolute distance, it looks at the number of intervening opportunities one must pass before reaching a desired destination.

It is predicated on the assumption that tourists are rational consumers who, given the choice of similar opportunities, will choose the most proximate one. Essentially, market access argues that more proximate destinations have a competitive advantage over destinations offering similar products that are less proximate (McKercher, 1998), especially among those who value time as a resource. This supposition is based on the argument that destination choice is often influenced by convenience and that, given a choice between similar destinations, the tourist will tend to choose the more convenient one. Market access can be measured by the relative difference in the time, cost, distance, or effort required to access different destinations (Pearce, 1989).

There are three main differences between market access and distance decay. The first is that it is a relative term, where absolute distance is less important than relative distance between competing destinations. As such, second, it is product or tourism type specific. The third issue is that distance decay has a consumer orientation by asking "how far do I want to travel?", while market access has a destination orientation by considering "how many destinations must the prospective consumer pass before reaching the desired one?"

Main difference between market access and distance decay

Distance decay and market access represent opposite sides of the same situation, as shown in Figure 5.8. Distance decay examines aggregate outbound travel patterns while market access is activity specific. In effect, distance decay represents the sum of individual demand and supply curves as defined by market access. Market access is a relative term whereby a distant destination may have strong market access if few if any intervening opportunities exist. Yet in aggregate for all types of travel, cumulative demand will likely be lower.

The main difference between market access and distance decay can be shown by considering the *x* axis of a graph. Distance represents the *x* axis in a distance decay curve, while intervening opportunities represent the *x* axis in market access curve.

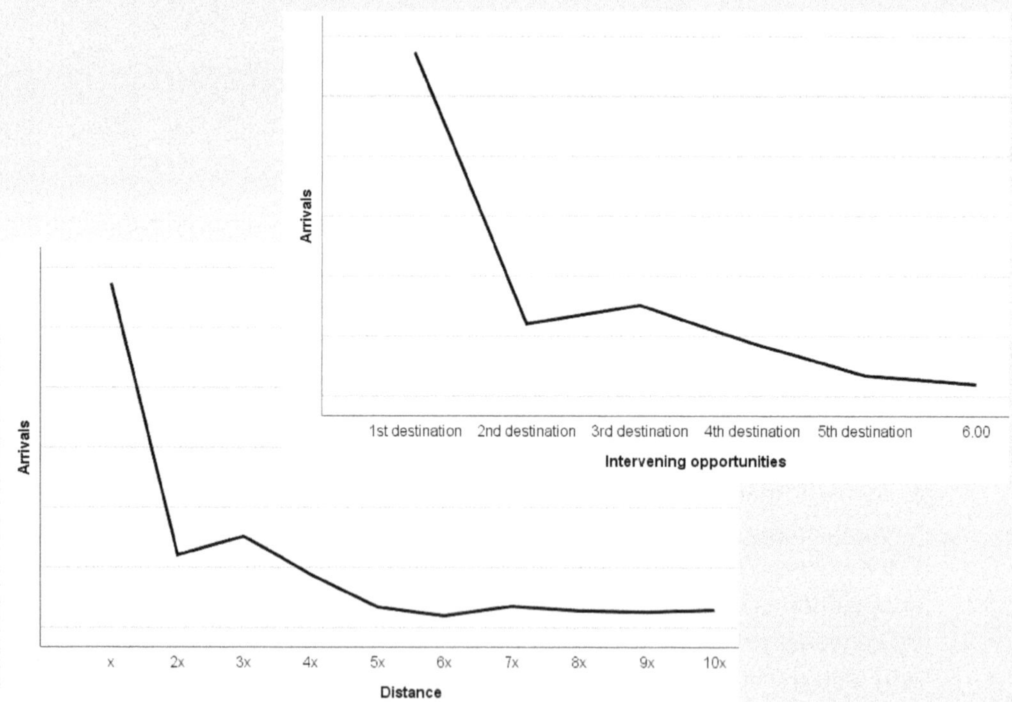

Figure 5.8: Difference between distance decay and market access

An example of the relative nature of market access will help explain the concept better. Consider a person who lives in Sydney, Australia. That person has a choice of close to ten beaches in the greater Sydney area including world-famous beaches like Bondi and Manly. A beach located 50 km away would be seen to have very weak market access because the consumer would have to pass by literally dozens of opportunities offering similar experience before reaching this beach. Now, consider the same person who wants to go skiing. The nearest ski fields to Sydney are located in Kosciusko National Park, some 500 km away, with the next nearest ski field located in the State of Victoria some 750 km from Sydney. Thus, Kosciusko ski fields enjoy strong market access because they represent the most proximate skiing opportunity. Relatively speaking, therefore, a beach 50 km away from the source market may be competitively disadvantaged due to the number of intervening opportunities, yet the ski field located 500 km away will have a strong competitive advantage because of the very lack of intervening opportunities.

One of the few studies of this type found destinations with both strong and weak market access had their own competitive advantages, for the impact of market access is influenced by time availability and, to a lesser extent, travel party type (McKercher, 1998). Short break tourists with the least amount of time were far more likely to select destinations with strong market access. By contrast, those with larger time budgets were more likely to bypass nearby destinations and head further afield, where they could escape the crowds. Likewise, families, especially those with young children chose destinations with strong market access, while couples tended to travel to places with weaker market access.

Market access has a number of development implications. Destinations with stronger market access are more likely to experience weekly cycles of weekend crowding and midweek slumps but are less likely to be affected by huge seasonal variations in visitation. Those with weaker market access are more likely to be affected by seasonality, but because of the distance involved are less likely to be affected by the weekly cyclical variations during their high seasons.

While market access may be a powerful source of competitive advantage it is also highly vulnerable, for the development of a new tourism product or destination area close to the source market can fundamentally change market access from being strong to weak. If the competitive advantage is based solely on lack of intervening opportunities, the creation of an intervening opportunity can eliminate that one source of competitive advantage.

Can market access be modified?

While the impact of distance is more or less universal, the impact of market access is variable. As such, market access can be modified by a number of factors. These include:

- *Removing artificial or political barriers to entry*. For example, Assaf and Josiassen's (2012) global studies revealed strong visa restrictions have a negative impact of the flow of travelers, while Balli, Balli and Cebecci (2013), with Lawson and Roychoudry (2013) suggest the elimination of visas can result in the tripling of arrivals.
- *Changing travel time*. The advent of low cost carriers has fundamentally transformed Spain's tourism sector, with Rebello and Baidal (2009), noting both quantitative and qualitative changes in tourism activity. The volume of tourists has increased, and importantly, more changes have been seen in the rise of independent travel, changes in accommodation choices and changes in visitor profile.
- *Changing price*. Changes in Spain's tourism sector have also been affected by the low price charged for air travel. It is now cheaper to fly to Spain than it is to take a train to Paris, effectively enhancing Spain's competitive position.
- *Changing speed*. The advent of the high speed rail network in China has opened up many regional destinations to new tourism opportunities, with visitor arrivals growing by at least 10% in the first year of operation (Wang et al., 2012, Wang et al., 2014).
- *Changing effort*. All one has to do is look at ski resorts. The advent of ski lifts has transformed skiing from an activity that only young and fit people could participate in to a mass activity available to all, young and old. The lift overcomes the challenge of having to climb a mountain. Instead, one can let a machine do the work that you would otherwise be expected to do, and you can enjoy the effect of gravity to take you down the mountain.

Itinerary models

Modelling inter-destination itinerary movements was a popular topic in the late 1980s and 1990s (Flognfeldt, 1999; Lue, Crompton and Fesenmaier, 1993; Mings and McHugh, 1992; Oppermann, 1995), but seems to have fallen out of fashion. Mings and McHugh (1992) identified four types of touring routes taken by domestic American tourists. These models include a 'direct route' itinerary taking the most direct path to and from the destination; the 'partial orbit' which consisted of a long transit leg followed by a short circular tour; the 'full orbit' itinerary which involved a circular tour and not taking the same route twice; and the 'fly-drive' itinerary which is a modification of the partial orbit except the transit leg was taken by air. Lue, Crompton and Fessenmaier (1993) identified four types of itineraries: the 'en-route' model, where tourists could make a number of stops on their way to a main destination; the 'base camp' model where they based in one place and then did a number of trips from there; the 'regional tour', involving a number of stops on a tour and the 'trip-changing' pattern that involved a multi-foci, touring trip visiting a number of destinations without overlapping any leg of the trip. Oppermann (1995), focusing on international travel, identified seven possible itinerary types, including the 'open jaw loop' model which applies to tourists who enter a country through one gateway and leave through another and the 'multiple destination areas loop' which basically consists of all other destinations mashed together.

Indeed, more than two dozen different inter-destination itinerary styles were identified. Yet, on closer inspection, they fall into four main types (McKercher and Zoltan 2014). They are shown in Figure 5.9 below and are:

- Single destination, there-and-back trip that may or may not include side trips.
- Transit leg to the destination area, followed by a circle tour, stopping overnight at different places.
- Circle tour with or without multiple access and egress points.
- Hub and spoke itineraries may be evident where tourists base themselves in a destination area and take side trips to other destinations

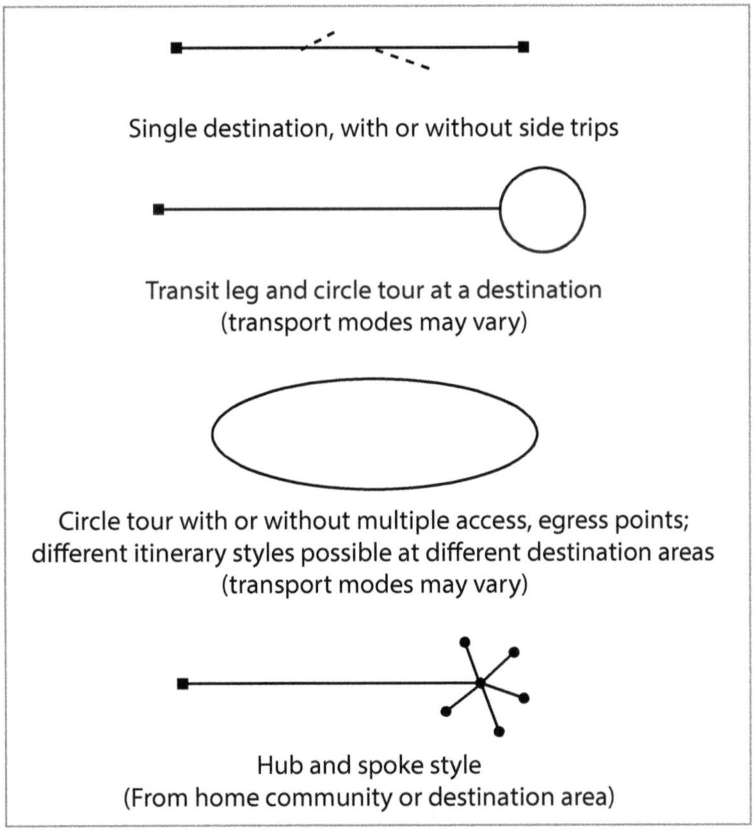

Figure 5.9: Inter-destination itinerary types. *Source*: McKercher and Zoltan, 2014.

All tourist movement patterns involve both a transit and a destination touring component as shown in Figure 5.10. The different types of transit and destination elements are shown below. The various tour combinations identified result from different mixing and matching of transit and touring elements.

- An outbound and inbound transit leg following the same route is implied in the single destination, base camp, stopover/en-route and the regional tour/partial orbit or destination area loop models.
- Multiple transit legs are needed for the various loop tours identified.
- In rare instances, tourists may embark on a single transit leg and then have an extended return tour home.

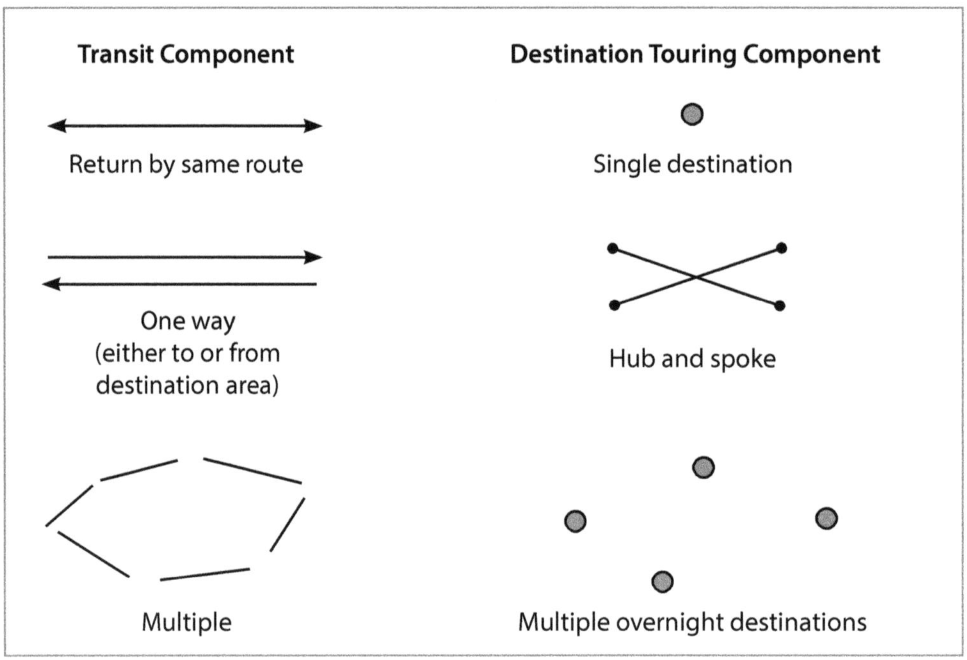

Figure 5.10: Transit and destination components of itinerary model. *Source:* McKercher and Zoltan, 2014.

Conclusion

The impact of distance on tourism is often undervalued, yet it exerts a profound impact on all aspects of travel from defining the size of the market, to reflecting different motives, to influencing behaviour and ultimately tourists' experiences. This chapter has explored a range of issues related to tourism flows and how the valuation of time, distance decay and market access influences tourism both for inter-destination and intra-destination travel.

■ Discussion points / Exercises

1. Discuss which groups of tourists are more likely to see travel time as a commodity or see it as a resource, and evaluate what impact that may have on their travel patterns.
2. Given that distance decay is a universal construct and that distance acts as a filter for certain segments, consider the Chinese and Indian outbound markets and analyse which destinations are going to be faced with a large influx of tourists as these markets grow and which are likely to attract the high end more mature market.
3. Identify an ETEZ near your location and examine the effect it has on tourism.
4. Conduct a market access analysis of three special interest segments of your choice.

Points to consider for revision

- How much time people have and how they value travel time plays a key role in tourist movements. People who see travel time as a commodity see value in the journey itself as much as being at the destination. Those who see it as a resource see travel time as a cost that should be minimised.
- Distance decay is a universal construct in geography and when tested empirically is also found to apply to tourism. More than half of all international travel is to land neighbours, while once distance passes a threshold of 4,000 km or more, source markets generally contribute little to overall demand.
- The shape of the distance decay curve is influenced by the positioning of the Effective Tourism Exclusion Zone, a zone where little or no tourism exists. If located near the source market, it will likely push demand further than if no ETEZ existed. A moderate distance from the source market, will likely accentuate the peak and then create a further secondary peak some distance away.
- Distance acts as a filter, advantaging some segments and disadvantaging others. As a result of this filtering effect, differences are noted in the profile, motives and in-destination behaviour as distance changes.
- Market access is like distance decay, but it is a relative term that is consumer specific. It evaluates the number of intervening opportunities that must be passed before reaching the destination. It too shows a typical decaying curve. However, market access can be modified.

- Itinerary models are influenced by both touring and destination components. How these two components interact leads to the various inter-destination itinerary models that have been developed.

Suggested additional readings

Lew A. & McKercher, B. (2006) Modeling tourist movement: a local destination analysis, *Annals of Tourism Research*, 33(2), 403 – 423.

McKercher, B. & Mak, D. (2019) The impact of distance on international tourism demand, *Tourism Management Perspectives*, 31, 340-347.

References

Assaf, A.G. & Josiassen, A. (2011). Identifying and ranking the determinants of tourism performance: a global investigation, *Journal of Travel Research*, 51(4), 388-399.

Balli, F., Balli, H. O. & Cebeci, K. (2013) Impacts of exported Turkish soap operas and visa-free policy on inbound tourism to Turkey, *Tourism Management*, 37, 186-192.

Becken, S. & Wilson, J. (2013) The impacts of weather on tourist travel, *Tourism Geographies*, 15(4), 620-639

Botti, L., Peypoch, N. & Solonandrasana, B. (2008) Time and tourism attraction. *Tourism Management*, 29(3), 594-596.

Bull, A. (1991) *The Economics of Travel and Tourism*, Pitman, Melbourne.

Butler, R. (2008) Modern tourism and its development in a post modern age. In Babu, S., Mishra, S., Parida, B (Eds), *Tourism Development Revisited: Concepts, Issues and Paradigms*. Response: London, pp 54-63.

Chavas J. P., Stoll, J. & Sellar, C. (1989) On the commodity value of travel time in recreational activities, *Applied Economics*, 21, 711-722.

Croes, R., Ridderstaat, J. & Rivera, M. (2018) Asymmetric business cycle effects and tourism demand, *Journal of Travel Research*, 57(4), 419-436.

Cvelbar, L., Mayr, M. & Vavpotic, D. (2018) Geographical mapping of visitor flow in tourism: A user-generated content approach. *Tourism Economics*, 24(6), 701-719.

Decrop A. & Snelders, D. (2005) A grounded typology of vacation decision-making. *Tourism Management*, 26(2), 121-132.

Eagles, P., Johnson, P., Potwarka, L. & Parent, C. (2015) Travel distance classes for tourism destinations: a proposal from Ontario Provincial Park camping, *Journal of Ecotourism*, 14(1), 64-84, DOI: 10.1080/14724049.2015.1071829

Eldridge D. & Jones J. P. (1991) Warped space: a geography of distance decay, *Professional Geographer*, 43(4), 500-511.

Flognfeldt, T. (1999) Traveler geographic origin and market segmentation: The multi trips destination case, *Journal of Travel and Tourism Marketing*, 8(1), 111-124.

Fotheringham, A. S. (1981) Spatial structure and distance decay parameters, *Annals of the American Association of Geographers*, 71(3), 425-436.

Gössling, S., Scott, D. & Hall. M. (2018) Global trends in length of stay: implications for destination management and climate change, *Journal of Sustainable Tourism*, 26(12), 2087-2101, DOI: 10.1080/09669582.2018.1529771

Graburn, N. (1983) The anthropology of tourism, *Annals of Tourism Research*, 10,9-33.

Graburn, N. (2018) Secular pilgrimage: A general theory of tourism. In Gmelch, S. & Kaul, A. (eds.) *Tourists and Tourism: A general reader 3rd Ed.* Long Grove: Waveland, pp. 17–28.

Greer, T. & Wall, G. (1979) Recreational hinterlands: A theoretical and empirical analysis in G. Wall (ed.) *Recreational Land Use in Southern Ontario* (Department of Geography Publication series #14), Waterloo University, Canada, pp 227-246.

Hall, C. (2005) Time, space, tourism and social physics, *Tourism Recreation Research*, 30(1), 93-98, DOI: 10.1080/02508281.2005.11081236

Hanink D. M. & White, K. (1999) Distance effects in the demand for wildland recreational services: the case of national parks in the United States. *Environmental and Planning Annals*, 31, 477-492.

Hooper, J. (2015) A destination too far? Modelling destination accessibility and distance decay in tourism, *GeoJournal*, 80, 33-46.

Hwang Y. & Fesenmaier, D. (2003) Multidestination pleasure travel patterns: Empirical evidence from the American Travel Survey, *Journal of Travel Research*, 32, 166 171.

Lawson, R. & Roychoudhury, S. (2013) Do travel visa requirements impede tourist travel? *Journal of Economics and Finance*, 40(4), 817-828.

Lee, H. A., Deniczi-Guillet, B. D., Law, R. & Leung, R. (2011). Robustness of distance decay for international pleasure travelers: a longitudinal approach. *International Journal of Tourism Research*, 14(5), 409-420.

Leiper, N. (1990) *Tourism Systems: An interdisciplinary perspective.* Department of Management Systems (Occasional Paper # 1). Palmerston North: Massey University

Lorde, T., Li, G. & Airey, D. (2016) Modeling Caribbean tourism demand: An augmented gravity approach, *Journal of Travel Research*, 55(7), 946-956.

Lue C., Crompton, J. & Fessenmaier, D. (1993) Conceptualization of multidestination pleasure trips, *Annals of Tourism Research*, 20(2), 289-301.

McKean J., Johnson, D. & Walsh, R. (1995) Valuing time in travel cost demand analysis: An empirical investigation, *Land Economics*, 71(1), 96-105.

McKercher, B. (1998) The effect of market access on destination choice, *Journal of Travel Research*, 37, 39-47.

McKercher, B. (2008) Segment transformation in urban tourism. *Tourism Management*, 29, 1215-1225.

McKercher, B. & Lew, A. (2003) Distance decay and the impact of effective tourism exclusion zones on in international travel flows, *Journal of Travel Research*, 42(2), 159-165.

McKercher, B. & Mak, D. (2019) The impact of distance on international tourism demand, *Tourism Management Perspectives*, 31, 340-347.

McKercher, B. & Zoltan, J. (2014) Tourist flows and spatial behaviour, in Hall, M. Williams A. and Lew (eds) *A Companion to Tourism, 2nd Ed* Blackwell, pp 34-44.

5

McKercher, B. Chan, A. & Lam, C. (2008) The impact of distance on international tourist movements, *Journal of Travel Research*, 47(2), 208-224.

Mings, R. C. & McHugh, K (1992) The spatial configuration of travel to Yellowstone National Park, *Journal of Travel Research*, 30, 38-46.

Nyaupane G. & Graefe, A. (2008) Travel distance: a tool for nature-based tourism market segmentation, *Journal of Travel & Tourism Marketing*, 25(3-4), 355-366, DOI: 10.1080/10548400802508457

Oppermann, M. (1995) A model of travel itineraries. *Journal of Travel Research*, 33(4), 57-61.

Pearce, D (1998) Tourism time budgets, *Annals of Tourism Research*, 15, 106-121.

Pearce, D. (1989) *Tourist Development 2nd Ed.* Harlow: Longman.

Plog, S. (1974) Why destination areas rise and fall in popularity, *Cornell Hotel and Restaurant Administration Quarterly*, 14(4), 55-58.

Rebello, J. & Baidal, J. (2009) Spread of low-cost carriers: Tourism and regional policy effects in Spain, *Regional Studies*, 43(4), 559-570.

Sirakaya, E & Woodside, A. (2005) Building and testing theories of decision making by travelers. *Tourism Management*, 26(6), 815-832.

Sirirak, W. & Pitakaso, R. (2018) Marketplace location decision making and tourism route planning, *Administrative Sciences*, 8(4), 72.

Tideswell, C. & Faulkner, B. (1999) Multidestination travel patterns of international visitors to Queensland, *Journal of Travel Research*, 37, 364-374.

Tobler, W. R. (1970) A computer movie simulating urban growth in the Detroit region, *Economic Geography*, 46, 234–240.

Walsh, R., Sanders, L. & McKean, J. (1990) The consumptive value of travel time *Journal of Travel Research*, Summer, 17-24.

Wang, Z., Huang, S., Zou, T. & Yan, H. (2012) Effects of the high speed rail network on China's regional tourism development, *Tourism Management Perspectives*, 1, 34-38.

Wang, D., Qian, J., Chen, T., Zhao, M. & Zhang, Y. (2014) Influence of the high-speed rail on the spatial pattern of regional tourism – Taken Beijing–Shanghai High-Speed Rail of China as example, *Asia Pacific Journal of Tourism Research*, 19(8), 890-912, DOI: 10.1080/10941665.2013.818049

Waters, N. (2017) Tobler's First Law of Geography. In Richardson, D., Castree, N., Goodchild, M., Kobayashi, A., Liu, W. & Martson, R. (Eds) *The International Encyclopedia of Geography*. Chichester: John Wiley and Sons. http://onlinelibrary.wiley.com/doi/10.1002/9781118786352.wbieg1011/pdf <downloaded Oct. 17, 2017>.

Wong, I., Fong, L. & Law, R (2016) A longitudinal multilevel model of outbound travel behavior and the dual cycle model, *Journal of Travel Research*, 55(7), 957-970.

Yeoman, I. & Lederer, P. (2005) Scottish tourism: Scenarios and vision, *Journal of Vacation Marketing*, 11(1), 71-87.

6 International Tourism

By the end of this chapter, the reader will be able to:

- Describe factors that influence bilateral tourism flows
- Identify why tourism is a politically selective activity
- Analyse the role of Bilateral Air Service Agreements in international tourism
- Describe the Freedoms of the Air
- Critique the role of the World Trade Organization in tourism

Introduction

International tourism is influenced by both small and big 'P' politics. Political factors influence who can visit a country and who is not welcome. As Artal-Tur et al. (2015) note, the impact of diplomatic relations on tourism flows is immense. Political relationships influence international air service agreements and through the General Agreement of Trade in Services, controlled by the World Trade Organization (WTO), influence the internationalization of tourism in all areas. To understand international tourism, then, you need a general understanding of the international trade dimensions that guide it and the politics behind some of them.

Bilateral tourism flows

Before starting though, it is worth appreciating that a number of conditions must exist before international tourism can occur. First, one country must recognize the existence of the other. Travel by Israelis is very difficult to many countries in the Islamic world and virtually impossible to many more due to the fact that they have never recognized Israel's existence. The same holds true for some breakaway republics in the former Soviet Union. Second, there must be diplomatic relations between countries. Third, commerce or trade agreements must be signed. While tourism may be a recreational activity pursued by individuals, at a national level, it is a highly sought after trade commodity. Then acceptable entry arrangements must be agreed upon that outline how people can visit, under what conditions and how long they can stay. These conditions include such features as the need for passports or visas as entry requirements and other entry conditions. They vary from country to country and are negotiated by each country as part of their diplomatic relations. Finally, a range of other barriers must be removed. As Prideaux (2005: 781) writes

> The quantum of the flow between country pairs is a function of a matrix of interrelated factors that includes the public and private sector structures supporting flows, diplomatic relations, and economic and noneconomic factors. Countries that have removed barriers to citizens undertaking foreign tourism have limited capability to regulate outbound flows, but still retain significant capacity to increase inbound flows through measures designed to enhance destination competitiveness and encourage citizens to substitute domestic for international tourism. Policies designed to alter the level of inbound and outbound flows through non-regulatory measures may become important where countries facing a balance of payments crisis decide to adopt policies designed to increase exports and reduce imports of tourism.

It is for this reason that Webster and Ivanov (2016) contend that one of the major barriers to allow people to travel is not the cost but political regulation of borders. Karaman (2016) adds that tourism flows are affected by political, economic and cultural factors whereby security conditions and international relations have a direct influence on mobility.

Politics of visas

While those of us in the Western world may take travel for granted, nowhere in international law does the right to enter foreign spaces exist (Neumayer, 2010). Instead, it is a privilege granted by both the generating and receiving country.

Either or both can grant easier access to facilitate movements or can create barriers to restrict movements (Liu and McKercher, 2014). The Universal Declaration of Human Rights, in principle, gives individuals the right to exit and enter their own country but not a right to enter foreign spaces. Indeed, it is estimated that approximately two-thirds of the world's population require a visa prior to departure, 15% can acquire a visa on arrival in country and only 18% are able to enter destinations on a visa-free basis (Bangwayo-Skeete and Skeete, 2017).

The issues of entry into different countries and terms of stay have received remarkably little attention within the tourism literature. Yet, they are fundamental to the understanding of tourism for they can encourage or discourage the free movements of people. As much as tourism is portrayed as a democratic activity accessible to most, in reality international tourism is highly politicized where certain types of undesirables are actively discouraged from entering a country while desirables are encouraged. Mau et al. (2015) comment that border selectivity is a central reaction to the opportunities and risks stemming from globalization. They further note that visa policies are seen as a central instrument in mobility restriction as a form of remote control or preemptive action. In essence whether governments impose or remove restrictions on cross-border mobility is a reflection of the friendliness between countries and the presence or absence of reciprocal trade agreements (Whyte, 2008). Webster and Ivanov (2016) indicate that states generally have policies that favour open borders, unless they are states that tend to face some type of real or perceived external threat.

People can enter a country in any number of ways. They may migrate on a permanent basis, come in through various professional and investor migration schemes, enter on student visas and perhaps be allowed to stay afterwards, come in as foreign workers, be identified as quality migrants, come in as part of a family reunion scheme and many other ways, including visiting as a tourist. The easiest way to gain access is as a tourist, especially since many people can gain visa-free entry. As a result it is also the easiest to abuse. Indeed, the vast majority of illegal immigrants enter a country legally as tourists and then overstay (Neumayer, 2006). It is because of this fact that many countries impose visa requirements as a means of creating a hurdle for visitors, as well as providing a means of tracking them once in the country (Whyte, 2008).

Countries face a difficult balancing act of, on the one hand, encouraging the free flow of people for political, social, economic and other beneficial reasons while, on the other hand, deterring people in order to maintain immigration control and uphold security (Neumayer, 2006). Visa policies fulfil this dual role for they act as a powerful screening mechanism that manages demand (Beenstock et al., 2015). But as Neumayer (2010) notes, visas are a crude mechanism to manage this trade off. Entry can be denied for a number of reasons, as shown in Table 6.1.

Table 6.1: Reasons why entry to a country can be denied

Does not meet the entry requirements
Has committed fraud, deception or misrepresentation in this or a previous application
Has a criminal record, has been arrested
Is a threat to national security
Does not have a good moral character
Has previous visa/immigration violations
Cannot prove to have strong ties to their current country of nationality and/or residence
Intends to reside or work permanently in the country
Fails to provide sufficient evidence/documents to prove eligibility for the visa sought after
Does not have a legitimate reason for the journey
Has no visible means of sustenance
Does not have health/travel insurance valid for the destination and the duration of stay
Is a citizen of a country to which the destination country is hostile
Has previously visited, or intends to visit, a country to which the destination country is hostile
Has a communicable disease, such as COVID-19
Has a passport that expires soon

Four types of visa regimes exist (Karaman, 2016):

- Visa exemption programs (for those seen as being from friendly countries or who pose no threat);
- E-visa applied prior to the trip;
- Visa on arrival; and
- The formal visa application process that requires attendance or submission of documentation to relevant consulates, as a form of pre-emptive mobility governance (Mau et al., 2015).

Visas typically state the terms of entry (i.e. no work permitted if a tourist), maximum duration of stay and other conditions. To receive one, applicants must often provide proof of a return ticket, evidence of sufficient funds to maintain themselves during the visit, evidence of immunization against certain diseases if needed, confirmation of hotel booking, employer's letter and, in some cases, may need to attend an interview in person (Prideaux, 2005). In general, nations with similar economic and political conditions, nearness and membership of international groupings share similar travel documentation requirements (Lee, Song and Bendel, 2010).

The granting of visa waiver conditions for residents of an entire country signifies that the country is friendly and that its residents are unlikely to pose a security threat, overstay or commit crimes in the country. The waiving of visa restrictions signifies that the receiving countries do not fear illegal immigration or entrance of unwanted individuals, wish to encourage international trade, foreign investment and other contacts from desirable economies or for other political reasons. Poor countries may also have an incentive to waive or modify visa requirements in the hope of encouraging foreign investment, while major tourist destinations have an incentive not to impose visa restrictions in order to remain competitive (Mau et al., 2015; Neumayer, 2010).

The system is incredibly unequal, though. Citizens of high income, politically stable nations enjoy far more liberal access than residents of poor countries. Luedtke et al. (2010) identified a range of factors and determined that level of political and civil liberties, per capita income, degree of internal insecurity, presence or absence of health epidemics and geographic location all play a role. Neumayer (2006) reports that passport holders from 25 countries facing the smallest number visa restrictions come from Western, high income OECD nations, with only one or two exceptions. Bangwayo-Skeete and Skeete (2017) found that an increase in the political stability index for a country resulted in an additional 13 countries granting visa-free access. They also found that citizens of collaborative multinational organizations, such as the European Union, also tend to enjoy greater visa-free access.

By contrast residents of poor, politically unstable countries or countries where the West seems to feel they pose a threat to global security face the most daunting visa restrictions (Webster and Ivanov, 2016). Members of the Francophonie, usually located in Africa, and the Arab League have less visa-free travel than most other countries (Bangwayo-Skeete and Skeete, 2017). Neumayer (2006) adds that countries which have the least visa-free access have a history of violent political conflict, strict autocratic regimes, are very poor or have some combination of these conditions.

A number of studies have documented both the negative impacts of entry restrictions on tourism and the positive implications of liberalized visa requirements. Neumayer (2010) suggests the imposition of visa restrictions can restrict bilateral tourism by up to 60% while others have questioned this figure and suggest its impact is more in the neighbourhood of a 25% to 35% reduction in arrivals (Atral-Tur et al., 2016; Lawson and Roychodhury. 2010). A number of other studies have also shown how the imposition of visa restrictions in response to unique events can deter tourism, at least over the short term (Li and Song, 2013 Song et al., 2012).

Other studies indicate that liberalizing visa policies can result in significant increases in visitor numbers (Balli, Ballie and Cebecci, 2013), with Lawson and Roychodhury (2013) suggesting that the elimination of visas can result in the tripling of arrivals. Visa-free travel for Koreans visiting Japan produced a 12%

increase in arrivals in the first year and 25% increase in the second year (Lee, Song and Bendel, 2010). Likewise, Hu (2013) notes that the implementation of the visa waiver program in the United States has encouraged an additional 1.8 to 2.7 million pleasure or business trips, generating an additional US$6.9 to US$10 billion in direct spending. A joint UN World Tourism Organization and World Travel & Tourism Council study on APEC countries (Asia-Pacific Economic Cooperation) concluded positive visa changes can result in a 5% to 25% increase in arrivals per year over a three-year period (UNWTO, 2013).

International Air Transport (Bilateral Air Service Agreements)

If you do not understand how international aviation works, then you did not understand how international tourism works. While it is easy to purchase an airline ticket, say between Europe and Australia, the actual ability of airlines to fly between these places involves a whole series of bilateral trade agreements signed between the government where the airline is based and every country or territory the airline flies through, lands, drops off or picks up passengers. First and foremost, it must be appreciated that aviation is a form of international trade governed by the same type of trade agreements between countries that also govern the trade in goods and services such as automobiles, manufactured goods, oil and even foreign direct investment. Governments own the airspace above their country or territory and therefore can negotiate permission to fly over or through that airspace (Fayed & Westlake, 2002). Because of this fact, they can control which countries they permit to fly their aircraft through that space, which airlines can fly, the capacity, costs, and to some extent even prices charged to consumers. Air rights are also highly valuable commodities that can be tied into other trade goods.

The rules for international air travel were set at the 1944 Chicago Convention on Civil Aviation when 53 governments met to establish an orderly arrangement for post war air travel (Lissitzin, 1964). The convention established the framework for the development of bilateral agreements between countries to ensure that the carriage of passengers, mail and freight would normally be restricted to the carriers of these countries. The 1944 Chicago Convention set the legal tone for international aviation to follow. The preamble states that the signing countries have agreed to certain principles and arrangements that will promote the development of international air transport on the basis of equality and opportunity. Article 44 largely eliminated open competition and, instead ensured that government owned carriers could protect their rights, restrict competition and operate profitably (DTRD, 1998).

There were two main outcomes of the Chicago Convention. The first was the assurance that international aviation would be strictly controlled as part of bilateral trade between countries. In doing so countries could dictate which airlines and which nations could trade with them through the aviation sector. The second key outcome was the established the so-called 'freedoms of the air', which will be discussed in greater detail later.

Remember a few critical factors when the Chicago Convention was signed. Most international air carriers were owned and operated by governments, not the private sector. Air travel was the domain of a very few modernised countries in the late 1940s. Canada, Australia, the US, the UK and some western European countries dominated international travel. Most other countries were not modernized enough to develop their own international carriers and were therefore dependent on other counties for international air travel. Moreover, aircraft technology limited the distance planes could fly, and so a series of agreements was needed to enable aircraft to land and refuel in order to complete long haul flights.

Air Service Agreements

The first international aviation agreement was the Bermuda Agreement signed in 1946 between the UK and the USA. It set the framework for all air service agreements up to and including the present. Some 3,000 bilateral air service agreements are now in force. They are, or should be lodged with ICAO, the International Civil Aviation Organization, based in Montreal, Canada. In recent years multilateral agreements that involve more than two states have also become popular. Typically, the traditional bilateral aviation agreements included the following elements (Aus, 2019; OECD, 2014):

- Routes to be flown, including gateways, which identify which cities are to be flown to and where stopovers are to be included. In this manner, governments can stimulate the expansion or creation of new gateway centres, promote one destination over another, or otherwise attempt to influence the flow of tourists to an area. Often the routes are identified due to market forces. In other instances, governments can lead.
- The capacity to be operated, either in terms of number of flights or number of passengers. Airlines are not obligated to fly that number of flights, instead it represents a maximum number that they can operate. Greater capacity than current demand is usually negotiated, as a means of encouraging growth of certain routes. This capacity is the key factor for the airlines, for it will determine the potential profitability of certain air routes as well as the frequency of flights. The other key factor to consider about capacity is that it includes ramp space or slots at airports, potentially enabling one country and its designated airline to dominate an airport and exclude competition.

- The designation of the air carrier(s). Before an airline can take up rights negotiated under a bilateral agreement, it must first be designated by its government as an international air carrier to the other government concerned. The other government can reject that designation if it is not satisfied that the carrier is substantially owned and effectively controlled by citizens of the designating country.
- How the capacity will be divided between countries. Normally it is split evenly, however, a number of other factors may affect the final division of capacity. Some of these factors relate to the ability of one country to deliver the capacity, while another may be related to trade-offs being made for preferential services elsewhere.
- Tariffs or the prices airlines can charge.
- Other issues including non-scheduled or charter operations, and 'cabotage'. Cabotage is a term that allows foreign carriers to fly domestic routes.

These types of agreements are incredibly restrictive in nature. They may work in favour of the country but work against the best interests of the airlines and the flying public for they restrict the market from playing any significant role in aviation. In recent years, pressure has been placed on countries to liberalize their air service agreements or open up their skies to more market forces. Liberalization does not mean the elimination of air service agreements, for they are still negotiated country to country (EU, 2013). What it does, though, is reduce the artificial constraints that have inhibited the development of the aviation sector for open skies agreements seek to provide broad opportunities for the free market to establish prices, routes and capacity as well as to remove restrictions on the number of carriers that can provide service (OECD, 2014). But, the degree of liberalization is still influenced by the existence of bilateral or multilateral agreements.

Freedoms of the Air

A second outcome of the Chicago Convention was the designation of so-called 'Freedoms of the Air', which form the basis for the types of conditions negotiated by government. Five formal freedoms exist, but over the years, ICAO has recognised the existence of four others (ICAO, 2016; 2019), as shown in Figure 6.1.

The first four freedoms are relatively straightforward. The first allows a country's carriers to overfly another country's airspace. For example, Russia would grant first freedom rights to a country like China in order to let it fly to France. Overflying fees are charged. Second freedom rights allow a plane to land for technical purposes, primarily to refuel for long haul flights. Second freedom rights are becoming less important as technological advances in airplane design enable them to fly much further.

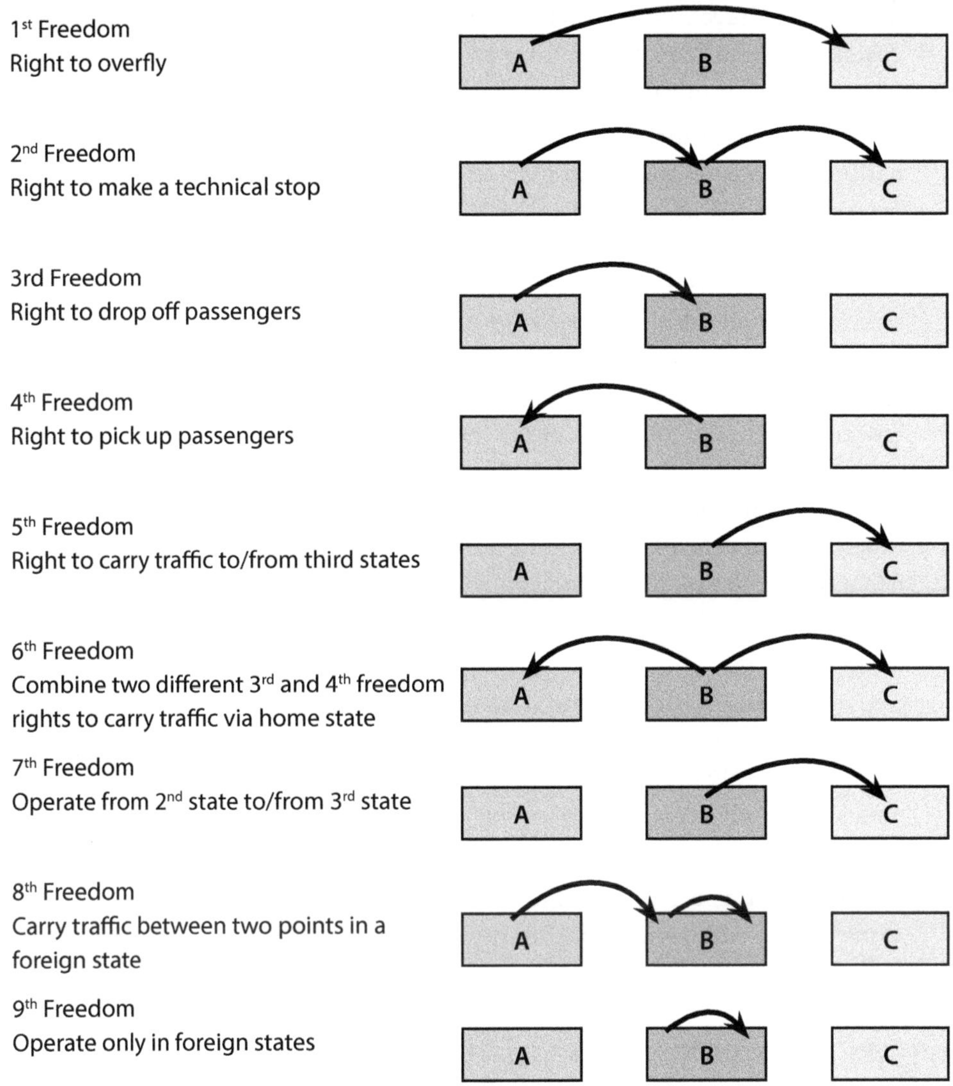

Figure 6.1: Freedoms of the Air. *Source*: ICAO (2016)

Third and fourth freedom rights are the basis of most agreements between countries, for they enable a country to drop off passengers in one destination and pick up others for the return trip. No country would negotiate third freedoms without also negotiating fourth freedoms for it would mean the aircraft would have to ferry back to its home port empty.

Fifth freedoms or beyond rights are the right of an airline of one country to carry traffic between two foreign countries as long as the flight originates and terminates in its own country. The granting of fifth freedom rights enables a carrier from a third country to compete on an international route not directly associated with that country. They were designed initially to enable countries that did not have international air carriers to benefit from international travel. It basically permits another country to enter into the international travel sector of two different countries.

The growth of international aviation and the entry of new nations with international air carriers has also created a de facto fifth freedom right, which have been recognized as 'sixth freedom', where intermediate countries can combine multiple third and fourth freedom agreements to enter into aviation markets that they would otherwise be excluded. For example, Country A would negotiate a separate 3rd and 4th freedom bilateral agreement with a country like Australia. It would then negotiate a similar 3rd and 4th freedom agreement with another country, like the UK. In doing so, it would create a one-stop service between the UK and Australia, and thus enter the market, even though neither the UK nor Australia may have negotiated additional capacity with each other. Thus by astute negotiation, nations can dramatically benefit from tourism to third countries. As a result, host or originating nations may not be benefiting from international tourism as much as they would normally expect to.

The premise behind the first six freedoms is that all flights must originate and terminate in the home country. The seventh freedoms enable third party carriers to fly between countries without necessarily having to start or stop in their home country. As an example, a German airline could potentially fly an international route exclusively between Italy and Spain. The seventh freedoms are rare and are most commonly seen in open markets like the European Union.

The eighth and ninth freedoms allow foreign airlines to enter the domestic market either as a leg of an international flight (eighth freedom) or as a standalone de facto domestic carrier (ninth freedom).

Tourism as international trade in services

Finally, most international trade today is governed by rules developed by the World Trade Organization (WTO). The WTO is a member-driven organization comprised of 164 member and 22 observer governments, as of 2019 (WTO, 2019a). It is estimated that members account for 98% or world trade and about 97% of global GDP. Its mandate is to provide the legal ground rules for international commerce that bind governments to keep their trade policies within agreed limits. The international trading system has an enormous effect on tourism, not only as far as transport is concerned, but also with the procurement of goods and services directly and indirectly (UNWTO/WTO/WTTC, 2018). Direct procurement relates to the provision of hotel and tour operations, while indirect procurement relates to the goods purchased to enable the delivery of services, such as the import of vehicles, staff and the like. It is for this reason that Dwyer (2015) writes about how international bodies such as the WTO and the International Monetary Fund have facilitated globalization by reducing trade barriers.

Currently, about 60 agreements and decisions have been developed. While each agreement is complex, they all share the same common, fundamental principles to varying degrees: removing discrimination between trading partners (most-favoured nation status); treating locally produced and foreign goods equally after they have entered a market (national treatment); gradually working toward freer trade by removing trade barriers; ensuring predictability, stability and transparency through legally binding commitments; promoting fair competition; and aiding economic development (WTO, 2019b). WTO agreements cover three broad areas of trade: goods, governed under the General Agreement on Tariffs and Trade (GATT); services governed under the General Agreement on the Trade in Services (GATS); and intellectual property rights, governed by the Trade-Related Aspects on International Property Rights (TRIPS). Each of these agreements states broad principles, while GATT and GATS also include extra agreements and annexes, lists of binding agreements and mechanisms to ensure transparency and resolve disputes

Some aspects of tourism are addressed within the terms and conditions stipulated by the General Agreement on Trade in Services. GATS covers all internationally-traded services, with one notable exception – services affecting air traffic rights as well as those directly related to the exercise of such rights. At present, more than 133 member countries or territories have made tourism commitments, more than any other service sector (George and Henthorne, 2007; WTO, 2018). These commitments relate to four aspects of tourism: hotels and restaurants; travel agencies and tour operators; tourist guide services; and 'other' (Hoad, 2003). It is noted that:

> "this level is greater than for any other sector, and indicates the desire of most members to expand their tourism sectors and to increase inward foreign direct investment as part of efforts to promote economic growth. Nonetheless, the existing level of GATS commitments does not imply the full liberalization of international trade in tourism services, especially when tourism-characteristic services are considered" (WTO, 2009: 14)

Basically, according to the UNWTO (ITC/UNWTO 2015:7):

> "Like the transactions in other services sectors, tourism transactions are therefore ruled by the provisions and commitments in the WTO's General Agreement on Trade in Services (GATS). Whenever tourism providers within the value chain import or export goods or services, however, tariffs or non-tariff measures falling under provisions in other WTO or regional agreements may become relevant... In addition, GATS

> commitments in services other than tourism and travel are also highly relevant for unlocking the potential of the sector. Commitments under financial services affect how easily tourists can access local currency or their domestic bank accounts when they wish to make purchases. Commitments under insurance services will affect where and how travel insurance is contracted."

The report goes on to mention that tourism has four ways of being traded under the GATS, even if tourists crossing borders are the most common form of international trade in tourism services. Each of these four modes of supply affects tourism and travel-related services:

- Mode 1 – Cross-border supply: travel agents, hotels, and travel guides can be accessed electronically from importing countries.
- Mode 2 – Consumption abroad: the mode of supply most typically associated with tourism.
- Mode 3 – Commercial presence: chains of hotels and tour operators frequently set up hotels in multiple locations.
- Mode 4 – Presence of natural persons: tour operators and hotels often employ foreign staff with skill sets that are not readily available in their country of operation.

Conclusion

This chapter has provided a brief introduction of some of the political dimensions of international tourism. Understanding these dimensions helps understand the varying nature of time and distance on tourism flows and also explains limitations to destination choice, to a great extent.

Discussion points / Exercises

1. What factors influence bilateral tourism flows? Select two countries that are seen to be antagonistic to each other and analyse the factors that prohibit international tourism activity.
2. Why is it that some countries impose strict visa requirements on some African and Islamic states?
3. What conditions must exist for visa-free access to be granted to residents of a country wishing to visit another country?
4. Discuss why airlines want liberalised Air Service Agreements, while some countries may be opposed to such liberalisation.

Points to consider for revision

- Granting the rights to travel internationally is fundamentally a political decision that is influenced by a range of factors.
- People can enter a country by any number of legal means. Tourist visas are the easiest way to enter a country and, therefore, are also the easiest to exploit.
- Countries impose a range of restrictions on who can enter a country, how long they can stay, and set other conditions to reduce the risk of illegal immigration.
- International air transport is a form of trade guided by the formation of bilateral air service agreements between states.
- The Freedoms of the Air set the rules for much international aviation activity.
- The General Agreement on Trade in Services (GATS) dictates much of the internationalisation of tourism activity, other than air service agreements.

■ Additional readings

Neumayer, E (2010) Visa restrictions and bilateral travel. *The Professional Geographer*, 62(2), 171-181.

Prideaux, B. (2005) Factors affecting bilateral tourism flows. *Annals of Tourism Research*, 32(3), 780-801.

UNWTO (2013) *The Impact of Visa Facilitation in APEC Economies*, UN World Tourism Organization and World Travel & Tourism Council. UNWTO, Madrid. https://www.e-unwto.org/doi/pdf/10.18111/9789284415960 <downloaded Jun 3, 2019>

References

Artal-Tur, A., Pallardó-López, V. & Requena-Silvente, F. (2016). Examining the impact of visa restrictions on international tourist flows using panel data, *Estudios de Economía*, 43(2), 265-279.

Aus (2019) The Bilateral System—how international air services work. https://www.infrastructure.gov.au/aviation/international/bilateral_system.aspx.

Balli, F., Balli, H. O. & Cebeci, K. (2013). Impacts of exported Turkish soap operas and visa-free policy on inbound tourism to Turkey. *Tourism Management*, 37, 186-192.

Bangwayo-Skeete, P. & Skeete, R. (2017) Who travels visa-free? insights into tourist hassle-free travel. *Journal of Travel Research* 56(3), 407-418.

Beenstock, M., Felsenstein, D. & Rubin, Z. (2015) Visa waivers, multilateral resistance and international tourism: some evidence from Israel. *Letters in Spatial and Resources Science* 8(3), 357-371.

DTRD (1998) *Submission by the Department of Transport and Regional Development to the Inquiry by the Industry Commission into International Air Services*, Canberra, March.

Dwyer, L. (2015) Globalization of tourism: Drivers and outcomes, *Tourism Recreation Research*, 40(3), 326-339, DOI: 10.1080/02508281.2015.1075723.

EU (2013) *An Overview of the Air Service Agreements Concluded by the EU*. European Union. http://www.europarl.europa.eu/RegData/etudes/note/join/2013/495849/IPOL-TRAN_NT(2013)495849_EN.pdf

Fayed, H. & Westlake, J. (2002) Globalization of Air Transport: The Challenges of the GATS. *Tourism Economics*, 8(4), 431–455.https://doi.org/10.5367/000000002101298214

George, B. & Henthorne, T. (2007) Tourism and the general agreement on trade in services sustainability and other developmental concerns. *International Journal of Social Economics*, 34, 136–146.

Hoad, D. (2003) The General Agreement on Trade in Services and the impact of trade liberalisation on tourism and sustainability. *Tourism and Hospitality Research*, 4(3), 213-227.

Hu, X. (2013) Economic benefits associated with the visa waiver program – a difference-indifference approach. *Global Journal of Business Research*. 7(1), 81-89.

ICAO (2016) *Manual on the Regulation of International Air Transport (Doc 9626) (3rd Ed.)*. Montreal: International Civil Aviation Authority. https://www.icao.int/Meetings/a39/Documents/Provisional_Doc_9626.pdf <downloaded June 3, 2019.>

ICAO (2019) *Freedoms of the Air*. Montreal: International Civil Aviation Authority. https://www.icao.int/pages/freedomsair.aspx <downloaded June 3, 2019>.

ITC/UNWTO (2015) *Tourism and trade: A global agenda for sustainable development*, ITC, Geneva: International Trade Centre and World Tourism Organization http://www.intracen.org/uploadedFiles/intracenorg/Content/Publications/Tourism_and_Trade__low%20res_2014-2015-335.pdf

Karaman, A. (2016) The pernicious impact of visa restrictions on inbound tourism: the case of Turkey, *Turkish Studies*, 17(3), 502-524. DOI: 10.1080/14683849.2016.1170602.

Lawson, R. and Roychoudhury, S. (2013) Do travel visa requirements impede tourist travel? *Journal of Economics and Finance*, 40(4), 817-828.

Lee, C., Song, H. & Bendel, L. (2010) The impact of visa-free entry on outbound tourism: a case study of South Korean travelers visiting Japan *Tourism Geographies*, 12(2), 302-323.

Li S. & Song, H. (2013) Economic impacts of visa restrictions on Tourism: A case of two events in China. *Annals of Tourism Research*, 43, 251-271.

Lissitzyn, O. (1964) Bilateral Agreements on Air Transport, *Journal of Air Law and Commerce*, 30(3), 248-263.

Luedtke, A., Byrd, D. G., & Alexander, K. P. (2010) The politics of visas. *The Whitehead Journal of Diplomatic and International Relations*, 11(1), 147-160.

Liu, A. & B. McKercher (2015) The impact of visa liberalization on tourist behaviors: The case of China outbound market visiting Hong Kong. *Journal of Travel Research*, 5595), 603-611. DOI: 10.1177/0047287514564599.

Mau, S., Gülzau, F., Laube, L. & Zaun, N. (2015) The global mobility divide: How visa policies have evolved over time, *Journal of Ethnic and Migration Studies*, 41(8), 1192-1213. DOI: 10.1080/1369183X.2015.1005007.

Neumaeyer, E. (2006) Unequal access to foreign spaces: how states use visa restrictions to regulate mobility in a globalised world. *Transactions of the British Institute of Geographers*, 31(1), 72-84.

Neumayer, E (2010) Visa restrictions and bilateral travel. *The Professional Geographer*, 62(2), 171-181.

OECD (2014) *Air Service Agreement Liberalisation and Airline Alliances*, https://www.itf-oecd.org/sites/default/files/docs/14airserviceagreements.pdf <downloaded June 3, 2019>.

Prideaux, B. (2005) Factors affecting bilateral tourism flows. *Annals of Tourism Research*, 32(3), 780-801.

Song, H., Gartner, W. & Tasci, A. (2012) Visa restrictions and their adverse economic and marketing implications – Evidence from China. *Tourism Management*, 33(2), 397-412.

UNWTO (2013) *The Impact of Visa Facilitation in APEC Economies*, UN World Tourism Organization and World Travel & Tourism Council. UNWTO, Madrid. https://www.e-unwto.org/doi/pdf/10.18111/9789284415960 <downloaded Jun 3, 2019>.

UNWTO/WTO/WTTC (2018) *Tourism, Trade and the WTO: A Joint Communication from UNWTO, WTO, ITC and WTTC*. UN World Tourism Organization. http://www2.unwto.org/ru/node/51274 <downloaded June 1, 2019>.

Webster, C. & Ivanov, S. (2016) The ideologies of national security and tourist visa restrictions. *International Journal of Tourism Policy*, 6(3-4), 171-190.

Whyte, B. (2008) Visa-free travel privileges: an exploratory geographical analysis, *Tourism Geographies*, 10(2), 127-149.

WTO (2018) *Tourism and Travel-related Services.* World Trade Organization. https://www.wto.org/english/tratop_e/serv_e/tourism_e/tourism_e.htm <downloaded June 1, 2019>

WTO (2019a) The WTO. World Trade Organisation. https://www.wto.org/english/thewto_e/thewto_e.htm <downloaded June 1, 2019>.

WTO (2019b) *General Agreement on Trade in Services.* World Trade Organization. https://www.wto.org/english/docs_e/legal_e/26-gats_01_e.htm <downloaded June 1, 2019>.

7 The Political Economy of Tourism and Approaches to Forecasting

By the end of this chapter, the reader will be able to:

- Describe the difference between political economy and economics
- Identify and describe the differences among Keynesian economics, Neoliberalism and Marxist perspectives
- Explain the difference between quantitative and qualitative forecasting models.

Introduction

While there is a vast literature on issues related to political economy, tourism scholars have generally paid little attention to this area of research. This chapter discusses some of the underlying debates related to political economy to provide a window into the fundamental role that economic forces play in the operation, sustainability and profitability of the global tourism system. In the near future, the need to transition to a carbon-neutral production system (see Chapter 15), and the immense difficulties that will be faced in reining in our increasing use of diminishing resources, will generate considerable discussion on the merits of various production systems such as Neoliberalism, socialism, neo-Marxism and post-Keynesian.

This chapter also considers a range of models used to forecast tourism activity, a task that is an important element in future tourism investment and planning activities.

The current production system is firmly embedded in the market driven capitalist system of economic production and for this reason, the following discussion will focus on capitalism. This chapter provides a neo-Marxist perspective to demonstrate a counter view to the predominate interpretation of capitalism through a neoliberal lens. Degrowth (see Chapter 15) also offers another vision for organizing a future economic system of production. Understanding how we arrived at our system of economic production is important and provides insights into the next stage of economic evolution, which will most likely to be based on a carbon-neutral economic production system.

So what is the difference between political economy and economics? Political economy is concerned with the process of production and its impact on national income and wealth, and the conflicts, inequalities and problems that arise in trade, law and government (Bianchi, 2018; Williams, 2004). Scholars generally approach the study of political economy from perspectives that include Liberalism, Keynesianism, Marxism, Socialism and Neoliberalism (and the many sub-branches of these schools). Economics on the other hand is concerned with production, distribution and consumption of goods and services and leaves aside political and social considerations although as Bianchi (2018) observed, the distinction between political economy and economics has become blurred because of the political nature of significant areas of neoliberal thought.

■ Capitalism

Fletcher, Fyall, Gilbert and Wanhill (2018: 615) define capitalism as "a social system based on individual rights where goods and services are produced and exchanged with minimal government interference". This simple definition encapsulates three important elements: a social system, exchange, and minimal government interference. Obviously, this definition is not a universal. The understanding of the term 'capitalism' has changed over time, is contested and continues to evolve. Views on the role that government should play in marketplace operations constitute the key points of difference in the standpoints adopted by various schools of economists. Li, Liu and Song (2019) identify three standpoints, the first being those who believe that the market is able to operate efficiently with little or no government intervention. This viewpoint is typified by the laissez-faire approach advocated by Adam Smith in the 18th Century and more recently by the neoclassical school that rose to prominence in the 1980s. The second group is broadly identified as the Keynesian school that believe that government intervention is essential for stabilising economic activity. The most recent standpoint is characterised by new Keynesian economics which models macroeconomic operations with micro foundations. In their analysis, Li et al. (2019) ignored the neo-Marxist school which takes a contrasting view of economic organisation based on the role of labour and capital.

Leaving aside issues related to the structure of 'social systems' outlined in Fletcher et al.'s (2018) definition of capitalism, many economies operate with a mix of State Owned Enterprises (SOE) and Privately Owned Enterprises (POE) and with varying levels of government intervention in the economy. The Chinese version of capitalism for example features both SOEs and POEs, both of which are subject to high levels of state oversight. Other countries such as the USA and the UK practice a version of capitalism that lies much closer to the principles of classical laissez-faire capitalism where labour markets are largely deregulated, and the state has largely withdrawn from operating state-owned enterprises. From a tourism perspective, an understanding of the form of capitalism adopted by specific countries is important. In China for example, POEs generally have close connections with state agencies and actors and are sometimes assigned roles in carrying out government policy (Milhaupt and Zheng, 2015). This is not the normal practice in western economies.

Laissez-faire economics

Adam Smith, an 18th Century scholar often referred to as the father of economics, postulated that competition and rational self-interest can lead to prosperity. His theories advocated minimal intervention by governments in the market, in the belief that removal of government controls would lead to ever-increasing prosperity. His theories underpinned the emergence of the capitalist economic system in the late 18th Century and economic liberalism, described as an economic and political philosophy that advocated support for private property and the market economy. The contributions by Smith and other scholars including David Ricardo, John Stuart Mills and Thomas Malthus led to the emergence of classical economics where the so-called natural law of production and exchange lead to a largely self-regulating economic system such that government intervention is not required. By extension, classical economics is the antithesis of non-capitalist economic systems of production such as planned economies and socialism.

The problem with this form of economic organisation is that crisis events (such as wars, major natural disaster, political upheaval, monopolies and pandemics) can severely disrupt self-regulating economic systems forcing government intervention. The COVID-19 pandemic provides an example of how a self-regulating economy is unable to deal effectively with crisis events where the solutions involved public health considerations rather than economic considerations (Prideaux, Thompson and Pabel, 2020). During the COVID-19 pandemic many governments including Japan, USA, UK and Australia funded large stimulus programs to prop up national economies. In Japan, the size of fiscal measures introduced to combat the economic impacts of the COVID-19 pandemic was equal to 21.1% of national income while in the US the size of fiscal measures were equal to 14.7% of its national income (Emmerson and Stockton, 2020). Interventions of this type have parallels with the economic policies introduced by governments during the latter years of the Great Depression (1929 to late 1930s) and the 2007/08 Global Financial Crisis (GFC).

Keynesian economics

Keynesian economics is based on the view that aggregate demand is unstable and volatile in market economies, and to avoid economic downturns of the type that occurred during the Great Depression, governments need to intervene in the market with both monetary and fiscal policy to stabilise the economy. In general, Keynes favoured a managed market economy based on government intervention when required to smooth out the business cycle. The approach advocated by Keynes contrasts to the earlier aggregate supply side focus that was a characteristic of classical economics. During the period from the latter part of the Great Depression to the mid-1970s Keynesian economics was viewed as the standard approach to economic policy development. During this period states managed economic downturns through macroeconomic interventions and welfare policies (Williams, 2004).

One of the enduring legacies of Keynesian economics was the formalisation of the concept of the multiplier effect (Rusu, 2011). Within a tourism context the multiplier effect refers to the direct, indirect and induced effects of tourism spending and is often used to measure success in terms of job creation and income generation. Direct and secondary (indirect and inducted) spending in other areas of the economy that are generated by tourism expenditure include income, investment, exports and government spending (Archer, 1989).

Another important concept introduced by Keynes (1936) was the idea of uncertainty. In periods of uncertainty, including during the GFC and the COVID-19 pandemic, consumers may postpone or cancel basic spending resulting in a fall in national economic activity (Gozgor and Ongan, 2017). Tourism is one economic sector that is likely to be affected by economic uncertainly, often leading to a fall in overall travel (Demir and Gozgor, 2018).

Keynesian and later post-Keynesian economic approaches were replaced by Neoliberalism economics from the early 1980s. This development has been attributed to the failure of Keynesian macroeconomics to manage stagflation during the 1970s, the collapse of socialist planning systems in the 1980s and the Latin American debt crisis of the 1970s (Lin, 2012). One of the major responses to these perceived failures of Keynesian economics was the growth of globalism and the subsequent weakening of controls that could be exercised by states over their domestic economies. More recently, the failures of Neoliberalism typified by the need for significant government interventions in the economy during the GFC have led to renewed interest in Keynesian economics and the use of public spending to stimulate growth and employment creation (Bramwell and Lane, 2009).

Neoliberalism

Neoliberalism is the most recent incarnation of capitalism and occupies a dominant position in the global political economy. However, its meaning is contested, widely debated and subject to constant evolution as the circumstances of national economies and the global economy change. For example, in a discussion on the evolution of the term over a number of decades, Boas and Gans-Morse (2009: 137) observed that "Neoliberalism has undergone a striking transformation, from a positive label coined by the German Freiberg School to denote a moderate renovation of classic liberalism, to a normatively negative term associated with radical economic reforms in Pinochet's Chile".

Peck, Brenner and Theodore (2018: 6) describe Neoliberalism as "an idealised vision of market rule and liberal freedoms, combining a utilitarian conception of market rationality and competitive individualism with deep antipathies to social redistribution and solidarity". From a similar perspective, Wearing, McDonald, Taylor and Ronen (2019) describe Neoliberalism as a process that aims to privilege free-market forces by reducing the role of government decision making in all areas of life.

Neoliberalism is strongly supported by major international organisations including the World Trade Organization, the International Monetary Fund (IMF) and the World Bank (Wearing et al., 2019). Contemporary Neoliberalism is associated with the views of influential economist Milton Friedman of the Chicago School of Economics. Friedman, a critic of government intervention, placed considerable emphasis on the efficiency of the market versus the distortions that he argued are commonly associated with government intervention and market regulation. Liberalism in this context has a direct lineage to laissez-faire economics which emphasises the role of the market (Palley, 2004).

In its current form, Neoliberalism is broadly characterised by laissez-faire style economic liberalism and free market capitalism and its defining features include deregulation, privatisation, globalisation, free trade, austerity and an increasing role for the private sector in the economy and society. As Jones, Parker and Ten Bos (2005) note, Neoliberalism as a set of economic ideals rose to prominence from the mid to late 1970s and is associated with economic policies introduced President Ronald Regan in the United States and Prime Minister Margaret Thatcher in the

UK. Jones et al. (2005:100) describe Neoliberalism as an "updated version of older ideas about 'liberal economics' which has long argued that markets should be free from intervention by the state. In its simplest version, it reads: markets good, government bad".

Neoliberal economic policies were enshrined in the so called 'Washington Consensus', a set of ten policy settings advocated by the US Treasury, the World Bank and the IMF that were prescribed for countries facing economic crisis. As Wearing et al. (2019) observed, the Washington Consensus represents a consensus between North American and European politicians, multinational corporations, corporate executives, regulatory agencies and media owners about the promotion of free-market ideologies as a superior mechanism for managing economies and equitable allocation of resources. One of the contradictions of Neoliberalism is the call by developed nations in Europe and the US for developing countries to adopt neoliberal policies based on opening their domestic markets to international trade while continuing to retain high levels of protection (subsidies) on sectors of their own economy, particularly agriculture (Jones et al., 2005).

Adoption of neoliberal policies saw a wave of deregulation and privatisation of government owned tourism related infrastructure including railways and airports. In the UK, the government privatised its railway system however the claimed benefits remain a matter of controversy. In Cairns, Australia, the state government sold the Cairns international airport to a private consortium resulting in a shift from local control to corporate control by a private company that has no local investors or board members who are resident in the region that the airport serves. Prior to its 2008 sale, the airport was owned by the Cairns Port Authority, a state government owned entity but with a local board of directors. The result of the sale is that the Cairns community has no direct representation in decisions that affect its most important tourism infrastructure asset. Globally, many destinations share a similar circumstance where decisions that affect the local tourism industry are made in distant board rooms and where considerations such as profits and rates of return on investment often have a higher priority than impacts on the local tourism industry or community. Issues of this type can create significant problems for communities attempting to plan the future course of their tourism sector.

In their critique of global tourism that is based on neoliberal practices, Wearing et al. (2019) describe a situation where global tourism has emerged as a consumer culture that has led to unsustainable forms of tourism production and consumption, which has begun to have adverse effects on local communities (over-tourism) and the ecosystem. The loss of control of local resources to multinational companies such as Airbnb is a symptom of the impacts of unregulated neoliberal tourism production.

In a critique of Neoliberalism Higgins-Desbiolles (2006: 1195) makes the point that "the mantra of tourism is an 'industry' that is subject only to the rules of the marketplace has been repeated so frequently that to think otherwise is almost viewed as nonsensical". In a later critique, Higgins-Desbiolles and Whyte (2015: 106) highlight the pressures that may be placed on communities to enter the market on terms that may be unfair and make them choose between "ecological integrity, social and community wellbeing" and economic growth, a consequence of Neoliberalism's focus on "unfettered growth".

■ A neo-Marxist perspective

Bianchi (2011: 17) stated that Marxist political economy has "made a profound contribution" to how we understand and critique ideas about the capitalist political economy. Given the previous critiques about the dominant role of Neoliberalism, it is useful to consider the contribution that neo-Marxist scholars can make to our understanding of the operation of global tourism. Observing that there were significant contradictions in laissez-faire economics that led to the exploitation of the working class, Marx postulated that under capitalism, profit was the value added by workers but not paid out in wages. The expropriation of profits by the owners of capital enabled them to gain control over the means of production and exploit workers. The end result of appropriating the surplus value of labour, where workers who are paid less than the full value of their labour, will lead to overproduction, over-accumulation and the ultimate self-destruction of the system. His ideas and those of his colleague Engels were incorporated into Marxist-Leninist philosophies adopted in part by the USSR. While the experiment failed, many of his insights into the contradictions of capitalism have been adopted by neo-Marxist scholars and continue to provide a useful critique of modern capitalism including Neoliberalism.

From a neo-Marxist perspective, the rise of the gig economy where work is casualized is a classic case of capital exploiting the worker. Removal of workers' rights to benefits, such as fixed hours of work, sick leave and annual holidays, enables firms to maximise profits without the obligation of returning some of the extra profit gained from productivity increases to workers. In the tourism industry, the trend to casualisation of the workforce is a trend that has been gaining pace for many years.

Bianchi (2011: 23) observed that "Neoliberalism and globalisation have stimulated a resurgence of the power of capital (or 'shareholder value') and thereby accentuated the tendencies towards concentration and centralisation of capital emphasised by Marx". One example of the concentration of capital cited by Bianchi is the cruise industry where two corporations handle nearly 80% of the Caribbean's passenger capacity.

Neo-Marxist thinkers have described tourism as a form of capitalism (Bianchi, 2011; Fletcher, 2011) where tourism growth provides a mechanism to alleviate the problems of overproduction predicted by Marx. From this perspective, excess capital is able to be reabsorbed into the economy as investment in tourism activity thereby providing a 'fix' to forestall a crisis of overproduction (Harvey, 1989: Fletcher, 2011). Fletcher (2011), building on the views of O'Connor (1988) that capitalism's 'second contradiction' is one of overproduction, argues that sustainable tourism, including ecotourism, provides a means to negate the crisis of capitalism's need to continually expand to survive. By providing a mechanism where finite natural resources can be used in a 'sustainable manner', the danger of natural resource exhaustion through over-extraction can be hopefully averted. From a neo-Marxist perspective, the recent emergence of concerns about overtourism can be interpreted as an example of overproduction that leads to a crisis, temporarily forestalled by the COVID-19 Pandemic.

Forecasting

Dwyer, Forsyth and Dwyer (2020: 159) describe forecasting as being "concerned with predicting the future, with estimating what will happen at some future time". From a tourism perspective, forecasting is important for supporting business cases for private sector investment and public sector infrastructure planning and development. The opportunity cost of failing to use forecasting techniques to estimate future demand may be high and result in over or under estimation of demand and poor planning (Calantone, Benedetto and Bofanic, 1987). For these reasons forecasting plays an important part in many aspects of tourism research and is extensively used by both the private and public sectors. In the public sector, forecasting is commonly used to evaluate how changes in government policy as well as externalities (such as international relations) may affect demand, employment and tax revenue. The private sector also relies on forecasting to determine changes in revenue, returns on investment and marketing. The importance of forecasting in the private sector is illustrated by annual 20-year forecast of growth in travel and the demand for aircraft undertaken by Boeing (2019), the results of which are used to inform aircraft design and future passenger demand.

Forecasting is also an important element in long-term planning. However, the ability to predict future economic activity, particularly during periods of uncertainty, is difficult because of unanticipated events.

Main categories

The main categories of forecasting are usually described as exploratory, speculative, normative/explicative and integrative (Van Doorn, 1984; Calantone, Benedetto and Bofanic, 1987).

- Exploratory forecasting relies on the extrapolation of past trends to predict what may happen in the future, based on the identification of relationships between independent and outcome variables. The major models used in exploratory research include time-series, casual methods, historical analogy, morphological analysis and projective scenarios).
- Speculative forecasting uses expert judgement to estimate what may happen in the future and use methods such as panel consensus, individual expert opinion, Delphi, expert opinion and brainstorming.
- Normative/explicative forecasting is based on identifying strategies that may lead to the attainment of a desired situation in a future time-space. Key models include Bayesian statistics, subjective probabilistic forecasting, prospective scenarios and pattern identification.
- Finally, integrative forecasting is an attempt to identify underling relationships between variables to achieve a convergence of results. Models include input-output analysis, multimethod models, cross impact analysis and dynamic systems models.

The main focus of forecasting research in the area of tourism has been in demand forecasting. Other scholars (Song, Qui and Park, 2019) classify forecasting models as time-series, econometric, AI-based models and judgement models.

In a review of the accuracy of forecasting used to evaluate tourist arrivals including short -, medium- and long-term arrivals in selected European countries between 2000 and 2013, Hassani, Silva, Antonakakis, Filis and Gupta (2017) found that no single model consistently outperformed all other models. The implication of this finding is that customised model building should be based on selection of models according to time horizons and the circumstances of the individual countries involved.

Scenarios

Demand forecasting is often subject to unanticipated disruptions that distort quantitative forecasting models and, in these circumstances, qualitative methods of forecasting such as scenarios may be appropriate (Prideaux, Laws and

Faulkner, 2003). Scenarios have emerged as an alternative to demand forecasting and are increasingly used as policy tools to inform policy development (Bradfield et al., 2005). From this perspective, scenarios are a method of attempting to understand the future and for providing guidance on the possible outcomes of alternative policy formulation. Whereas demand forecasting generally relies on known relationships between variables such as past demand growth, trends in GDP and exchange rates, scenarios provide opportunities for exploring alternative futures based on selective combinations of knowns (e.g. past demand factors) and alternative futures.

There are a number of approaches to scenario building and these range from the alternative stories of the future (Yeoman, Munro and McMahon-Beattie, 2006) to empirical scenarios of the type employed by the IPCC to build models of future climates. Peeters and Dubois (2010) describe four groups of scenarios based on combinations of exploratory/ normative and quantitative/ qualitative projections. Exploratory (plausible) scenarios usually extrapolate trends and are sometimes used as a form of forecasting. Normative (desirable) scenarios are used to identify a desirable future and employ backcasting approaches to identify pathways to achieve this future (Prideaux et al., 2003: 476).

Scenarios are extensively used by the Intergovernmental Panel on Climate Change (IPCC) and other bodies to examine climate futures under varying combinations of greenhouse gas emissions. The IPCC (2007:145) defines scenarios of this type as "a coherent, internally consistent and plausible description of a possible future state of the world. Scenarios are not predictions or forecasts but are alternative images without ascribed likelihoods of how the future might unfold". IPCC scenarios are extensively used in the tourism literature to develop an understanding of what the future may look like in terms of temperature, precipitation and sea levels and then apply this knowledge to inform policy and planning. Given the significance of unanticipated changes to demand such as those that occurred from the 9/11 terrorist attack on the US and the COVID-19 pandemic, there is a strong case for a greater use of scenarios in tourism forecasting.

Conclusion

Tourism scholars have mostly ignored the insights that an understanding of political economy can bring to debates on the impacts that national economic organization can have on tourism demand and supply. As this chapter has pointed out, the manner in which national economies are organized has significant implications for tourism. For example, the casualization of large sections of the tourism workforce has occurred because national governments have adopted neoliberal economics.

The need to move to a new carbon neutral production system and to reduce the exploitation of the Earth's remaining stock of non-renewable resources is an issue of considerable importance to tourism and is likely to become of increasing interest to scholars in future years.

Discussion points / Exercises

1. What are the key differences between a capitalist and a Marxist approach to political economies?
2. Why were Keynesian economic approaches replaced by neoliberalist approaches?
3. What are the key differences among the various approaches to forecasting?

Points to consider for revision

- The key driver of tourism is the interplay between demand for travel and supply of services.
- A variety of political economy approaches have been adopted to facilitate this interplay, to greater or lesser success, depending on the system adopted.
- Both state-owned and privately-owned enterprises function within a tourism system, but the mix varies by economy.
- Forecasting involves more than predicting the future, for the outcome of forecasts influences government policy and private enterprise business decisions.

Additional readings

Britton, S. (1982) The political economy of tourism in the Third World, *Annals of Tourism Research*, 9, 331-359.

Martins, L., Gan, Y. & Ferreira-Lopes, A. (2017) An empirical analysis of the influence of macroeconomic determinants on World tourism demand, *Tourism Management*, 61, 248-260.

Mosedale, J. (2011) *Political Economy of Tourism: A Critical Perspective*, Routledge, Abingdon and New York

References

Archer, B. (1989) Tourism and island economics: Impact analyses, *Progress in Tourism and Hospitality Management*, 1, 125-134.

Bianchi, R. (2011) Tourism, capitalism and Marxist political economy, in Mosedale, J. (ed), *Political Economy of Tourism: A Critical Perspective*, Routledge, Abingdon and New York (pp. 17-37).

Bianchi, R. (2018) The political economy of tourism development: A critical review, *Annals of Tourism Research*, 70, 88-102.

Boas, T.C. & Gans-Morse, J. (2009) Neoliberalism: From new liberal philosophy to anti-liberal slogan. *Studies in Comparative International Development*, 44, 137–161. https://doi.org/10.1007/s12116-009-9040-5.

Boeing (2019) *Commercial Market Outlook 2019-2038*, Boeing, https://www.boeing.com/commercial/market/commercial-market-outlook/

Bradfield, R., Wright, G., Burt, G., Cairns, G. & van Der Heijden, K. (2005) The origins and evolution of scenario techniques in long range business planning, *Futures*, 37, 795–812.

Bramwell, B. & Lane, B. (2009) Economic cycles, times of change and sustainable tourism, *Journal of Sustainable Tourism*, 17(1), 1-4, DOI: 10.1080/09669580802608061.

Calantone, R., Di Benedetto, CA, & Bojanic, D. (1987) A comprehensive review of the tourism forecasting literature, *Journal of Travel Research*, Fall, 28-39.

Demir, E. & Gozgor, G. (2018) Does economic policy uncertainty affect tourism? *Annals of Tourism Research*, 69 (C), 15-17.

Dwyer, L., Forsyth, P. & Dwyer, W. (2020) *Tourism Economics and Policy*, 2nd edition, Channel View, Bristol.

Emmerson, C. & Stockton, I. (2020) How does the size of the UK's fiscal response to coronavirus compare with other countries? *Institute for Fiscal Studies*, 14 May, downloaded from https://www.ifs.org.uk/publications/14845.

Fletcher, J., Fyall, A., Gilbert, D. & Wanhill, S. (2018) *Tourism Principles and Practice* 6th edition, Pearson, Harlow.

Fletcher, R. (2011) Sustaining tourism, sustaining capitalism? The tourism industry's role in global capitalistic expansion, *Tourism Geographies*, 13(3), 443-461.

Gozgor, G. & Ongan, S. (2017) Policy uncertainty and tourism demand: Empirical evidence from the USA, *International Journal of Tourism Research*, 19, 99-106.

Harvey, D. (1989) *The Condition of Postmodernity: An Inquiry into the Origins of Cultural Change*, Basil Blackwell, Oxford.

Hassani, H., Silva, E., Antonakakis, N., Filis, G. & Gupta, R. (2017) Forecasting accuracy evaluation of tourist arrivals, *Annals of Tourism Research*, 63, 112-127.

Higgins-Desbiollles (2006) More than an "industry": The forgotten power of tourism as a social force, *Tourism Management*, 27, 1192-1208.

Higgins-Desbiolles, F. & Whyte, K. (2015) Tourism and human rights, in Hall, CM., Gössling, S. & Scott, D. (eds), *The Routledge Handbook of Tourism and Sustainability*, Routledge, London and New York.

7

IPCC (2007) *Climate Change 2007: Impacts, adaptation and vulnerability. Working Group II contribution to the Intergovernmental Panel on Climate Change. Fourth Assessment Report.* International Panel on Climate Change, Geneva.

Jones, C., Parker, M. & Ten Bos, R. (2005), *Business Ethics, a Critical Approach*, Routledge, Abingdon.

Keynes, J.M. (1936) *The General Theory of Interest, Employment and Money*, MacMillan, London.

Li, S., Liu, A. & Song, H. (2019) Does tourism support supply-side structural reform in China? *Tourism Management*, 71, 305-314.

Lin, J. (2012), *New Structural Economics: A Framework for Rethinking Development and Policy*, World Bank Researcher Observer, 26, World Bank, Washington, DC.

Marx, K. (1974) *Capital: A Critical Analysis of Capitalist Production*, Volume 1, Lawrence and Wishart, London (First English edition published 1887).

Milhaupt, C. & Zheng, W. (2015) Beyond ownership: State capitalism and the Chinese firm, 103 *Georgetown Law Journal* 665, available at http://scholarship.law.ufl.edu/facultypub/696.

O'Conner, J. (1988) Capitalism, nature, socialism: A theoretical introduction, *Capitalism, Nature, Socialism*, 1(1), 11-38.

Palley, T. (2004) From Keynesianism to Neoliberalism: Shifting paradigms in economics. *Foreign Policy in Focus*, May 5th, retrieved from https://fpif.org/from_keynesianism_to_neoliberalism_shifting_paradigms_in_economics/

Peck, J., Brenner, N., & Theodore, N. (2018) Actually existing neoliberalism. In D. Cahill, M. Cooper, M. Konings, & D. Primrose (eds), *The Sage Handbook of Neoliberalism* (pp. 3–15). London: Sage.

Peeters, P. & Dubois, G. (2010) Tourism travel under climate change mitigation constraints, *Journal of Transport Geography*, 18, 447 – 457.

Prideaux, B., Laws, E. & Faulkner, B. (2003) Events in Indonesia: Exploring the limits to formal tourism trends forecasting methods in complex crisis situations, *Tourism Management*, 24(4), 511-520.

Prideaux, B., Thompson, M. & Pabel, A. (2020) Lessons from COVID-19 can prepare global tourism for the economic transformation needed to combat climate change, *Tourism Geographies*, 22(3), 667-678.

Rusu, S. (2011) Tourism multiplier effect, *Journal of Economics and Business Research*, 1, 70-76.

Song, H., Qiu, R. & Park, J. (2019) A review of research on tourism demand forecasting: Launching the Annals of Tourism Research Curated Collection on tourism demand forecasting, *Annals of Tourism Research*, 75, 338-362.

Van Doorm, J. (1984) Tourism forecasting and the policy-maker – criteria of usefulness, *Tourism Management*, 5(1), 24-39.

Wearing, S., McDonald, M., Taylor, G. & Ronen, T. (2019) Neoliberalism and global tourism in Timothy, D. (ed) *Handbook of Globalisation and Tourism*, Edward Elgar, pp. 27-43.

Williams, A. (2004) Towards a political economy of tourism, in A. Lew, CM. Hall and A. Williams (eds.) *A Companion to Touris*m, Blackwell, Oxford, pp. 61-73.

Yeoman, I., Munro, C. & McMahon-Beattie, U. (2006) Tomorrow's: World, consumer and tourist, *Journal of Vacation Marketing*, 12(2), 174-190.

Part 3 – Evolution of Tourism

Tourism changes tourism! The question is how? This part of the book examines the three major evolutionary/revolutionary and lifecycle models that have shaped our thinking of how tourism, in general, and destinations, in particular, change over time.

Chapter 8 compares and contrasts Butler's and Plog's complementary models and engages in a discussion about which one better reflects how destinations change. Butler adopts a supply-side perspective with carrying capacity as the deterministic factor that influences the likelihood of a destination going into decline. Plog, on the other hand, adopts a supply-side perspective and argues that destinations attract a progressively softer type of tourist. If not managed carefully, destinations are at risk of sowing the seeds of their own destruction.

Chapter 9 takes a different perspective and introduces the concept of complexity theory to describe how destinations evolve rapidly over time. In spite of its broad acceptance in other fields of study, complexity theory has drawn relatively little attention in tourism studies, even though it better reflects the real world situation.

8 Lifecycle Models

By the end of this chapter, the reader will be able to:

- Describe the biological basis of destination lifecycle models
- Evaluate the validity of Butler's and Plog's lifecycle models
- Analyse the strengths and limitations of each model
- Compare and contrast the two models and describe how they integrate closely
- Explain the role of carrying capacity

Introduction

Two tourism lifecycle models developed by Butler (1980) and Plog (1974, 2001) more than 40 years ago have dominated the conceptualisation of this topic. Everybody is or should be familiar with the broad themes discussed in each model. But, while the Plog and Butler models are well known, they are not necessarily known well, for the original sources are somewhat dated, and in the case of Butler's work, obscure. Indeed, one wonders how many people have actually read the original source? Instead, they are so common that knowledge generally comes from secondary sources which invariably summarize the manuscripts into a few sentences or short paragraphs, and in doing so, simplify the complex and subtle issues raised about the nature of destinations and how they change over time. At worst, the works are condensed to such an extent that the only thing shown is the respective graphic and subheadings. It is for this reason that the reader is encouraged to read each paper as a supplement to this chapter.

Most readers should also be aware of the basic strengths and criticisms inherent in each model. For example, the devotion to the product metaphor is the main reason why Butler's work is simultaneously praised and criticized – praised because of its elegance, simplicity and intuitive logic, and criticized because it is difficult to prove empirically (Tooman, 1997; Weaver, 2000) and is deficient as a predictive tool (Agarwal, 2002; Hovinen, 2002; Moss et al., 2003). Likewise Plog's work has been both praised and criticized – praised as an early attempt to segment the tourist market by psychographic profile (Chandler and Costello, 2002; Harrill and Potts, 2002; McGuiggan, 2003) and criticized for not accurately depicting lifecycles (McKercher, 2005a).

Each had humble origins as each was written first as a conference paper and only later converted into a journal article. They are also a product of their times. Tourism in the late 1970s was undergoing rapid expansion as the development of jet aircraft led to the massing of tourism for the first time. Many remote economies identified tourism as their only economic development option and pursued a path of large scale development under the premise that the economic benefits of tourism would outweigh any adverse social and environmental impacts. They were unaware of many of the realities of tourism, and so it took authors such as Budowksi (1976), Doxey (1975), Jafari (1974), Wenkman (1975) and many, many others to write cautionary tales about spontaneous development and the inherent risks involved. Butler, in particular focused on the social and physical environmental impacts of tourism when developing his lifecycle model, and used carrying capacity as the deterministic variable to define the health of destinations. Plog took a supply-side perspective and wanted to understand reluctance to fly commercially.

Biological origins

Both Butler's and Plog's models, and indeed the broader product lifecycle and diffusion models on which they are based, have their origins in biological theory (Osland, 1981). Three complementary models influenced the development of the marketing or geographic origins of the lifecycle model.

The first and simplest analogy is the lifecycle that all higher organisms progress through (Crawford, 1984). As Tellis and Crawford (1981) note, individual biological specimens live predictable patterns of birth, growth, maturity, decline and death. Polli and Cook (1969) cite the usefulness of this analogy for the identification of each stage permits the evaluation of a series of tactical and strategic considerations that can be taken to either progress to the next stage or stem decline. However, this idea is overly simplistic for the biological model applies only to individual specimens, such as a person, a frog or a leaf (Crawford, 1984). It does not apply when groups of things are considered.

Instead, Osland (1991) suggests we need to consider studies of populations of species. This modification forms the second model. Here Osland (1991), summarising the literature notes the collective shape of the lifecycle curve is influenced by a range of biological factors, including food supply, competitors and the carrying capacity of the environment where the entity lives. Polli and Cook (1969) found it related well to the theory of the diffusion of products, innovation and the overarching lifecycle collections like entities evolve through. As they note, a product finds initial resistance to widespread acceptance and is purchased by only a limited segment of the buying population. Later, a larger segment of buyers adopt it, and sales begin to increase but eventually, the rate of growth decreases as the proportion of adopters gets closer and closer to a maximum. The rate of adoption remains constant throughout the maturity phase and diminishes in the decline phase. This approach makes more intuitive sense, but has been often omitted in favour of the more simplistic individual lifecycle model.

However, both the specimen and population models are static. Instead, Tellis and Crawford (1981) support an evolutionary model of the lifecycle. As they state "in biology the evolutionary cycle is used to explain the phenomenon of the growth and proliferation of species, groups of biological specimens that can breed within but not outside each group. Indeed, while the life-death cycle is a rigid, highly predictable one telling the story of each biological specimen (the individual horse, dog, etc.), the evolutionary cycle is a dynamic, open-ended one telling an even more exciting story of the origin, growth, and proliferation of entire species" (Tellis and Crawford, 1981: 127). Later Crawford (1984:6, 7) argues that an evolutionary approach is more suitable to consider for it contains four features of products. They are:

- **Cumulative,** where each change in the object under study builds on previous changes. This condition has been well documented in the Darwinian concept of evolution;
- **Directional**, where each change moves the object to greater complexity, efficiency and diversity. First, the species becomes more diverse, with more variations in the group. Second, the variations permit greater efficiency wherever the group needs it and third, the evolution introduces greater complexity. The evolution of the telephone, the car, even the cake mix, documents all three of these aspects of direction in change;
- **Motivated**, which refers to the source of change or what is causing it. In biology there are three such sources. The first is internal to the specimens or its gene sets. Second is the environment where living conditions favour certain mutations or certain behaviour changes. Third, there is mediation, which is

unique to humans, where we try to eliminate undesirable species (the measles virus), develop useful ones (hybrids of corn), and to control or maintain rare ones (zoo animals);

- **Patterned**, in that there is a clear pattern of change that closely corresponds to what we can call the traditional lifecycle pattern of: divergence or the start of new clans or species; development through growth through increased numbers; differentiation by increasing variation, including subspecies; stabilization with long periods of stability or stagnation, when there is no force for change; and perhaps, demise of the loss of a species.

Butler's model

Butler's model has become the industry standard (Wang et al., 2016). It is often quoted, but not particularly accurately. In particular, the model has often been described as a 'lifecycle' model that is assumed to borrow heavily from the product lifecycle concept. But, right from the start, there is a misconception, for Butler's original paper is titled *The Concept of a Tourist Area Cycle of Evolution* and not as it has been summarised recently as the TALC (tourism area lifecycle) model. The difference is profound. Lifecycle models imply a finite cycle with a beginning, middle and end. Evolution, on the other is an ongoing process of continuous change from a simple to a more complex form with change occurring in response to environmental conditions by natural selection of randomly appearing mutations. Evolution, therefore, implies an open ended, infinite process where change is part of the normal order of events. Moreover, it relates to how entire systems collectively interact to influence change. Murphy (1985) was one of the first to liken destinations to an ecosystem in his discussion of community tourism. Over time, evolution can produce systems that are almost completely different from those found during the initial stages although one can logically map the progression or change over time.

The opening paragraph of his paper states unequivocally

> "There can be little doubt that tourist areas are dynamic, that they evolve and change over time. This evolution is brought about by a variety of factors including changes in the preferences and needs of visitors, the gradual deterioration and possible replacement of physical plant and facilities, and the change (or even disappearance) of the original natural and cultural attractions which were responsible for the initial popularity of the area. In some cases, while these attractions remain, they may be utilized for different purposes or come to be regarded as less significant in comparison with imported attractions." (Butler 1980:5)

So, he argues from the outset that tourism changes tourism and that destinations evolve, rather than have a classic fixed lifecycle. Where the 'lifecycle' confusion comes in though, is that he argues the evolution can be mapped along the classic product lifecycle curve for convenience sake, more so than for absolute accuracy. But, note from Figure 8.1, the phases intentionally do not correspond with product phases.

His model is predicated on a number of assumptions:

- Demand changes as destinations evolve, with the type of tourist becoming increasingly 'softer' as destinations mature.
- Supply of products and services changes, usually in response to changing consumer demand, but occasionally in a proactive manner to lead demand change. Again, the change is to the creation of 'softer' products – better quality accommodation, better services and more built attractions that develop the extrinsic appeal of the destination, rather than relying on the intrinsic appeal of its original cultural and heritage landscape.
- The limits of growth are determined by the destination's carrying capacity and once it is reached or exceeded, the destination is at risk of entering decline. At risk, but not pre-determined to do so!

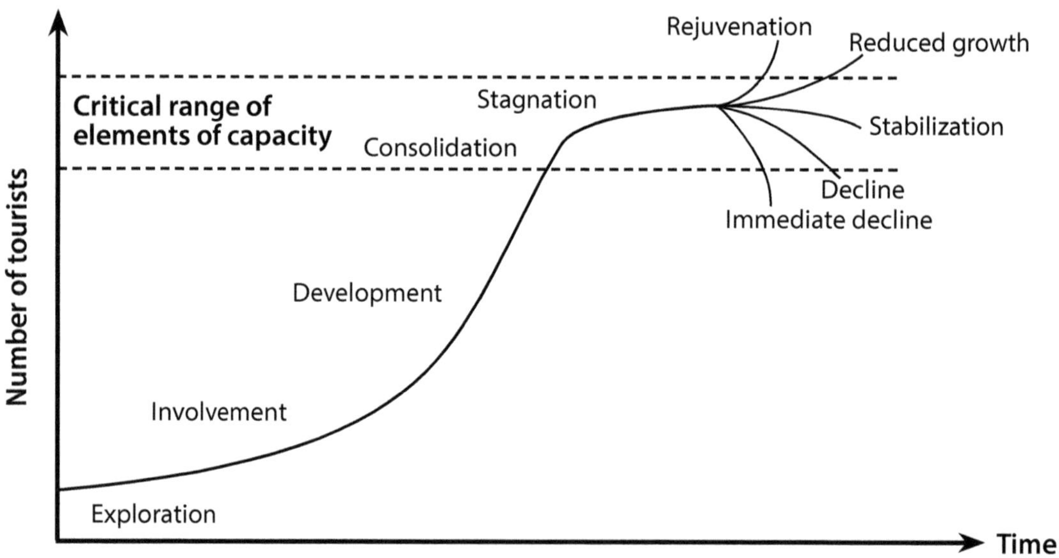

Figure 8.1: Butler's lifecycle. *Source*: Butler, 1980.

The model has five main stages before a series of post-stagnation options emerge. The key features are shown in Table 8.1 below and summarised briefly below.

Table 8.1: Stages in Butler's Lifecycle

Stage	Characteristic
Exploration	Few explorers or drifters visiting sites with no public facilities Visitors attracted to the destination by a natural physical feature Specific visitor type of a select nature
Involvement	Limited interaction between local residents and the developing tourism industry leads to the provision of basic services Increased advertising induces a definable pattern of seasonal variation Definite markets begin to emerge
Development	Development of additional tourist facilities and increased promotional efforts Greater control of the tourist trade by outsiders Number of tourists at peak periods far outweighs the size of the resident population, inducing rising antagonism by the latter towards the former
Consolidation	Tourism has become a major part of the local economy, but growth rates have begun to level off A well-delineated business district has taken shape Some of the older deteriorating facilities become less attractive to visitors Local efforts are made to extend the tourist season
Stagnation	Peak numbers of tourists and capacity levels are reached The destination has a well-established image, but it is no longer in fashion
Post-stagnation	Five possibilities, reflecting a range of options that may be followed, depending partly on the success of local management decisions. At the extremes are rejuvenation and decline.

Exploration represents the first stage for a destination. Small numbers of independent tourists arrive. There is little existing tourism infrastructure and visitors tend to be attracted by the natural or cultural assets of that destination. They use local facilities. Tourism tends to be small scale and locally owned. Although relationships between hosts and guests may be intense, overall impacts are small because numbers are small. Usually leakages from the local economic system are small, since most of the tourists' needs are produced locally.

As the number of visitors increases, the destination enters the **Involvement** stage. At this point, a formal tourism industry begins to appear, dominated almost exclusively by local residents. A distinct tourist season may begin to appear and also some formal travel arrangements for tour operations can emerge. Contact between local residents and tourists remains high. Relationships between hosts and guests, although a little more formalized, are still personal and the scale of tourism and its growth rate are still manageable. Economic benefits per tourist can be quite large, since leakages remain small.

The **Development** stage reflects the period where large scale tourism begins. The area begins to develop an identity as a tourism destination, large numbers of tourists begin to arrive, and large scale development begins. Local involvement in tourism declines and outside investment increases. Purpose built attractions begin to replace natural or cultural attractions. The type of tourist will change from someone seeking to explore the destination to someone looking for more of a rest and relaxation holiday. It is during this stage that adverse social and environmental impacts tend to be noted first, as local control over tourism diminishes and residents must begin to adjust their lifestyles to cater to the needs of the growing numbers of tourists. Congestion, traffic problems, increased competition for scarce resources, inflation and disruption of local ways of life are noted.

Consolidation occurs when the growth rate slows, but the total number of tourists continues to increase. Tourism is now a major economic activity with the industry dominated by national or multinational organizations. Butler suggests the destination takes on the characteristics of a product which is branded and marketed as a single entity. Because of the sheer number of tourists, much money enters the economy, however, leakages are also large as the amenities these tourists look for are likely to be imported. Antipathy to tourism may reach its peak because of serious social and environmental impacts, while the economic benefits may be dispersed in an uneven manner.

Stagnation occurs once the peak number of arrivals has been reached and parts of the physical, social, psychological and ecological environment have reached or exceeded their carrying capacity. Tourist enclaves may emerge whereby tourists and locals are separated, mostly because locals have been pushed out of tourism development zones. As a result, contact between tourists and hosts is highly institutionalised. The area is now well known, but has fallen out of favour and instead of generating new arrivals will become increasingly reliant on repeat visitors. Local involvement in tourism starts to return as national and multi-national firms begin to abandon the place.

After stagnation comes decline. Right? Wrong! In classic product lifecycle literature, decline normally follows stagnation, but not here. Butler identifies five possible options, of which only two relate to decline. These options are:

- **Decline** (slow or rapid) where the area enters a rapid or slow period of decline. The destination area falls out of favour with tourists.
- **Stability** where the absolute number of tourists stays relatively stable or declines only slowly over time.
- **Rejuvenation** (slow or rapid) where the destination area reinvents itself.

Yes, three options are quite positive. On the one hand, the destination can maintain a level of maturity and remain stable indefinitely. But, equally, as well, Butler illustrates that it can enter a period of renewed growth. Admittedly, rejuvenation will involve a conscious strategic decision on behalf of the local community, either by adding new and different built attractions and activities to target a new market, or by taking advantage or previously untapped natural resources, again to attract new visitors.

His model is not without its critics as well as supporters, as acknowledged by Butler (2011) in a reflective commentary. Many have argued that it lacks utility because it is not deterministic and predictive. One cannot show when one phase ends and the next begins. Butler (2011:6) counters this by stating "the purpose of the model was to draw attention to the dynamic nature of destinations and propose a generalized process of development and potential decline which could be avoided by appropriate interventions (of planning, management and development), or as suggested in the title of the article, the management of resources."

■ Carrying capacity

Perhaps, though, the biggest criticism of Butler's model is the idea that carrying capacity is a fixed concept. Butler argues that the limits of growth are defined by the carrying capacity of the resort area and once exceeded, a range of negative and sometimes irreversible impacts are likely to occur (Butler, 1999). The UN World Tourism Organization (2018:3) defines tourist carrying capacity as "the maximum number of people that may visit a tourist destination at the same time, without causing destruction of the physical, economic and sociocultural environment and an unacceptable decrease in the quality of visitors' satisfaction." The idea has been broadened in the literature to include many more concepts, such as: *experiential/psychological* carrying capacity – the lowest degree of enjoyment tourists are prepared to accept before they start seeking alternative sites or destination; *economic* carrying capacity the ability to absorb tourist functions without squeezing out desirable local activities; *political* carrying capacity at which point political support for tourism diminishes; and *social* carrying capacity, which is the tolerance of the host community for tourism.

The concept of carrying capacity was recognized well when the original paper was written, but since then has been subjected to a great deal of criticism since the real world does not fit well into an idealized hypothetical world (McCool and Lime, 2001). Butler himself even acknowledges this point (Butler, 2011). The biggest challenge is the focus on numbers as a determinant of carrying capacity, with the assumption that large numbers equate to negative impacts (Koens et al., 2018). The second issue is that somehow it is an objective measure. But, McCool and Lime (2001) ask the question of from whose perspective is carrying capacity measured – the local or the tourist? Tourists have a much higher threshold than locals. Likewise, Jurado et al. (2013) highlight the challenge of which aspect of carrying capacity to measure: social, physical, psychological, economic or political? They may all be quite different. A third issue is the inherent seasonality of tourism, and with it comes the cost decision of what level of infrastructure is needed. Overcrowding may be an issue during part of the year in highly seasonal destinations but may not be a problem at other times. This observation leads to the fourth concern about carrying capacity and that is, from a social perspective, it tends to forget that people are resilient and adjust to change. Social disruption is transitional, whereby eventually communities adapt to these changes over time (Perdue, Long and Kang, 1999).

Perhaps the biggest issue is that carrying capacity is a malleable, subjective and complex concept that is more of a social construct than an ecological absolute (Lindberg, McCool and Stankey, 1997; Weaver, 2000). As such, it can be altered through new product development and site hardening. Romeril (1989) feels that while carrying capacity is simple in theory, the management application is complicated by strong value dimensions. Some users will adopt a liberal attitude, accepting significant modification of the environment as an acceptable outcome for greater use of a resource, while others will perceive that any modification as representing unacceptable degradation.

Sites can be hardened and the carrying capacity increased. Indeed, changes in carrying capacity are implicit in the development of the Butler framework, for the process of touristification is the process of carrying capacity enhancement. Building small, locally owned properties enhances the carrying capacity of a destination somewhat, while building larger resorts and hotel complexes increases it a lot. Likewise, the shift from natural and cultural attractions to built attractions is in essence an exercise in carrying capacity expansion. The same can be said for most infrastructure development from water and sewage provision to electricity provision and enhancement of the transport infrastructure.

■ The unrecognized subtlety in Butler

Most people tend to look at the S-shaped curve as a reflection of the entire evolutionary process of the destination. They fail to appreciate the most subtle and exciting condition in Butler's lifecycle. It is open ended. The x axis, reflecting time, and the y axis, reflecting the number of tourists, have arrows at the end implying that destination evolution is not bounded by either time or the volume of visitors (McKercher, 2005b). Instead it is an open ended process that extends beyond the narrow confines of the framework shown. The open-ended nature of the lifecycle model is also reflected in what happens post stagnation, where the outcome is also open ended.

In reality, the framework shown by Butler does not reflect the destination's total lifecycle from beginning to end, but rather individual phases in an on-going lifecycle pattern. Choy (1992) recognized this pattern explicitly in his work on the island destinations, while Prideaux's (2000) model, discussed elsewhere, recognizes this implicitly. It is also explicit in the complexity theory approach to destination change.

Figure 8.2 shows how each of these individual lifecycle phases can be aggregated to reflect the overall evolution of a destination. Destinations may undergo multiple stages of exploration, involvement, development, consolidation and stagnation as the nature of the destination changes. Moreover, the critical range of capacity that defines the consolidation and stagnation phase is also malleable and can change as the nature of the destination changes. The trend may be upward, as shown in Figure 8.2, downward, in the case of destinations in long term decline, or highly variable in destinations undergoing rapid evolution. Not all destinations necessarily evolve through all of Butler's stages. Some will go directly from post stagnation to rebirth, while others will enter a period of decline before entering a recovery period.

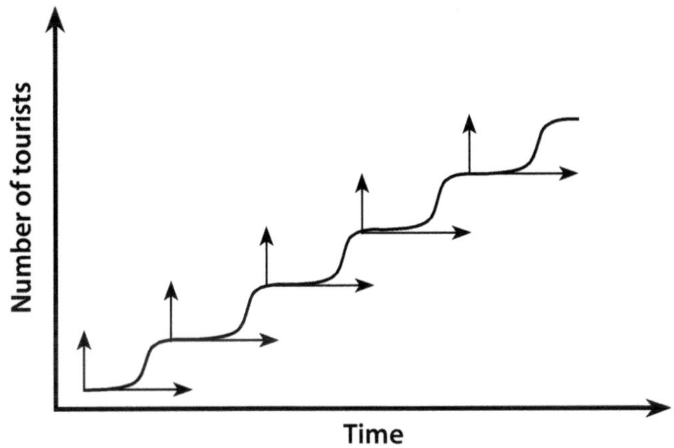

Figure 8.2: Lifecycle phases aggregate to create the evolution of a destination

Plog's model

The other tourism model referred to in the literature was developed by the social psychologist Stanley Plog. Plog's lifecycle model is based on the underlying assumption that as a destination changes it attracts a progressively less adventuresome tourist. In doing so, these tourists require more and more entertainment oriented attractions that reflect, increasingly, their home environment. As he states (Plog 1974:55) "any approach to understanding why destination or resort areas rise and fall in popularity inevitably must involve an understanding of the psychology of people who travel."

Plog's work as a lifecycle model has been largely forgotten in recent years. Instead the main contribution is seen as being one of the first attempts to define market segments by psychographic characteristics or by the tourists' underlying psychological profile. But, as with Butler, this reinterpretation of his work is a misconception, for the original paper was titled "*Why Destination Areas Rise and Fall in Popularity*". His work looks at how destinations change as they attract initially the adventuresome tourist who discovers the place, through growth as the market base broadens and then finally to decline and demise as the market base shrinks.

His paper states unequivocally (Plog, 1974: 58) "the introduction of each new type of traveler has meant a larger population base from which to draw travelers... [but] when the appeal of the resort passes the magic mid-point in the population curve of travelers... it begins to draw on a smaller number of travelers." Elsewhere he adds (Plog, 1974: 58) "thus, we can visualize a destination moving across a spectrum, however gradually or slowly, but far too often inexorably towards the potential of its own demise. Destinations carry with them the potential seeds of their own destruction as they allow themselves to be more commercialized and lose qualities which originally attracted tourists."

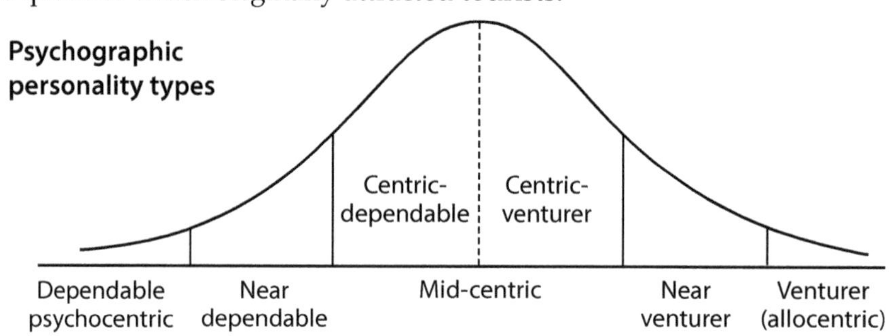

Figure 8.3: Plog's Lifecycle model. Source: Plog, 2001.

Most of us are familiar with his model (Figure 8.3) and with the descriptors he developed initially to label each group. Two features make this model unique from virtually every other lifecycle model. First, it is read from right to left. While it may be confusing to some, the reason is actually very simple, for the groupings are based on responses to a series of Likert scales questions. People who were more likely to answer '5' tend to be Allocentric in nature, whereas those who answer closer to the '1' end of the scale are more Psychocentric. The second feature is that the categories are quantitatively based. Plog feels all populations are normally distributed, including the tourist population. As such, they can be segmented based on deviation from the norm. The Midcentric groups capture all those people whose responses were within one standard deviation from the average score, while the other groups are based on scores of between one and two standard deviations and greater than two standard deviations from the mean, respectively. It is for this reason that the mid group represents just under two thirds of the population, while the two 'Near' groups represent about one-third of the population and the two groups at the extreme end represent the rest.

The labels have changed over time as first he tried to soften the negative connotations applied to people at the psychocentric end of the spectrum and later as the model was commercialized (see www.besttripchoices.com). Table 8.2 summarizes some of the key features of each group, along with the different labels attached to them, while each is described briefly below.

Allocentrics or **Venturers** are the first type of tourist attracted to a destination. The term refers to the root word 'allo' which means very informed, or a person with varied interests and 'centric' meaning the focus of his/her patterns. An Allocentric tourist is somebody who is interested in a sense of discovery and likes to be immersed in new activities.

As they begin to talk to their friends about the vacations that they have recently discovered, their friends are attracted to destinations. The **Near Allocentric/ Pioneer** sector are not discoverers, but are tourists who can still withstand a fair amount of strangeness. However, they do require hotels, restaurants and tourist services. The advent of this group really represents the beginning of the tourism lifecycle.

As destinations become more popular, the **Midcentric** market begins to arrive. Plog initially identified only one group of Midcentrics who represented two-thirds of all tourists. He has since divided them into two groups based on whether their inclination is to the Allocentric or Psychocentric end of the spectrum. The **Voyagers** share more in common with Allocentrics, in that they are somewhat adventuresome and want to engage in wanderlust activities. The **Journeyers** share more in common with the Pyschocentrics. They still enjoy travel but are more cautious. Both groups represent the typical tourist who requires a large number of services and facilities and who travels primarily for rest and recreation. An influx of Midcentric tourists is thought to displace the other groups. The destination has now reached its maximum potential because it is an attracting the broadest audience possible.

As the appeal of the resort passes a certain point, it begins to attract **Sightseers / Near Psychocentric / Near Authentic** who are rather constrained in their travel. They travel less and prefer at home style of accommodations and experiences. The **Psychocentric / Dependable / Authentic / Traditional** tourist is the opposite of the Allocentric. The root of the word is psyche, which means self. These tourists are self-focussed, fearful of change and the outside world. They travel little and when they do prefer as little change from their home environment as possible. Plog argues that destinations move along the continuum from Allocentric to Pyschocentrics, and in doing so, inevitably must enter a period of decline.

Table 8.2: Plog's classification

Current Segment name*	Traditionals	Sightseers	Journeyers	Voyagers	Pioneers	Venturers
Initial name	Pyschocentrics	Near Psychocentric	Midcentric	Midcentric	Near Allocentrics	Allocentric
Other names	Authentic / Dependables	Near Authentic				
Approximate size of group	About 3%	17%	30%	30%	17%	about 4%
Importance and frequency of travel	Travel unimportant Take fewer trips and stay away for fewer nights than all others	Travel not very important and travel less than others	Travel somewhat important and travel an average amount	Travel important and travel an average amount	Travel important and travel often	Very important part of their life and travel the most
Preferred types of travel and destinations	Nearby well established destinations Major hotel chains and entertainment facilities not available at home Prefer domestic travel and familiar atmosphere Like group tours	Prefer popular places with lots of sunshine Repeat visitor Prefer domestic travel Cruises are popular Package tours if travelling internationally	Like crowds Travel for fun, recreation, and to spend time with family Prefer domestic travel Cruises are popular Package tours if travelling internationally	Prefer a modest selection of good hotels Repeat visitors to favourite places Prefer independent travel but will buy packages Prefer to be active Wanderlust	Like to travel to foreign destinations and seek new experiences Plan trips and set an itinerary of places they want to visit Wanderlust Soft adventure Travel internationally to culturally different places	Like to visit relatively unknown, uncommon and non-tourist destinations before others Prefer longer duration trips Want to explore and discover Unique cultures are especially attractive Meet people of different backgrounds Prefer high activity and hard adventure Happy to tough it

* (www.besttripchoices.com)

Questioning Plog

McKercher's (2005a) critique highlights three assumptions that form the basis of the model that may not be valid.

1) Can destinations exist simultaneously at different places in the continuum?

The model defines lifecycle stage from the perspective of a single market. In the original case lifecycle seemed to be assessed from the perspective of a New Yorker. Thus, Asian and African destinations were seen as appealing to the Allocentric or Venturer tourist. But, from the perspective of someone living in Asia, a place like Japan may not be all that foreign and may be seen as lying on the mid-point of the scale. How a destination is classified, therefore, depends on the perspective of the market doing the classification. In other words, perceived lifecycle stage is market specific, rather than destination specific. Each market has a unique perception shaped by geographical proximity, political orientation, stage of economic development, similarities or differences in culture, language and shared histories and a host of other factors.

A few researchers have acknowledged the destinations can be seen to exist at multiple stages in their lifecycle simultaneously. Prideaux's (2000, 2004) resort development spectrum model identifies a series of growth phases that resorts can progress through as their market base broadens. The shift from one stage of development to another (i.e. from regional to national or national to international) is induced by a broadening in demand producing or resulting in a concomitant broadening of supply. If a destination can exist simultaneously at multiple points along its lifecycle continuum, depending on the perspective of different markets visiting it, then it cannot be said any destination is at a definitive lifecycle stage.

8

2) Can markets evolve fully through their lifecycles?

Plog posits that destination lifecycle is similar to the diffusion model of products. It begins with innovators (Allocentric / Venturer segment) discovering a place, experiences growth as early adaptors (Near Allocentrics / Pioneers) visit and shows signs of maturity as Midcentrics visit. It then enters terminal declines when the destination loses favour with all but the rump group of laggard or Psychocentric / Dependable / Authentic / Traditionals tourists.

This proposition is predicated on the validity of the assumption that each market can evolve fully through this lifecycle pattern. To do so, each market must be a source of Allocentric / Venturer, Midcentric and Psychocentric/Dependable tourists. This proposition ignores the significant mitigating impacts that physical and cultural distance will have on the type of tourist who is willing to visit. Smith and Xie (2003) illustrate the impact of cultural distance in their work examining tourist flows from America to Canada. Their comparisons of the volume of visitors entering Canada as opposed to travelling within the United States showed that the act of crossing an international border added the equivalent of an average of 1,650 km to a journey and that crossing from English-speaking Vermont to French-speaking Quebec added the equivalent of 3,700 km. Therefore, not all markets can evolve. Some must start as Allocentric and can never progress beyond Midcentric. Others start as Psychocentric.

3) Must psychocentric destinations be in decline?

Lastly, the assertion that Psychocentric destinations must be in decline needs to be questioned. This model suggests that tourism destroys tourism and yet there is very little hard evidence to support this proposition. Instead, tourism changes tourism. Many so-called Psychocentric destinations located in the pleasure periphery of large cities have undergone a renaissance over the last few decades as destination marketing organizations and local business people have responded to the growing market demand for short break holidays, spa vacations, and the growth in the meeting and convention market. True, the nature of tourism demand may have changed as these destinations evolve to cope with cyclical demand, peaking on weekends and troughing during the week. But, in doing so, many have also become four season resort areas.

Does Plog work empirically?

In principle, Plog's ideas make a lot of sense for they match closely to Cohen's (1972, 1979) ideas of strangeness vs familiarity and also his five modes of tourist experience that will be discussed in Chapter 11. The emotionally and psychologically restrained Psychocentric tourist can cope with little change and therefore needs a large environmental bubble if travelling far from his or her comfort zone. This profile matches the Diversionary mode of experience, for the person will travel to escape, but not to re-create. The two Midcentric groups of tourists (Voyagers and Journeyers) who travel for fun, escape and to spend time with family and friends match with Cohen's Recreational tourist experience mode. They prefer simple to consume experiences with a large environmental bubble, typified by Cohen's institutionalized tourists. By contrast, the adventuresome Allocentric or Venturer type of tourist travels specifically to escape from an environmental bubble to experience the authentic destination. This person corresponds to Cohen's individual Drifter or Explorer and would take trips associated with the Experimental, Experiential and Existential modes of travel.

In theory! Reality, though is different, for many tests show Plog's model is good at identifying tourists' ideal destinations, but not where they actually travel. Figure 8.4, shows the difference between the ideal destination and the actual destination visited. In reality, more people will visit so called Psychocentric or Traditional destinations, regardless of their profile.

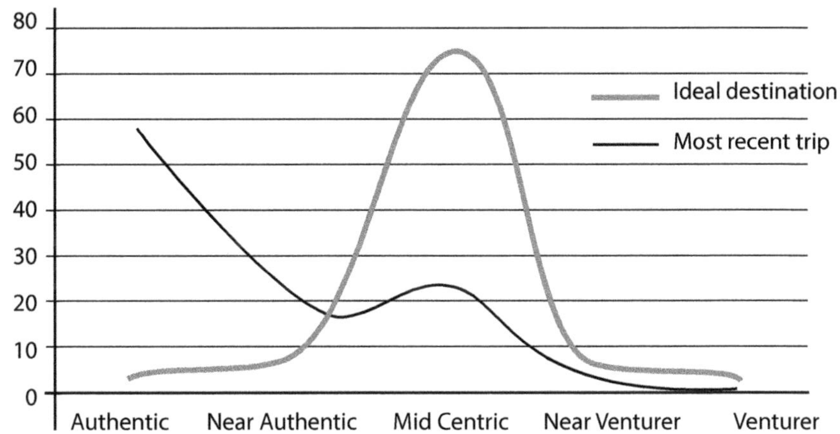

Figure 8.4: Actual vs desired destination choice. *Source*: from Litvin, 2006.

Table 8.3 shows how even Venturers (Allocentrics) prefer Traditional (Psychocentric) destinations. Litvin (2006) revealed that ideal destinations fit nicely along Plog's proposed continuum but that actually places visited leaned heavily toward the Psychocentric end of the continuum. A follow-up study by Litvin and Smith (2016), as shown in Table 8.3, revealed that almost 90% of all travel could be considered to Psychocentric destinations instead of the predicted 20% (including both Psychocentric and Near Psychocentric destinations). Moreover, instead of two-thirds of travel being to Midcentric destinations, the actual figure was closer to ten percent. Likewise, little travel was taken to Allocentric destinations.

Table 8.3: Destination choice by type of tourist

	Psychocentric / (Traditional / Sightseers)	Midcentric (Journeyers / Voyagers)	Allocentric (Pioneers / Allocentrics)	All tourists
Allocentric destinations	0.4%	0.9%	1.6%	0.9%
Midcentric destinations	7.8%	10.9%	15.6%	11.2%
Psychocentric destinations	91.8%	88.2%	82.8%	87.9%

Source: Litvin and Smith, 2016.

Look closely at the 'most recent trip' line shown in Figure 8.4. It looks an awful lot like the classic distance decay curve, further reinforcing the fact that distance decay seems to be a dominant law in tourism. Why is it that actual trip patterns reflect a distance decay curve and not the predicted Plog-like line? Time availability, cost travel party size, motivation and a range of other factors dictate that most travel is to nearby destinations. There is another factor as well. Plog notes that the Allocentric tourist is a very active tourist, taking many trips a year, while those who lean towards the Allocentric side of the Midcentric continuum are also active travellers. It is unrealistic to assume that everyone of them will take a typical Allocentric type of trip, but as active travellers they will likely take more short haul trips to nearby destinations, as well as the possibility of a longer haul trip that reflects their underlying personality traits. This observation further challenges the belief that Psychocentric destinations must be in decline.

These findings also raise a further two key question about Plog's work. First, he does not consider the role differing motivations may play in individual trip choice. Instead, he gives blanket descriptors and assumes all travel by that person fits the generic descriptor. But, one's underlying psychological make up is not necessarily a deterministic factor that defines the motivations for a trip. You may have an Allocentric personality, but simply want to get away from your job for a weekend, or spend some time with family and young children. Your motives may dictate that you take a Pyschocentric or Midcentric type of trip. Having said this though, it is unlikely that someone with a Psychocentric personality would want to climb Mt Everest.

In fact, an alternative interpretation of Plog is that one's personality traits determine the upper limit of the type of trip an individual is comfortable with unless a range of special considerations are made to reduce the level of strangeness to an acceptable level. This relationship is shown in Figure 8.5.

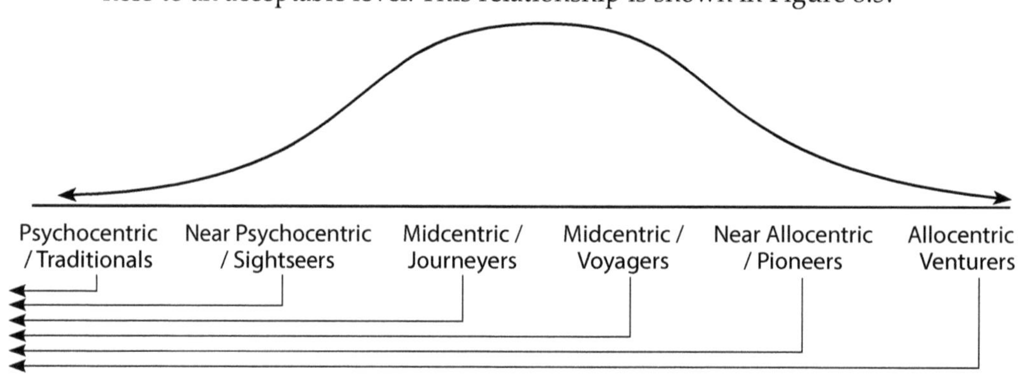

Figure 8.5: Plog's psychographic characteristics determine the 'maximum' type of trip

Much like as seen with distance decay, where anybody who can travel can travel a short distance, the same situation applies to Plog. Anybody who can travel can choose a Psychocentric destination or type of holiday and probably be quite satisfied with that experience if that type of trip reflects their desired motives, regardless of their underlying personality characteristics. A Midcentric type of personality would likely be satisfied with a type of trip that reflects a Psychocentric were Midcentric experience if that was their motive and the Allocentric tourist would be satisfied with all types of trips up to and including an Allocentric trip.

But, because of his or her underlying personality traits, the Psychocentric type of tourist would be unlikely to take a more adventuresome type of trip unless a huge environmental bubble was created, as is often done when either travelling to visit friends (high degree of familiarity) or travelling in a package tour (high degree of safety). The Midcentric tourist would also likely not be interested in a more venturesome type of trip, unless significant measures were taken to reduce the level of strangeness to an acceptable level.

The second feature is that Plog assumes destination populations are exclusive instead of cumulative. Here, he suggests that 'softer' tourists progressively displaces the hardier tourists until only a rump of the population remains. But Prideaux (2000) suggests otherwise, as do Litvin's various studies.

The complementarity of Butler and Plog

Butler and Plog are essentially looking at destination lifecycles from different sides of the same coin. Butler adopts the supply-side development process in response to changing demand and, rarely, in anticipation of changing demand, while Plog adopts a demand-side perspective and examines how the supply of facilities and services for tourists changes as the profile of tourists change. Indeed, Butler (1980) makes a number of specific references to Plog indicating that the Allocentric tourist can be found during the exploration stage of the lifecycle, the Midcentric tourist is typical in the development stage and the Psychocentric tourist can be found in the stagnation stage of the lifecycle.

Table 8.4 compares the features of Butler's and Plog's lifecycle stages to illustrate just how complementary they both are. Small numbers of tourists who are satisfied with limited facilities, want intense contact with the local community and are attracted initially by the intrinsic natural and cultural assets of the destination area reflect the desires of the prototypical Allocentric tourist. The Near Allocentric tourist shares much in common with the Allocentric tourist but is not quite as hardy and, therefore, wants some better services and facilities. This segment is up to six times larger than the Allocentric segment which then stimulates the initial development of purpose built tourist facilities often by local entrepreneurs who see a business opportunity to cater to this growing market. It is at this stage of the lifecycle that tourism development begins to occur rapidly.

The rapid growth development phase corresponds with the Midcentric tourist discovering and popularizing the destination area. Remember that according to Plog's quantitative measures, Midcentrics represent two-thirds of the travelling population. The Midcentrics who lean towards Allocentrism want to explore and discover but they want to do so from the comfort of good facilities and services.

Table 8.4: Complementarity of Butler and Plog

Butler	Plog
Exploration Small numbers initially Restricted by lack of access, facilities Limited local knowledge Use of local facilities Intense contact with local residents Appeal rests with natural and cultural attractions	**Allocentric / Venturer** Prefer non-touristy areas First to 'discover' new places Can cope with limited facilities Prefer heavy interaction with local residents Independent tourists Travel a high priority
Involvement Visitor numbers begin to grow Initial development of tourist facilities Contact with locals high Emergence of a tourist season	**Near Allocentric / Pioneer** Popularise places Want a comfortable bed Want tourist facilities including resorts, hotels and activities Travel a high priority
Development Well defined tourist area Local involvement in development declines as outside organisations establish facilities Development of built attractions Formalisation and commercialisation of natural and cultural attractions Peak season crowding begins to emerge	**Midcentric / Voyageurs & Journeyers** Rapid growth continues Tourists want better quality and variety in facilities, accommodation and activities Want to be isolated from local community and happy to be in tourist enclaves
Consolidation Growth continues but rate of increase slows National and multinational firms are evident Less growth in new developments Well defined tourism nodes	**Near Pyschocentrics / Sightseers** Want well developed facilities Prefer built attractions Want known brands Prefer domestic travel Travel not a high priority
Stagnation Peak arrivals reached Area well known but beginning to lose its appeal Artificial, built attractions dominate Heavy reliance on repeat visitation	**Pyschocentrics / Traditionals** Prefer familiar destinations Heavy repeat travel Prefer highly developed places Prefer familiar environment including known brands Low activity level Travel a low priority

Those who begin to lean towards the Psychocentric end of the spectrum are starting to become less active and more demanding of facilities that represent the extrinsic appeal of the destination. Because of these factors well-defined tourist areas need to be developed and the essential character of the destination changes.

The consolidation and stagnation stages correspond closely to the Psychocentric end of the spectrum where the destination area has lost its unique character and broad appeal and instead begins to attract a heavily repeat market of low activity visitors. The Near Psychocentric tourist is not particularly active, wants well-developed facilities and known attractions while the Psychocentric tourist is the least active of all groups of tourists.

Conclusions

These two models still resonate today. They are found in most texts and referred to frequently in the literature. But, neither asks the key question of 'do destinations have lifecycles or simply progress through a series of lifecycle phases?' We know destinations evolve. Plog says they evolve from a demand perspective as an increasingly softer group of tourists force change to the destination. Butler takes a more supply-side approach and also sees how the provision of goods and services changes as destinations evolve. Both also assert that in a single point in time it is possible to depict the destination as existing at some point on their various continua.

But, neither answers the question of whether destinations have lifecycles or simply progress through a series of lifecycle phases. Plog in fact asserts their lifecycles are finite, while Butler's work is interpreted by many as also depicting a finite process, in spite of the model's open-ended nature. Indeed, Plog (2001) says a destination needs an earthquake to revise its fortunes. Butler (1980) argues places can be revitalised, but also presents a rather pessimistic view that revitalisation involves targeting a specialist interest group of tourists who may provide a short term period of growth before the place loses its competitiveness once more.

Where is the empirical evidence of such a finite lifecycle process? Few if any destinations have truly lost their flavour and disappeared, unless they have been victims of suburban encroachment or some natural disaster. Indeed, we may need a formal rethink of the entire lifecycle model. Food for thought.

■ Discussion points / Exercises

1. Butler and Plog adopt different biological models when developing their lifecycles. Butler supports an evolutionary model whereas Plog supports a diffusion model. Both have gained acceptance in the literature. Which one do you support and why?

2. People seem to support either Butler's or Plog's idea, but rarely both. Yet, each model has its merits. Select which model you support and then argue why the other model also has merits.
3. What factors would propel a destination into a new lifecycle phase as suggested in Figure 8.2?
4. Each of these models has its strengths and its limitations. Can you think of another model that captures the dynamic nature of destinations better?
5. What factors would cause an Allocentric tourist to select a Psychocentric destination?
6. Visit http://besttripchoices.com/travel-personalities/quiz and take the quiz to see what types of holiday suits you. Do you agree with this assessment?

Points to consider for revision

- Two competing lifecycle models have dominated the literature over the past 40 years. Both have strong biological roots.
- Butler's model is predicated on the assumption that carrying capacity is fixed and once the carrying capacity of a destination is exceeded, it could enter a period of decline, but is not mandated to do so.
- The subtlety of Butler's model, and in particular the recognition that it is open ended suggests destinations can evolve through multiple lifecycles.
- Plog's model is based on the diffusion or innovation of products.
- Plog posits destinations tend to have a closed lifecycle and that once they pass the magic midpoint of their market curve, the seeds of their demise are sewn.
- Empirical testing of Plog's model does not support the absoluteness of his lifecycle, primarily because he does not see motivations as a key determinant of a trip.
- Plog's model probably represents the maximum extent of change each segment is willing to undergo unless they wrap themselves in a bubble of familiarity.
- Butler's and Plog's models are highly complementary and overlap in many places.

Suggested additional readings

Butler R. W. (1980) The concept of a tourist area cycle of evolution: implications for management of resources *Canadian Geographer*, 24(1), 5-12

Butler, R. W. (2011) *Tourism Area Lifecycle*. Contemporary Tourism Reviews. Goodfellow.

Litvin, S. & Smith, W. (2016) A new perspective on Plog psychographic system. *Journal of Vacation Marketing*, 22(2), 89-97.

Plog, S. (1974) Why destination areas rise and fall in popularity *Cornell Hotel and Restaurant Administration Quarterly*, 14(4), 55-58.

References

Agarwal, S. (2002) Restructuring seaside tourism: The resort lifecycle *Annals of Tourism Research*, 29(1), 25-55.

Budowski, G. (1976) Tourism and conservation: Conflict, coexistence or symbiosis? *Parks*, 4, 3-6.

Butler R. W. (1980) The concept of a tourist area cycle of evolution: implications for management of resources *Canadian Geographer*, 24(1), 5-12.

Butler, R. W. (1999) Sustainable tourism: A state-of-the-art review, *Tourism Geographies*, 1(1), 7-25, DOI: 10.1080/14616689908721291.

Butler, R. W. (2011) *Tourism Area Lifecycle*. Contemporary Tourism Reviews. Goodfellow.

Chandler, J. & Costello, C. (2002) A profile of visitors as heritage tourism destinations in East Tennessee according to Plog's lifestyle and activity level preferences model. *Journal of Travel Research*, 41, 161-166.

Choy, D. (1992) Lifecycle models for Pacific Island destinations, *Journal of Travel Research*, 31(3), 26-31.

Cohen, E. (1972) Toward a sociology of international tourism. *Social Research*. 39(1), 164.

Cohen, E. (1979) A phenomenology of tourist experiences, *Sociology*, 13, 170-201.

Crawford, M. (1984) Business took the wrong lifecycle from biology. *Journal of Consumer Marketing*, 1(3), 5-11.

Doxey, G. V. (1975) A causation theory of visitor-resident irritants: methodology and research inferences. In *Travel and Tourism Research Associations Sixth Annual Conference Proceedings* (195-198), San Diego, September.

Harrill, R. & T. Potts (2002) Social psychological theories of tourism motivation: exploration, debate and transition. *Tourism Analysis*, 7(2), 105-114.

Hovinen, G. (2002) Revisiting the destination lifecycle model. *Annals of Tourism Research*, 29(1), 209-230.

Jafari, J. (1974) The socio-economic costs of tourism to developing countries. *Annals of Tourism Research*, 1(7), 227-262.

Jurado, N., Damian, I. & Fernandez-Morales, A. (2013) Carrying capacity model applied in coastal destinations. *Annals of Tourism Research*, 43,1-19.

Koens, K., Postma, A. & Papp, B. (2018) Is overtourism overused? Understanding the impact of tourism in a city context. *Sustainability*, 10(12), 15. https://www.mdpi.com/2071-1050/10/12/4384/htm <downloaded April 26, 2019>.

Lindberg, K., McCool, S. & Stankey, G. (1997) Rethinking carrying capacity. *Annals of Tourism Research* 461-465.

Litvin, S. (2006) Revisiting Plog's model of allocentricity and psychocentricity...one more time. *Cornell Hotel and Restaurant Administration Quarterly*. 245-253.

Litvin, S. & Smith, W. (2016) A new perspective on Plog psychographic system. *Journal of Vacation Marketing*, 22(2), 89-97.

McCool, S. & Lime, D. (2001) Tourism carrying capacity: tempting fantasy or useful reality? *Journal of Sustainable Tourism*, 9(5), 372-388, DOI: 10.1080/09669580108667409.

McGuiggan, R. (2003) A vacation choice model incorporating personality and leisure constraints theory. *Tourism Analysis*, 8(2), 183-186.

McKercher, B. (2005a) Are psychographics predictors of destination lifecycles? *Journal of Travel and Tourism Marketing*, 19(1), 49-55.

McKercher, B. (2005b) Destinations as products? A reflection on Butler's lifecycle. *Tourism Recreation Research*, 30(3).

Moss S., Ryan C. & Wagoner, C. (2003) An empirical test of Butler's resort product lifecycle: Forecasting casino winning. *Journal of Travel Research*, 41, 393-399.

Murphy, P. (1985) *Tourism: A Community Approach*, Routledge, London.

Osland, G. (1991) Origins and development of the Product Lifecycle Concept. *Scholarship and Professional Work - Business. 237.* http://digitalcommons.butler.edu/cob_papers/237 <downloaded Nov 5, 2019>

Perdue, R., Long, P. & Kang, Y. (1999) Boomtown tourism and resident quality of life: The marketing of gaming to host community residents. *Journal of Business Research*, 44(3), 165-167.

Plog, S. (1974) Why destination areas rise and fall in popularity. *Cornell Hotel and Restaurant Administration Quarterly*, 14(4), 55-58.

Plog, S. (2001) Why destination areas rise and fall in popularity: An update of a Cornell Quarterly Classic. *Cornell Hotel and Restaurant Administration Quarterly*, 42(3), 13-24.

Polli, R. & Cook, V. (1969) Validity of the product lifecycle. *The Journal of Business*, 42(4), 385-400.

Prideaux, B. (2000) The resort development spectrum – A new approach to modeling resort development. *Tourism Management*, 21, 22-240.

Prideaux, B. (2004) The resort development spectrum: The Case of the Gold Coast, Australia. *Tourism Geographies*, 6(1), 26-58.

Romeril, M. (1989) Tourism and the environment: Accord or discord? *Tourism Management*, September, 204-208.

Smith S. L. & Xie, P. F. (2003) Estimating the distance equivalence of the Canada-U.S. Border on U.S.-to-Canada visitor flows. *Journal of Travel Research*, 42(2), 191-194.

Tellis, G., & Crawford, C. M. (1981) An evolutionary approach to product growth theory. *Journal of Marketing*, 45(4), 125-132.

Tooman L. A. (1997) Applications of the tourism life-cycle model. *Annals of Tourism Research* 24(1): 214-234.

Wang, X., Weaver, D. Li, R. & Zhang Y. (2016) In Butler (1980) we trust? Typology of citer motivations. *Annals of Tourism Research*, 61, 216-218.

Weaver, D. (2000) A broad context model of destination development scenarios. *Tourism Management*, 21, 217-224.

Wenkman, R. (1975) The Pacific tourism blight. *Annals of Tourism Research*, 3(2), 68-77.

9 Tourism as a Complex System

By the end of the chapter, the reader will be able to:

- Describe why existing lifecycle models are deficient
- Appreciate that tourism, by its very nature, is inherently volatile
- Explain how and why tourism functions in a nonlinear manner
- Explain the key features of a complex system.

Introduction

The last chapter examined Butler's and Plog's lifecycle models. They suggest that tourism works as a linear, predictable manner, moving logically through a pre-determined lifecycle. While the speed of change may be variable, progression through the stages up to and possibly including decline seems inevitable. These models and the models identified earlier in this book share a number of features in common. They recognize that any tourism system begins with the tourist and that also any system needs some destination features and a linkage between the tourist and the destination. The models all have a number of advantages helping us to understand the constituent parts of tourism and some of the simple linkages between elements. They are also deficient in a number of areas, though, for they do not work in practice. This chapter takes an alternative view to the organization and evolution of tourism by looking at tourism from the perspective of complexity theory. In the last 20 years, complexity theory has made strong inroads into management disciplines, but has only recently gained limited interest in the tourism sector, notably from Rodolfo Baggio (Baggio, 2008; Scott, Baggio and Cooper, 2008; Sainaghi and Baggio, 2017).

Existing tourism models

The idea of examining tourism from the perspective of complexity theory was mooted first by Faulkner and Russell (1997, 1998) and McKercher (1999) who questioned the utility of existing models to explain how destinations change over time. They argued that most models proposed a reductionist approach to tourism, where one could understand how it worked by disaggregating it into its component parts, identifying the relationships between parts and then re-aggregating it. Such models make it easy to study tourism's components, but do not necessarily explain how tourism works. Moreover, they tend to be written from the perspective of an overarching public sector tourism organization with the implicit understanding that somehow tourism can be controlled from above by planners, as the World Tourism Organisation asserted 25 years ago (WTO 1994) and continues to assert.

And yet, they do not explain why it is so hard to control the genie of tourism once the development process starts and why, as Pearce (1989) stated, spontaneous, catalytic and extensive development are the most common types of development observed, especially in emerging economies. It also does not explain why, after almost 50 years of concern about and research into the social, cultural and environmental impacts of tourism, overtourism continues to remain a vexing problem today. If, indeed, tourism can be controlled, then one would have expected these issues could has been resolved a long time ago.

Why existing models do not work

Why do existing models not work? Models represent simplified versions of reality in order to help make sense of complicated ideas. By design, they are selective as to which elements they include and therefore which they exclude. They tend to focus narrowly on selected destination variables and argue that simple cause and effect relationships exist between these variables. As a result, most of the models cannot appreciate the independent, yet complex interrelationships that exist between and among the multitude of players involved in tourism, where any tourism business must both coexist with and compete fiercely against other businesses to survive.

The models further fail to reflect the dynamic nature of tourism, where hundreds or thousands of businesses, depending on the size of the destination, enter into and exit the marketplace, change ownership or reposition themselves radically each year. And so instead of representing a stable, closed system, the tourism system can be seen as an open system subject to constant dynamic interactions with a whole series of internal and external agents (McDonald, 2009; Brouder and Eriksson, 2013). As a result, a degree of instability is both inherent and essential.

From time to time apparently stable systems are thrown into a period of tremendous instability (Farrell and Twining-Ward, 2004) that rattle them to their core only to re-emerge in a different shape, stronger and more resilient. These perturbations can be caused by external factors such as COVID-19, war, terrorism or natural disaster; internal events involving new developments or financial failure of key players; or by technological innovation affecting transportation linkages in the tourism distribution system. Let's not forget how fickle tourists are as well, as they tend to overreact positively to the latest fad or negatively to bad news.

Instead, most models focus on the stability of systems, or their orderly linear change. They are predicated on predictability and central tendencies and the threat posed by outliers. Yet, outliers are often precursors of change. The models, thus, cannot accommodate the ongoing turbulence that is inherent even in mature, stable tourism communities. Nor can they explain the apparent organic resilience seen in many systems (Baggio, 2008; Cochrane, 2010).

In short, at their heart, the models do not appreciate that tourism operates in a non-linear manner that is akin more to a living ecosystem than to a machine (McKercher, 1999), with a clearly defined a hierarchy of dominant and subservient players and clear inter-relationships between entities. All other tourism activity revolves around these features, including the demand for accommodation, secondary attractions, amenities, services, shopping, other activities, improved access, and the interest in the travel trade to bring tourists to the area. Farrell and Twining-Ward (2004: 277) are more blunt when they state "the central problem is that tourism researchers schooled in a tradition of linear, specialized, predictable, deterministic, cause-and-effect science, are working in an area of study that is largely nonlinear, integrative, generally unpredictable, qualitative, and characterized by causes giving rise to multiple outcomes, quite out of proportion to initial input."

A new approach – complexity

Traditional scientific thought, or the Cartesian-Newtonian approach, views systems as machines (Faulkner and Russell, 1997). To understand how a machine works, all you have to do is break it into its component parts, understand how each part works and how they fit together. Moreover, any machine should work in a predictable manner, where a certain input produces the same output time and time again. Faulkner and Russell (1997, 1998) highlight some other features of the Cartesian-Newtonian approach. To begin, it is assumed that the failure to understand how a system works is due to lack of information and if enough information can be gathered, then any system can be understood fully. Systems also function by fairly simple scientific laws, whereby understanding the laws means that you can control the system. They are deterministic. They tend toward equilibrium and are driven by negative feedback, where individual differences, externalities or other changes noted are seen as being an exception or noise.

Complex systems are dramatically different. Chaos theory, and its companion model, complexity (Lewin 1993) emerged from the realization that many systems operate in a complex, non-linear, non-probabilistic, non-deterministic and dynamic systems manner and not as a machine (Gleick, 1987; Kellert, 1993, as per Overman, 1996). Instead of operating in a machine-like manner, these systems have been likened to living systems where a vast number of elements form an array of rich, subtle, complex and varied relationships. Baggio (2008:3) defines complexity as:

> "structures which depend only in the part on the nature of their constituents and whose global behaviours and characteristics cannot be deduced from the knowledge of their elementary building blocks. Complexity concerns the unforeseen adaptive capacities and the emergence of new properties in systems that arise as the quantity and the quality of the connections among individuals and organizations increase... A system is considered complex if its parts interact in a nonlinear manner. Simple cause and effect relationships among the elements rarely exist and instead a very little stimulus cause unpredictably large effects or no effect at all."

Sedarti, Santos and Pintassilgo (2018) assert destinations act as complex systems with numerous interactions between the sectors operating within destinations and multiple stakeholders with varying and conflicting interests. They say that the complexity of the problems that emerge in tourism systems and the dynamic interrelationships between and among stakeholders means that traditional linear thinking has to be discouraged.

At first glance, relationships in such systems may appear to be haphazard and random, but at a deeper level recognizable patterns emerge that help explain the system's functioning as a whole. Because the relationships are open and so complex and because this complexity engenders an innate level of instability, it is extremely difficult to predict accurately the future movement of the system. Indeed, one of the features of complex systems is their inherent adaptability (Stevenson, Airey and Miller, 2009), which is influenced by and in turn influences its environment.

Scott, Baggio and Cooper (2008:176) identify the key features of complex adaptive systems, including:

- **Nondeterminism** – it is impossible to anticipate precisely the behaviour of such a system even knowing the function of all of its elements.

- **Presence of feedback cycles** both positive and negative that influence the overall behaviour of the system.
- **Distributed nature** where many properties and functions cannot be precisely localized, for in many cases there are redundancies and overlaps.
- **Emergence and self-organization** where the system exhibits emergent properties that are not directly identifiable or predictable from the knowledge of single components, meaning the system continues to evolve in its complexity up to the next self organization level.
- **Limited decomposability** where the dynamic structure of the system has to be studied as a whole and it is difficult, if not impossible, to understand how it works by decomposing it into its component parts
- **Self-similarity** whereby the system will look like itself at different scales if magnified or made smaller. Thus complexity can be used to understand global tourism, as well as tourism at a national, regional, local and enterprise level.

Some other features must also be considered. Complex systems work on the principle of the strange attractor. In business, strange attractors have been likened to a common vision, sense of meaning, strategy or value system that drives people to achieve a common goal (Svyantek and DeShon, 1993; Covey, 1994; Zahra and Ryan, 2007). As a result, the system will manage itself, often in an unknowing manner, towards a common goal (McKercher, 1999). But because the system is non-probabilistic, because the relationships between elements are so rich, and because the system is open to external stimuli, it is impossible to predict accurately the future position of the over time. Instead, the rules will dictate where the system is likely to move within broad parameters. Yet, all systems will evolve and self-organize to ensure the survival of their components (McDonald, 2009).

The butterfly effect is probably the best known element of complex systems. It states that a butterfly flapping its wings in one location could trigger a weather event in a different, distant location. Of course, like most homilies, this one is overly simplistic. Indeed, the butterfly effect is a misnomer. The correct term is "sensitivity to initial conditions." This idea recognizes that small and apparently insignificant changes can precipitate a chain reaction that culminates in a fundamental shift in the structure of the system in one place, but the same action may have no effect somewhere else. Small changes may initiate positive feedback loops that feed on themselves, spinning more and more benefits. Thus it is possible for the initiative of a single entrepreneur to precipitate a major shift in the evolution of a tourism destination, with changes reverberating throughout the destination.

The corollary of the butterfly effect is the lock-in effect. Inertia within a system is also common, whereby an accident of history or initiative produces such a strong network of mutually reinforcing relationships that may endure long after the initial conditions have been superseded. The classic example is the computer keyboard. The keyboard we all use was designed for manual typewriters, with the letters organized in such a way that the keys would not jam. The advent of electronic keyboards overcomes the problem of jamming. Yet, the keyboard design remains unchanged.

Both the butterfly and lock-in effects relate to the feedback cycle of complexity. Linear systems theory is predicated on negative feedback where systems move to a stable state. (Stevenson et al., 2009). Yet, tourism functions as much by positive feedback as it does by negative feedback. There is a reason why the strong seem to get stronger and the weak seem to get weaker.

Phase shift and the associated idea of the edge of chaos recognize that from time to time otherwise stable systems receive such profound shocks that they are shifted from one state to another. Chaotic systems may have experienced prolonged stages of relative stasis, interrupted by (r)evolutionary change or phase shift. Indeed, a feature of large, interactive, dynamical systems is that they evolve naturally toward the edge of chaos (Lewin, 1993). Zahra and Ryan (2007: 855) note that contrary to the usual representation of the destination life cycle as presented by Butler's diagrammatic use of the S-shaped curve, tourism development is often characterized by nonlinear spurts and interventions that are significant. Like any open system, the tourism system can be exposed to exogenous shocks.

Because complex systems are self-organizing, they have tremendous adaptive ability and are remarkably resilient. It has been observed that living systems, when confronted with change, have the capacity to fall apart so that they can reorganize themselves to be better adapted to their current environment (Flower, 1993). Within a tourism context, Faulkner and Russell (1997) and McKercher (1999) note otherwise apparently stable systems may be upended from time to time and completely reorganised, resulting in a new, stronger and more resilient system emerging. Cochrane (2010) comments on how quickly a system can change depends on its adaptive capacity as a function of its ability to absorb disturbance and reorganise itself. It may emerge as fundamentally the same type of entity or may re-emerge in a substantially different form.

Components of a complex system

McKercher (1999) proposed an open-ended complex tourism system comprised of nine different sets of actors. It is reproduced in Figure 9.1. While the model is inelegant, it does identify key component and their roles. The actors include, and are described briefly below:

- The **Traveller**, who is the essential player in tourism, for without people travelling no tourism would occur;

- The **Communication** vectors used to connect the traveller to the destination;
- The **Considerations** or factors that influence the effectiveness of the Communication vectors used;
- The **Destination** or **Internal tourism community** consisting of all businesses involved in tourism at the destination;
- **External tourism agencies** (public and private sector) that try to influence tourism;
- Other **tourism-related externalities**, such as alternative tourism destinations that affect a destination's ability to attract travellers;
- **Non-tourism related externalities,** or macro-environmental forces, such as changing political, economic or social conditions, war, natural disaster, that affect people's ability to travel.
- **Outputs** from the system – both desired and undesired, and;
- **Rogues** or **chaos makers** who can push a system to the edge of chaos.

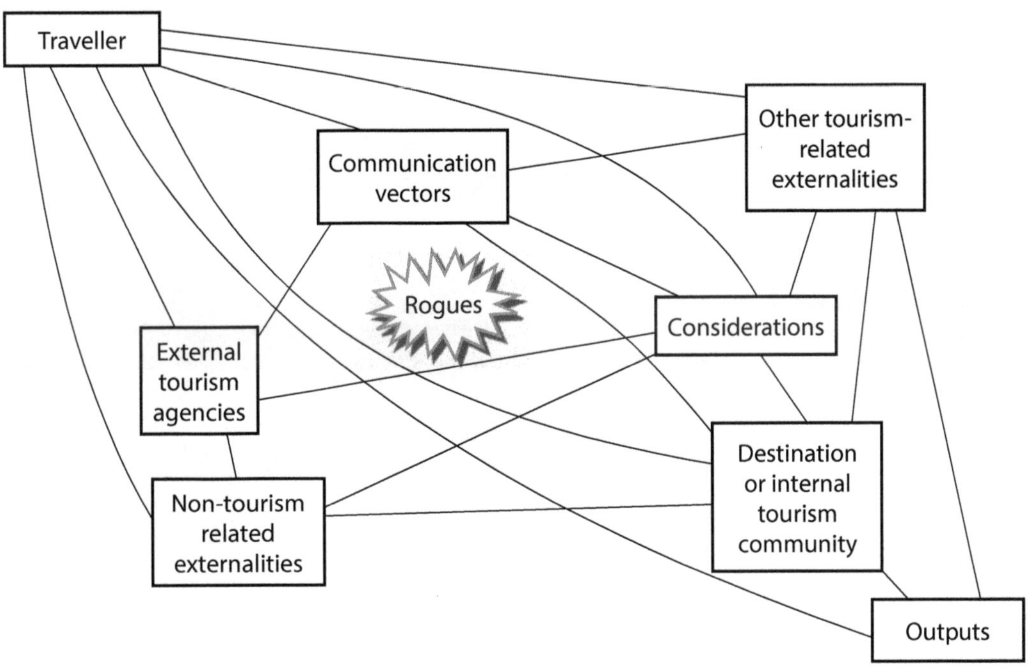

Figure 9.1: A complex tourism system. *Source*: McKercher (1999)

Essentially, tourism is about people travelling, with the business of tourism about linking tourists to destinations. For this reason, the traveller or tourist must be the starting point in the consideration of any tourism model. Markets are dynamic, erratic and non-linear (Faulkner, 1997; Pike, 2005), with tourism markets, in particular, being noted by their great volatility (Var, Kaplan and Yurt, 2006). If the key input does not behave in a rational manner, then the entire system cannot function in a deterministic manner.

Communication vectors connect the traveller with the destination and include both formal marketing activities as well as the dissemination of markers (discussed in Chapter 4). Communication channels are exceedingly complex (Pearce, Tan and Schott, 2004), involving all levels of the formal travel trade, other tourism retailers, the efforts of public sector marketing and promotion agencies and direct communication between businesses and the travelling public.

However, the effectiveness of different communications activities is influenced by a number of considerations. Considerations from a consumer's perspective, relate to getting the right information, while from the destination's perspective they relate to resource availability and expertise. Other factors such as distance, travel party composition and willingness/ability to respond positively also affect the likelihood of connecting the tourist to the destination.

Other tourism related externalities can have a dramatic impact on tourism activity. Lifou Island, Noumea has benefited from the development of South Pacific cruises originating in Australia. A vibrant community-based tourism sector has been developed to cater to day trippers from the cruise ships. Sometimes the net result can be negative, where the development of new destination areas may lead to the decline in other areas, especially if market access has changed.

The destination or the internal tourism community describes the businesses that exist within any destination area. The word 'community' is used in its ecological sense, where a community is described as a living, dynamic, self-organizing system with a group of organisms interacting with one another, cycling and recycling matter through the system. Anything crossing the boundary of that community is considered as either being an input or an output. Importantly, the community as a whole is reliant on the ongoing health of its keystone species.

External tourism agencies, those public and private sector agencies that function outside the commercial tourism world, can have an effect on most of the elements within this model. No product, destination, region or country operates in isolation. Its success depends on how well it competes against other tourism products and how well they compete against it.

Tourism initiatives that occur elsewhere can have a fundamental impact on the competitiveness of a destination, just as initiatives within the system can affect the competitiveness of other destinations. For this reason, a variety of other tourism related externalities need to be included in any tourism model.

Tourism is not immune from global events, such as war, terrorism, natural disasters or disease. A vast array of non-tourism related externalities can dramatically impact the sector. The Severe Acute Respiratory Syndrome affected Asian tourism dramatically in 2003 and 2004 as this photo of a near empty commercial airliner illustrates. The more recent COVID-19 pandemic has posed the greatest threat to the global economy in almost 80 years and has brought tourism to a standstill.

Two other elements that have the potential to plunge tourism into chaos: one is external to the tourism system, the other is internal. The role of non-tourism related externalities is well recognized in the tourism literature. These externalities occur rapidly and without warning. Previously stable tourism communities are plunged into a state of chaos, usually caused by a dramatic fall in visitors. What is less well recognized is the role of chaos makers, or rogues operating within the tourism system (Faulkner and Russell, 1998; Russell and Faulkner, 1999). These individuals can single-handedly transform an organization, destination, region or country.

Lastly, this model considers the outputs of tourism as an intrinsic aspect of any tourism system, for the impacts or perceived impacts of tourism at the host destination or other destinations are increasingly affecting its development. Governments support tourism out of the belief that it will provide a variety of positive outcomes, such as employment, enhancement of the tax base or the provision of foreign currency. In addition, tourism is seen as a powerful tool to help modernize societies. On the other hand, opponents of tourism citing its legacy of adverse impacts have successfully blocked tourism development proposals. Positive or negative, the impacts of tourism are playing a greater role in development decisions.

■ What does it mean?

Complexity theory helps explain the structure of tourism better as well as the evolution/revolution of destinations.

Structure of tourism

The model proposed by McKercher (1999) builds on the types of existing models identified in Chapter 3, but importantly adds externalities as a critical component of any tourism system. This model begins with the tourist, as do the models from Leiper (1990), Mill and Morrison (1995) and Goeldner and Ritchie (2006). Like the other models, it also appreciates some element of exchange. Hall's (1995) model explicitly recognises outputs, or as he states impacts as the key outcome of tourism, as does this model.

Where the complexity model differs, though, is in the recognition of the role other tourism related externalities, external tourism agencies and non-tourism related externalities play in structuring the tourism system. The actions of other destinations, destination marketing organisations and the like have a direct and profound impact on the development of any destination. Likewise the role of non-tourism related externalities can fundamentally alter the structure of a destination.

Destination evolution

The adoption of complexity theory means that destinations do not have lifecycles per se, nor do they evolve in a predictable, slow manner. Instead they tend to go through periods of relative stability before they are fundamentally shaken and transformed. Stability may include an increase, decrease or no change in arrivals and destinations may also enjoy a period of development in response to current market situations. But, at some point this period of stability ends, caused by either a conscious choice on behalf of the destination's managers and governing bodies, or by one or more external events that occur beyond the control of the destination. The resulting evolutionary pattern is similar to Choy's (1993) scalloped pattern discussed in the previous chapter, whereas the destination progresses through a series of mini lifecycles as it evolves and changes its character. In many ways, then, the 'earthquake' Plog (2001) mentioned is not a rare event, but a fundamental part of the life of any destination.

Surprisingly little research has been conducted examining this topic, mostly because it is difficult to conduct a sufficiently long duration study to document the impacts and their resultant consequences. It must be recognised, though, that the revolutionary events that push places to and beyond the edge of chaos are fundamentally different from the occasional shocks that disrupt systems over the short term, but leave no long lasting impact. Natural disasters and crises can rattle systems, but their impact is usually transitory. Instead, what we are talking about is transformative events that completely change the tourism system.

Prideaux (2000) integrated elements of Butler, Plog and chaos into his model of resort development (Figure 9.2). Whereas Butler and Plog regard destinations or a destination's markets as single entities, respectively, Prideaux recognizes their complexity. Indeed, he argues that the catalyst to destination evolution is the emergence of new markets which stimulate new development that in turn, transform the destination. He also identifies a number of elements of change that push destinations to the edge of criticality and initiate phase shift.

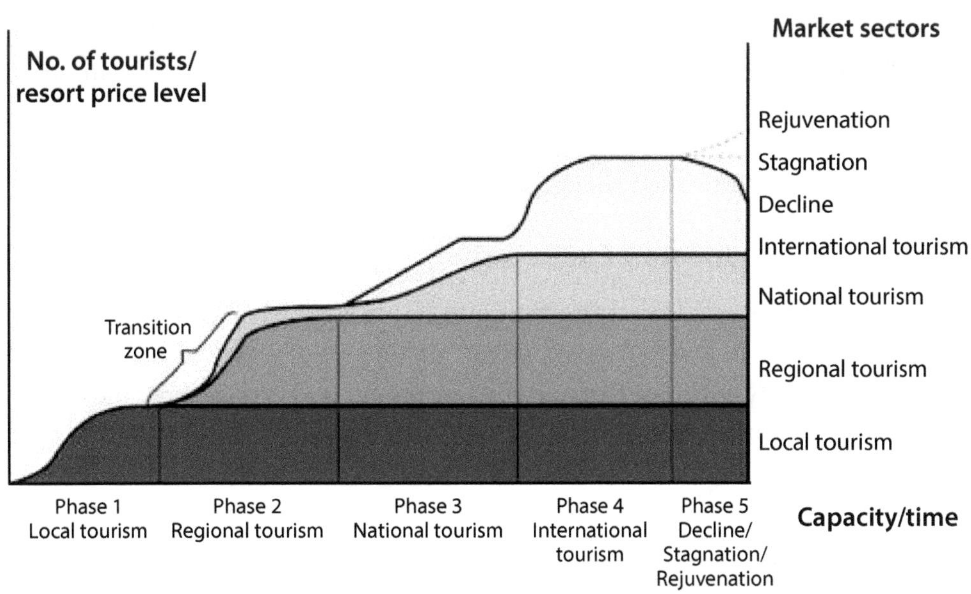

Figure 9.2: Prideaux's Lifecycle Model. *Source*: Prideaux, 2000.

Liu and McKercher (2015) and McKercher and Zhang (2017) wrote about how changes to the visa policy between the Hong Kong SAR and China fundamentally altered Hong Kong's tourism sector. Arrivals tripled in a period of 15 years or less, which in turn spurred the development. A total of 96 hotels offering 39,000 rooms were operating at the end of 2002, along with another 4800 guest house rooms. By the end of 2016, more than 260 hotels offered 75,000 guest rooms with guest houses providing another 12,500. Total overnight and same day visitor expenditure stood at HK$58 billion at the end of 2003, with visitors from mainland China contributing just under HK$38 billion. By 2015, total expenditure by overnight and same day visitors passed HK$300 billion, with Chinese tourists accounting for HK$237 billion of that sum. Moreover, the profile and nature of the tourist changed, whereby Hong Kong has become a short break main destination for people from nearby China. Such a profound change is different from the one-off shock of events like SARS (Severe Acute Respiratory Syndrome) that affected Asia in 2003. Arrivals collapsed for a period of three or four months, but recovered soon afterwards, leaving no long term impact.

Conclusion

Much has changed in our thinking of how tourism is structured and evolves over the last 40 years. Systems models and the Cartesian-Newtonian approach that try to make things ordered nicely have largely been questioned and disproved. Instead, we now appreciate that tourism is a far more complex activity involving a myriad of stakeholders at all levels who interact in a manner that is hard to model and predict accurately. The complexity model proposes an alternate structure for tourism that suggests we need to consider externalities as well as internalities when looking at the structure and evolution of tourism. It also suggests that the nature of destination change is more revolutionary than evolutionary.

Discussion points / Exercises

1. Compare and contrast Plog's and Butler's models with the complexity model described in this chapter. Where are they similar and where do they diverge?
2. Identify different chaos makers or rogues and discuss how they have shaken the tourism system.
3. Complexity looks at the longer term evolution of a destination. Can one-off crises alter a destination's lifecycle? If so, how?
4. Plog suggests the less robust tourist displaces the more robust one. Prideaux suggests the results are cumulative. Argue which case you feel is more applicable in tourism.

Points to consider for revision

- Tourism is a fundamentally unstable activity that is guided by the rules of complexity and chaos theory.
- Tourism systems function much life living ecosystems.
- Phase shift, the lock in effect and the butterfly effect are particularly relevant to tourism.

Additional readings

Baggio, R. (2008) Symptoms of complexity in a tourism system. *Tourism Analysis*, 13(1), 1-20.

Faulkner, B. & Russell, R. (1997) Chaos and complexity in tourism: in search of a new perspective. *Pacific Tourism Review*, 1, 93-102.

Prideaux, B. (2000) The resort development spectrum – A new approach to modeling resort development. *Tourism Management*, 21, 225-240.

References

Baggio, R. (2008) Symptoms of complexity in a tourism system. *Tourism Analysis*, 13(1), 1-20.

Brouder, P. & Eriksson, R. (2013) Tourism evolution: on the synergies of tourism studies and evolutionary economic geography. *Annals of Tourism Research*, 43, 370-389. http://dx.doi.org/10.1016/j.annals.2013.07.001.

Choy D. (1992) Life cycle models for Pacific Island destinations. *Journal of Travel Research*, 31(3), 26-31.

Cochrane, J. (2010) The sphere of tourism resilience. *Tourism Recreation Research*, 35(2), 173-185, DOI: 10.1080/02508281.2010.11081632

Covey, S. R. (1994) The strange attractor. *Executive Excellence*, 11(8), 5-6.

Farrell, B. & Twining-Ward, L. (2004) Reconceptualizing tourism. *Annals of Tourism Research*, 31(2), 274-295.

Faulkner, B. & Russell, R. (1997) Chaos and complexity in tourism: in search of a new perspective. *Pacific Tourism Review*, 1, 93-102.

Faulkner, B. & Russell, R. (1998) *Movers and shakers: The chaos makers on the Gold Coast.* Paper presented at the 1998 Australian National Tourism and Hospitality Research Conference, Gold Coast, February.

Flower, J. (1993) The power of chaos. *Healthcare Forum*, 36(5), 48-55.

Gleick, J. (1987) *Chaos: Making a New Science.* Abacus Books London.

Goeldner, C. R. & Ritchie, J. R. B (2006) *Tourism Principles, Practices and Philosophies* 9th Ed John Wiley and Sons.

Hall, M. (1995) *Introduction to Tourism* 2nd Edition, Longman Australia, Melbourne.

Leiper, N, (1990) *Tourism Systems; An Interdisciplinary perspective.* Department of Management Systems (Occasional Paper # 1). Palmerston North: Massey University.

Lewin, R. (1993) *Complexity: Life on the Edge of Chaos.* Phoenix, London.

Liu, A. & B. McKercher (2015) The impact of visa liberalization on tourist behaviors: The case of China outbound market visiting Hong Kong. *Journal of Travel Research,* 55(5), 603-611. DOI: 10.1177/0047287514564599.

McDonald, J. (2009) Complexity science: an alternative world view for understanding sustainable tourism development. *Journal of Sustainable Tourism,* 17(4), 455-471.

McKercher, B. (1999) A chaos approach to tourism. *Management,* 20(4), 425-434.

McKercher, B. & Zhang, R. (2017) Hong Kong as a supranational destination for Chinese tourists. *China Tourism Research.* 13(2): 178 – 102. DOI: 10.1080/19388160.2017.1350613

Mill R. C. & Morrison A. M. (1985) *The Tourism System: An introductory text,* Prentice Hall, NJ.

Overman, E. S. (1996) The new science of administration: Chaos and quantum theory. *Public Administration Review,* 56(5), 487-491.

Pearce, D. (1989) *Tourist Development 2nd Ed.* Harlow: Longman.

Pearce, D., Tan, R. & Schott, C. (2004) Tourism distribution channels in Wellington, New Zealand. *International Journal of Tourism Research,* 6(6), 397-410.

Pike, S (2005) Tourism destination branding complexity. *Journal of Product & Brand Management,* 14(4), 258-259. https://doi.org/10.1108/10610420510609267.

Plog. S, (2001) Why destination areas rise and fall in popularity: An update of a Cornell Quarterly Classic *Cornell Hotel and Restaurant Quarterly,* 42(3), 13-24.

Prideaux, B. (2000) The resort development spectrum – A new approach to modeling resort development. *Tourism Management,* 21, 225-240.

Russell, R. & Faulkner, B. (1999) Movers and shakers: Chaos makers in tourism development. *Tourism Management,* 20(4), 411-423.

Sainaghi, R. & Baggio, R (2017) Complexity traits and dynamics of tourism destinations. *Tourism Management,* 63, 368-382.

Scott, N., Baggio, R. & Cooper, C. (2008) *Network Analysis and Tourism: From Theory to Practice.* Clevedon: Channel View Publications

Sedarati, P. Santos, S. & Pintassilgo, P. (2018) System dynamics in tourism planning and development, *Tourism Planning & Development,* DOI:10.1080/21568316.2018.1436586.

Stevenson, N., Airey, D. & Miller, G. (2009) Complexity theory and tourism policy research. *International Journal of Tourism Policy,* 3(2), 206-220.

Svyantek D. J. & DeShon R. P (1993) Organisational attractors: a chaos theory explanation of why cultural change efforts often fail. *Public Administration Quarterly,* 17(3), 339-355.

Var, T., Kaplan, M. & Yurt, O. (2006) Challenges and opportunities of developing wine tourism in a small community in Turkey. *e-Review of Tourism Research,* 4(1), 9-18.

WTO (1994) *National and Regional Tourism Planning: Methodologies and Case Studies,* Routledge, London.

Zahra, A., & Ryan, C. (2007) From chaos to cohesion—Complexity in tourism structures: An analysis of New Zealand's regional tourism organizations. *Tourism Management,* 28(3), 854-862.

Part 4 – The Tourist

So far, the book has looked at defining tourism, understanding its constituent parts to see how tourism works and examining different lifecycle models. Now it is time to consider the central component of tourism – the tourist. The tourist has been the subject of much research by sociologists, anthropologists, psychologists, members of the leisure and recreation community and others for close to 60 years. The next few chapters review their thoughts.

Chapter 10 asks the important question of why do people travel? This chapter explores a range of progressively sophisticated models and theories that try to explain tourist motivation. It begins with a review of Maslow's hierarchy and concludes with a discussion of Pearce's current Travel Career Pattern.

Chapter 11 considers the thoughts of sociologists and anthropologists. It begins by delving into ideas that were prevalent in the 1960s and then considers such core concepts as the environmental bubble, tourism as a quest and the liminal nature of travel.

Chapter 12 focuses more on how the leisure and recreation community views tourism in general and issues relating to propensity and intensity in particular. This chapter examines how the evolving literature on barriers to travel translates into different attitudes toward travel and subsequent travel patterns, the confounding role of the leisure paradox and how generational cohort theory influences travel.

10 Why Do People Travel?

By the end of this chapter, the reader will be able to:

- Describe the strengths and weaknesses of Maslow's hierarchy of needs
- Analyse Dann's push-pull model
- Apply Crompton's motives
- Critique Iso-Ahola's optimal arousal model
- Understand Gutman's means-end model
- Apply Pearce's evolving travel career ladder/pattern.

Introduction

Understanding why people travel for pleasure has been the focus of tourism research since the earliest days of tourism academia. Iso-Ahola (1983: 50) states "while many questions about recreational travel remain to be raised and answered, perhaps the most fundamental of them all are the ones that deal with motivation." As Song and Bae (2018) note in their review of the literature, motivation influences destination choice, perceived benefits, satisfaction and on-site experiences.

A number of increasingly sophisticated models has been developed over the years that has attempted to develop a comprehensive theory of tourist motives. Initially models were adapted from consumer behaviour literature. Recently, though, tourism specific theories have been developed. The task has been challenging for the underlying reasons to travel are covert, in that they reflect individual's private needs and wants (Pearce, 2005). As a result, many of the earliest attempts to examine the issue conflated motives with activities pursued and trip purpose (see Lundberg, 1972), while other popular models tended to identify unmet needs as motives, even though they are different (Tasci and Ko, 2017).

Another challenge with trying to identify motives is that different disciplines adopt different approaches to explore this concept. Jamal and Lee (2003) argue, for example, social psychology and sociology represent the two major discipline areas that have influenced this area of research. Sub-disciplines within each host discipline, in turn direct studies. As a result, the study of motives is typical of Smith et al.'s (2013) type two theory where many legitimate, but competing models and/or theories exist. For example, Dann (1981), Page (2019) and Jamal and Lee (2003) identify a range of approaches that have been adopted over the years, including:

- Travel as a response to what is lacking yet desired;
- Destination pull in response to motivational push;
- Needs based approaches where pleasure travel is a means to satisfy unmet needs;
- Intrinsic motivations as reflected by a desire to obtain some reward for travelling;
- Values based approaches;
- Disequilibrium and the need for a break;
- Motivation as fantasy and a liminal break;
- Expectancy based approaches which look at achieving outcomes as a prime motivation;
- Quest for authenticity and a search for one's centre;
- Motivation as a trip purpose, etc.

An additional challenge in developing motivation theory was highlighted by Pearce (2011:43) who stated:

> The criteria for a sound motivational framework highlight some of the unique attributes of tourism ... its episodic, dynamic, relationship-dependent, future-oriented and varied character ... Good motivation theory [thus] needs to be multi-motive, dynamic, measurable and relatively easy to communicate. In addition, it is most essential that the approach functions as an organiser and synthesizer of existing information, with a desirable characteristic that it has a future-oriented or predictive capacity.

This chapter explores the evolving literature on motives from Maslow through to the Travel Career Pattern proposed by Pearce and his colleagues.

Before beginning though, it is important to define terms. Dann (1981) explains that etymologically 'motivation' is derived from the Latin word 'movere' to move. Pearce (2005:54) defines motives as the forces that drive travel behaviour. They can be biological and socio-cultural. Iso-Ahola (1982:257) defines them as internal factors that arouse, direct and integrate a person's behaviour. Moutinho (1987:16) identifies motives as a state of need, a condition that exerts a push on the individuals toward certain types of action that are likely to bring satisfaction.

In a similar manner Mill and Morrison (2012) suggest a distinction is needed between needs and wants. A *need* is an amorphous concept. People may only be partially aware of their needs. Some people say it is a latent, or hidden element. A *want* is the concrete definition or transformation of a need. The difference between a need and a want is awareness.

Maslow's Hierarchy of Needs

In many ways, Abraham Maslow is the father of motivation research and, moreover, his work has influenced tourism motivation research in the past and continues to do so to the present. He sought to understand why people are driven to act by trying to understand how particular needs can be met at particular times (Kotler et al., 2003). He posited that human needs can be classified in a hierarchical manner (Maslow, 1943) from the most pressing to the least pressing. His theory suggests lower order needs had to be met before higher order needs can be considered, but once met, they fade into the background as we strive to have higher order needs satisfied. Maslow (1943) also asserts that for most people the hierarchical nature of needs is relatively fixed. He does recognise, though, that needs are related to each other, being arranged in a hierarchy of prepotency. As he explains "the most prepotent goal will monopolize consciousness and will tend of itself to organize the recruitment of the various capacities of the organism. The less prepotent needs are minimized, even forgotten or denied (Maslow, 1943: 394, 395)."

His model (Figure 10.1) can be divided into deficiency needs and growth needs (McLeod, 2018), with the first four levels referring to deficiencies that need to be fulfilled before the top level growth need of self-actualization can be met. It is typically portrayed as a pyramid, with its dimensions described below:

- Physiological needs of food, water, sex, air, sleep, clothing;
- Safety and security needs encompassing freedom from threat or danger or a secure, orderly and predictable environment;
- Social needs like love and belonging needs that reflect feelings of connection to and affectionate relationships with family, friends, teammates or one's community;
- Esteem needs of self-respect, accomplishment, reputation, recognition and prestige; and
- Self-actualisation needs of fulfilling one's potential (Hsu et al., 2008).

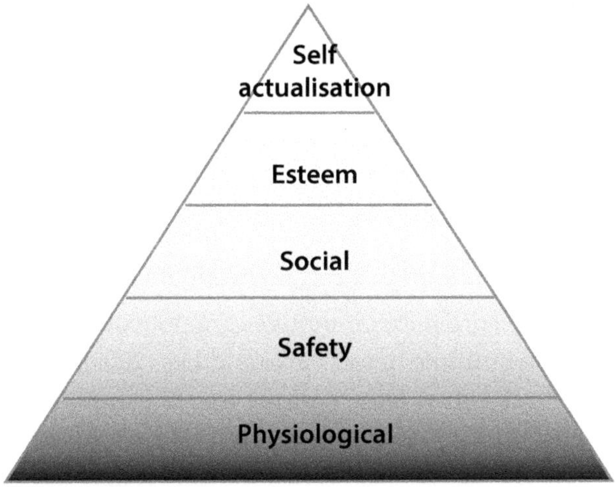

Figure 10.1: Maslow's Hierarchy of Needs

The model is intuitively appealing, in part because of its parsimoniousness and in part because we can all relate to its dimensions. It has been used so widely in tourism that a *Google.Scholar* search using the keywords *Maslow* and *Tourism* yielded over 20,000 results, covering every conceivable topic from destination marketing to food tourism consumption, motives for volunteer tourism, seniors' travel, impacts and residents' perceptions, tourism planning, quality of life, dental tourism and many, many more topics. Virtually every tourism marketing and consumer behaviour text along with most introductory tourism texts also refer to it. Yousaf et al. (2018) and others have also provided a range of examples of how destinations can provide goods and services that can satisfy each tier of needs.

There is only one problem with Maslow's model. There is very little evidence to support it. It is particularly difficult to apply Maslow's hierarchy in understanding why people travel, although it may be useful in understanding how tourism experiences are delivered. A number of criticisms of the theory have been raised in both the tourism and mainstream psychology literature. To begin, there is scant empirical evidence to support the conclusions (Kaur, 2013; Kay, 2003). The reason is that needs are not necessarily hierarchical nor are they singular (Denning, 2012). You may be able to seek social interaction without necessarily having physiological and security needs met. Indeed, that is the very basis of much commercial adventure tourism, where social interaction is a driver of participation and yet people willingly place themselves in some physical risk. Maslow suggests such actions are incompatible.

The observation that needs are not singular is also challenging. Again, using commercial adventure tourism as an example, the drivers of participation may be physiological, social and esteem related, and in some cases self-actualisation may come into play. All may be equally powerful in the participation decision, with no single item standing out. This issue is something that has vexed Pearce in the development of his travel career pattern. More on this later.

Moreover, from a tourism standpoint, the same action may satisfy many needs simultaneously, defying simple categorisation. Eating may satisfy a physiological need if the person is hungry, a relationship need if dining with friends and relatives, a self-esteem need if eating at a fine dining restaurant or perhaps even a self-actualisation need if the person is a culinary tourist and has always wanted to eat a certain type of food in a certain destination.

Likewise, some core tourism motives are not covered by the hierarchy or can be subsumed into multiple tiers. Pearce and Lee (2005), for example, identify novelty and escape/relaxation as two of the four motivational backbones behind all travel decisions, yet they do not easily fit into the traditional hierarchy of needs. Finally, the causal link between need and behaviour is not as absolute as desired.

Indeed, Pearce and Packer (2013: 389) conclude that "in some sixty years after its initial formation, it can be suggested that the value of Maslow's work lies not in the hierarchical system associated with his studies, but resides more simply in the diversity of motive forces he identified."

Plog

The same observation can be made about Plog's (1974, 2001) models. Good ideas, but not applicable in practice! Plog's ideas have been discussed in detail in Chapter 8 and need not be repeated here again, other than to acknowledge that his Allocentric/Psychocentric continuum has been used as a proxy to understand tourist motives. Critics of Plog's model as a motivational theory state that it neither explains tourist motivations nor predicts tourist behavior (Huang and Hsu, 2009). Litvin and Smith (2016), reviewing the literature, note many people have questioned how such a simplistic model that gives a person an absolute label can predict travel behaviour when tourists travel with different motivations on different occasions. Yoo et al (2018) also show little empirical support for his work. One other criticism is that he lumps two-thirds of the travelling public into the same Midcentric category, which proves to be of little utility when trying to segment the market.

What his works do, though, is recognise that tourists are not a homogeneous group and that a variety of factors influence their likelihood of travelling and the type of holiday they may prefer ideally. He also argues effectively that one's underlying social-psychological make-up may set the upper limits of the type of travel one is comfortable with, unless a large environmental bubble is created. It must be placed within the historical context of the lack of tourist motivation theory some almost 50 years ago. Early researchers, such as Lundberg (1972) were among the first to write about why people travel. But, looking at this work in hindsight, it is rather simplistic, for it conflates purpose (sports, adventure tourism), with activities (such as shopping), motives (ego enhancement, escape, personal growth) and other factors (socially acceptable behavior, acquiescing to cultural norms). Plog's work represented a paradigm shift in that he tried to associate underlying personality traits to travel preferences. It is just that this work has been superseded by more sophisticated research.

Dann's Push–Pull model

The late 1970s and early 1980s saw a flourishing of research on motives by such scholars as Dann (1977), Crompton (1979) and Iso-Ahola (1982). Much of this work is still relevant to this very day.

Dann's (1977) push-pull model is arguably the best known and most ubiquitous one relating to motivations (Crompton, 1979; Uysal et al., 2008; Prayag and Ryan, 2011). Its popularity lies in part because it provides a simple and intuitive approach to define tourist behaviour (Klenosky, 2002) and in part because it attempts to show a link between drivers of travel and destination choice (Yousaf et al., 2018). It also builds on and/or informs many other motivational works, including Maslow's hierarchy, Crompton's work and to some extent Pearce's travel career approach (Pearce, 2011; Prayag and Ryan, 2011). Moreover, Klenosky (2002) states push and pull factors have generally been characterized as relating to two separate decisions made at two separate points in time – one focusing on *whether* to go, the other on *where* to go. Below is a summary of the key points of this important work.

Essentially, Dann (1977) argues that push factors are those factors that predispose tourists to travel, while pull factors are the features of the destination that attract tourists to certain places. He illustrates that most of the research conducted to date (remember 1977) focussed on the pull factors of destinations, in much the same way that marketing theory on the time focussed on the pull factors of products (the 'if you build it they will come' syndrome). Push factors were largely ignored. Yet, he argues that while destinations may provide a range of products and services, the actual decision to visit a place is predicated on a prior need to travel. Push factors then are the antecedent considerations before pull or destination factors can be considered. Thus, he states the question of what makes people travel can only be answered by considering push factors, whereas the question of why do they choose certain destinations may involve both push and pull considerations.

Push factors drive people to travel, while pull features are those of the destination that serve to attract people and satisfy their needs. Graham Dann argued two factors, ego enhancement and anomie drive tourism. Ego enhancement, coupled with the desire to learn about exotic lost civilisations, drives many people to visit places like Angkor Wat in Cambodia. Alternately, anomie, or the desire to get away from it all is a powerful push factor to explore remote regions like Kyrgyzstan. In other instances, anomie may be the driving force behind seeking a sun, sand and sea resort holiday, where people simply need a short break from their everyday life.

He argues that the motivation to travel lies in the twin factors of anomie and ego-enhancement. Anomie refers to a state of disequilibrium from an anomic or socially disoriented society. Thus, one push factor is the desire to get away from it all or transcend the isolation of everyday life. Tourism provides an opportunity for stronger and more influential social interaction for those living in an isolated world, or conversely social isolation if living in a world where privacy is at a premium. Both needs can only be fulfilled when away from the normal environment.

Ego-enhancement too derives from personal needs, where from time to time we need to have our ego boosted. One strategy is to travel where a tourist can go to a place where his/her social position is unknown and, consequently, can feel superior on that basis on such anonymity. Moreover, status can be gained when the person returns from the trip.

These two push factors are really quite limited and do not explain all types of pleasure travel. Both Crompton's (1979) and Pearce's (2011) various studies expand the range of push factors to a far more comprehensive list.

Combined, Dann asserts that travel can represent the fulfilment of basic needs, while also offering an alternate world to that in which the tourist lives. It also lets the tourist behave in a way that is normally circumvented in their daily life, by enabling subterranean values to be released and allowing the person to reside, temporarily in a fantasy world. Such a fantasy world relieves the humdrum work of the anomic life, while providing the opportunity for ego-enhancement by acting out in an alien personality.

Pull factors, on the other hand represent those attributes of the destination that appeal to tourists. Fyall et al. (2019) characterise them as encompassing social and cultural features of the destination, its physical characteristics and the range of attractions that exist to satisfy needs. Indeed, Uysal et al. (2008) indicate the number of identified pull factors is limited only by the scope of the destination. It is therefore, easier to define and measure pull factors. Yet, they suggest that the attitude of a tourist toward a vacation destination may be a measure of that destination's ability to pull or attract the tourist. But, as Pearce (2011) indicates, pull factors are variable and as such are hard to isolate as key features that induce travel to a destination. He illustrates that one activity or destination attribute can be valued by different travelers for very different reasons. Nikjoo and Ketabi (2015) found pull factors varied depending on the motive of the tourist. Those who were driven by anomie and wanted to escape tended to select resort style vacations, while those who were travelling for ego enhancement chose more culturally oriented holidays.

Crompton's Motives

John Crompton's (1979) work titled *Motivations for Pleasure Vacation* is one of the most influential motivational studies published in the last 40 years. It has received over 4000 citations and is referred to in most texts. As Crompton notes (1979: 409), motivations are only one of the variables that may explain tourist behavior, but are considered to be the crucial variable because they are the impelling and compelling force behind all behavior. This relatively simple, though elegant study involved 39 unstructured interviews with a convenience sample of adults. While the study explored a range of socio-psychological variables, motives were a primary focus of the research. Interviews were recorded and content analyzed to identify patterns of responses which provided the basis of the conceptual issues identified in the study. The key features of the study are summarized below.

Essentially, Crompton's work is another form of push study. He began by identifying a four stage continuum in the travel decision process. Like Dann before him, and as strongly inferred by Jafari (1987) and Cohen (1979), the process begins with the recognition of a state of disequilibrium, which could best be satisfied by the second stage, the recognition of the need for a break from existing routine. Disequilibrium could be classified into two categories, short term, and long term. Short term disequilibrium occurs when a particular set of circumstances or events triggers disruption to homeostasis. It demands immediate attention and could be satisfied by a single vacation. In many ways, short term disequilibrium explains the popularity of short break vacations. On the other hand, long term disequilibrium has deeper causes. It is considered ever present but postponable. It could only be resolved by a series of pleasure vacations. In some ways, as well, short term disequilibrium is reflected in Cohen's Recreational tourist who travels to re-create and rejuvenate, while long term disequilibrium may be reflected by the type of vacation taken by Cohen's Diversionary tourist who travels simply to sooth a troubled soul.

Uysal et al. (2008) pointed out that disequilibrium in an individual's cultural, social and psychological needs can be a primary motivation for travel. Interestingly, Crompton notes a break from routine does not necessarily mean a change to doing different kinds of things or a change in lifestyle. But he does suggest some form of inversion as Graburn (1983, 2018) suggested, whereby the break from routine may occur in a different geographic or social setting. Activities could also be pursued at differing levels of intensity.

The third phase in the continuum is the consideration of behavioral alternatives, where people can choose to stay at home, travel for pleasure, to visit friends and relatives or go on a business trip. The fourth stage examines the range of motives that can help determine the nature of the trip and the choice of destination.

A total of nine motives emerged from this work. Seven could be classified as socio-psychological and two as cultural. Socio-psychological motives were overtly identified rarely by respondents and proved difficult to articulate. Crompton argues that such motives reflect the 'hidden' agenda that drives travel. The cultural motives were more concerned with the destination's attributes than with the social or psychological status of the individual. Importantly, they are not exclusive, but operate in tandem or in combination with other motives. Crompton thus concludes that a destination can attract customers whose motives are not homogeneous or necessarily compatible. The motives are summarised in Table 10.1.

Crompton also made another important observation. He determined that particular pull attributes of destinations are relatively unimportant if people traveled to have broadly based social psychological needs of regression, escape and relationship development met. Thus, almost any destination could satisfy these types of needs. This situation is especially true in multi-product urban destinations that represent a dynamic agglomeration of attractions, culture, events, landscapes and services that provide tourism space for consumption (Framke, 2002). This issue is discussed more fully in Chapter 4 on Attractions.

Table 10.1: Crompton's Motives

Socio-psychological motives

1 Escape from perceived mundane environment
- Temporary escape
- No single optimum environment
- Anticipation of the event can outlast the event

2 Exploration and evaluation of self
- Re-evaluate and rediscover more about oneself
- Liminal state aided self-discovery

3 Relaxation
- More mental relaxation than physical relaxation
- Pursue activities of interest

4 Prestige
- Rarely stated overtly

5 Regression
- Do things that were inconceivable in their normal lives, like adolescent or childlike behaviour

6 Kinship

7 Social interaction
- Meet new people
- Most likely to interact with tourists and not locals

Cultural motives

8 Novelty
- Curiosity, adventure, new and different experiences

9 Education
- Develop a rounded individual

John Crompton developed a broader set of push factors that drive people to travel, including relaxation and regression. Here one of the authors tries indoor skiing in Dubai, an activity he has not pursued for many years.

Iso-Ahola optimal arousal theory

Seppo Iso-Ahola (1982) argued that proponents of push-pull theories in general, and sociologists like Dann in particular, did not pay enough attention to psychological theory when considering motivation (Moscardo, 2014). He also suggested that deep insights into travel motivations could arise from examining the vast array of leisure research, since pleasure tourism is an extension of leisure (Iso-Ahola, 1983). He argued people pursue leisure as a potential satisfaction-producer for two main reasons: to provide certain intrinsic rewards such as feelings of mastery and competence; and to leave the routine environment behind (Iso-Ahola, 1982: 258). He categorised these two elements as an 'approach' or 'seeking' and 'avoidance' or 'escapist' dimensions. Both are present in leisure motivation in general and in tourism motivation in particular, although he does suggest the unique nature of tourism may place greater emphasis on the escapist motive.

People may be pushed to escape the personal world, such as personal problems, difficulties or the interpersonal world of co-workers, family members and relatives. And, they may seek personal rewards such as feelings of mastery, learning about other cultures and the like, or interpersonal rewards, such as deeper social interaction with family and friends, meeting new friends or socialising with members of a travel party. Together these four elements can be organised in a two by two matrix, where tourists can be placed in one of four cells (Figure 10.2). One group may travel primarily to escape their personal environment and seek personal rewards, while other groups may escape their personal environment and seek interpersonal rewards; escape their interpersonal environment and seek personal rewards; or escape their interpersonal environment and seek interpersonal rewards. Interestingly, while Iso-Ahola criticises Dann's work, Pearce (2011) asserts both the escaping and seeking dimensions represent a form of push motive.

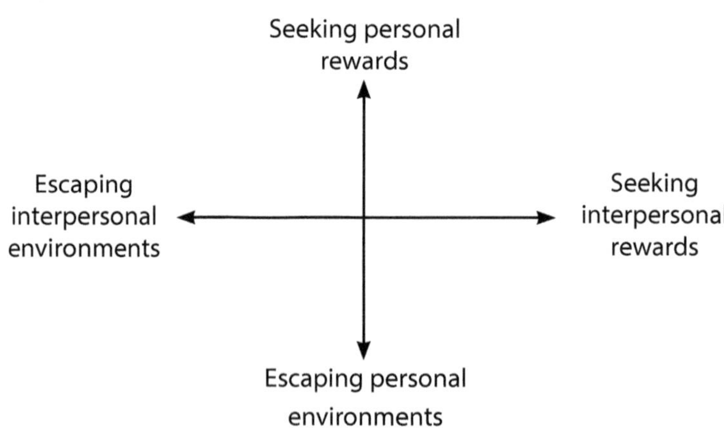

Figure 10.2: A social psychological model of tourism motivations. *Source*: Mannel and Iso-Ahola, 1987.

The optimal arousal theory (Iso-Ahola, 1983) that evolved from these two considerations, suggests that tourists must balance between novelty and security. The extent of escaping from and degree of seeking to represents the desired state of stimulation that tourists can accept (Pearce, 2011; Prayag and Ryan, 2011). As Iso-Ahola (1983: 53/54) states

> when the social environment provides too many inputs and stimuli for the individual to handle (i.e. the situation is too incongruous in relation to the stored information), the individual tends to withdraw from the environment. On the other hand, when the environmental inputs are overly similar to the coded and stored information and experience, the individual becomes bored and seeks situations which offer more incongruity, uncertainty, novelty, and complexity. Thus, this tendency to seek the optimal incongruity and arousal channels the individual's leisure behavior.

Gutman's means-end chain theory

Another model that has been used with some success in tourism is Gutman's (1982) means-end chain model. McDonald et al. (2008) note that it has been widely used in the marketing literature to identify values-based motivations behind consumption. It has been used sparingly in tourism, though, mostly because of its qualitative nature and the concomitant need for detailed data collection and an arduous analysis phase. Also, because it is situation specific, it is difficult to develop a series of generalizable motives that can be applied broadly. None the less, Klenosky (2002) asserts it provides a practical framework for examining the relationships between the pull attributes of a destination and the higher level motivational forces important to the individual traveler. Jiang et al. (2015) have used it to understand China outbound travel and Kim et al. (2016) have used it to examine pilgrimage tourism. Indeed, Jiang et al. (2015) argue it is particularly useful for understanding personal values.

Essentially, the model assumes that customers do not purchase a product for its own sake, but for the desired benefits that accrue, or alternately to minimise undesired consequences (Gutman, 1982). Thus, means-end theory is based on the idea that products, and the attributes they possess, represent the 'means' by which consumers obtain important consequences or benefits and reinforce important personal values or 'ends' (Gutman, 1982). Product features or attributes themselves have no inherent meaning to consumers and instead gain their meaning or importance through the consequences they are perceived to provide or help one to avoid (Klenosky, 2002).

Means-end theory encompasses three elements: the features or aspects of a product or service (physical or abstract); consequences (benefits accruing to users when they experience the product or service whether functional or psychosocial); and value as the highest level of end, and the most fundamental goal, whether instrumental or terminal (Kim et al., 2016). Jiang et al. (2016) thus say it describes the hierarchical relationships between product attributes (the means), the consequences for the consumer provided with these attributes (the benefits) and the personal values (the ends) these consequences reinforce.

Methodologically, means-end theory is used primarily as a qualitative research tool whereby through a series of laddering questions one can dig down into the deepest meaning behind behaviour. In this way the underlying emotions, consequences and personal values that drive consumer choice can be revealed (Woodall, 2013). Klenosky (2002), for example, illustrates that someone may visit an area because it has nice beaches (product), with the consequence that someone can get a tan, leading to a further consequence of looking good or feeling better. But as one digs deeper, the true value is one of self-esteem and feeling better about oneself.

Pearce from travel career ladder to travel career pattern

The current state of the art for tourist motivation is Pearce's travel career pattern model. It has been an evolving framework that was first proposed some 30 years ago as a travel career ladder, based on Maslow's hierarchy. It has since undergone a number of iterations, modifications and revisions until today the travel career pattern model bears little resemblance to the initial piece of work. Unlike the other models and type three and four theories identified above, there is no single piece of work that one can summarise. Instead, the following section summarises a number of studies that reflect the dynamic nature of the mode.

Pearce posited initially that people have travel careers. As he summarised in an early paper: "people have a career in their tourist behaviour. Like a career at work, people may start at different levels, they are likely to change levels during their life-cycle and they can be prevented from moving by money, health and other people. They may also retire from their career or not take holidays at all and therefore not be part of the system." (Pearce 1992:125)

The initial iteration proposed a travel career ladder (TCL) based on Maslow's hierarchy. Like Maslow, the needs of travelers were seen as organized into a hierarchy or ladder with relaxation needs being at the lowest level, followed in order by safety/security needs, relationships needs, self-esteem and development needs, and finally, at the highest level, fulfilment needs (Pearce and Lee, 2005:

227). The premise was that people's motivations would change as their level of experience grew. And, as they grew, they would seek to have higher needs met. While he recognised that people's motives may be found at multiple levels, the initial model suggested one set of needs on the ladder would dominate, thus enabling someone to be positioned neatly at one place on the ladder.

Great idea, intuitively appealing, but like Maslow, not practical in practice. Ryan (1998) wrote an insightful critique of the model arguing it is too simplistic and not supported empirically. He also noted that empirical research conducted to that point has been cross sectional and, like a lot of work on lifecycle theory, can identify a situation at a single point in time, but does not necessarily show change over time. He then went on to raise six questions about the concept, including its lack of philosophical congruence with Maslow, the validity of measures used and the methods adopted in different studies, and possible dissonance between expressions of needs and the ability to satisfy needs. In addition, subsequent work by Pearce and Lee (2005) and McKercher and du Cros (2008) suggest the pyramid may actually be inverted, for inexperienced tourists consistently express views that their first trips reflect a degree of self-actualisation, but as tourists gain experience, their needs tend to move down the pyramid to have lower order needs of relationship, or safety needs met.

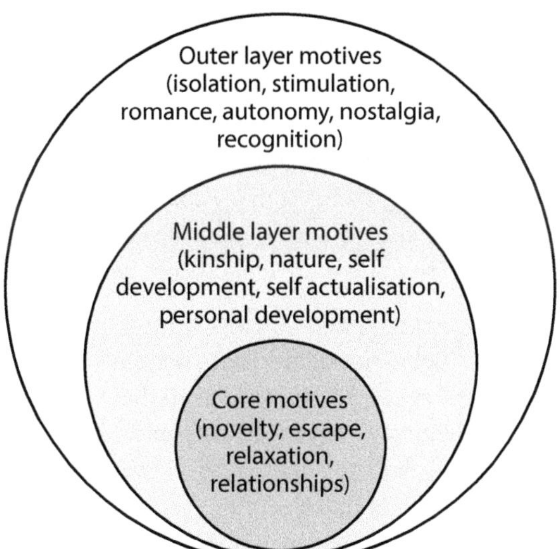

Figure 10.3: Pearce's travel career pattern

Pearce took these issues on board and over the next 20 years he revised the model. To begin, he modified the idea of a career 'ladder' to a career 'pattern.' Though this might seem like a semantically small measure, it was actually quite important. As Pearce and Lee (2005) and Pearce (2011) noted, the term ladder drew too much attention, with the concentration on either being at one step in the hierarchy at one single time, or the prospect that people could ascend or descend a ladder. Accordingly, the idea of a ladder was de-emphasised and replaced with a travel career pattern (TCP) approach, a more dynamic, multilevel motivational structure that is seen as critical in understanding travel motivation (Pearce and Lee, 2005: 227). This revised model retained the notion that travellers' motives changed with experience, but importantly, added that a pattern of motives was evident and not necessarily a single dominant one (Pearce, 2011).

Table 10.2: Disaggregating constituent motive variables from Pearce's travel career pattern

Factor	Constituent variables
Motive level: Core	
Novelty	Experiencing something different Feeling special atmosphere of the vacation destination Having fun Visiting places related to my personal interests
Escape/relax	Being away from daily routine Getting away from everyday physical stress/pressure Getting away from the usual demands of life Getting away from everyday psychological stress/pressure Giving my mind a rest Not worrying about time Resting and relaxing
Relationship	Being with others who enjoy the same things as I do Contacting with family/friends who live elsewhere Doing something with my family/friends Doing things with my companions Strengthening relationships with family/friends Strengthening relationships with my companions
Motive level: Middle	
Kinship (relationship-security)	Being near considerate people Being with others if I need them Being with respectful people Feeling personally safe and secure Feeling that I belong Meeting people with similar interests/values
Self-development through host-site involvement	Developing knowledge of the area Experiencing different cultures Following current events Learning new things Meeting new and varied people Meeting the locals Observing other people in the area

Nature	Being close to nature Being harmonious with nature Getting a better appreciation of nature Viewing scenery
Self-actualization	Being creative Experiencing the things I have always wanted to do Feeling inner harmony/peace Gaining a new perspective on life Understanding more about myself Working on my personal/spiritual values
Self-development – personal development	Develop my personal interests Developing my skills and abilities Gaining a sense of accomplishment Gaining a sense of self confidence Knowing what I am capable of Using my talents and skills
Motive level: Outer	
Isolation	Avoiding interpersonal stress and pressure Being away from crowds Enjoying isolation Experiencing open space Experiencing peace and calm
Stimulation	Being spontaneous Experiencing the risk involved Experiencing thrills Exploring the unknown Feeling excitement Having daring/adventuresome experiences Having unpredictable experiences
Romance	Being with people of the opposite sex Having romantic relationships
Autonomy	Being independent Being obligated to no-one Doing things my own way
Nostalgia	Reflecting on past memories Thinking about good times I've had in the past
Recognition	Being recognized by other people Having others know I have been there Leading others Sharing skill and knowledge with others Showing others I can do it

Source: Pearce, 2005.

The most recent version (Pearce, 2005, 2011) argues that trip decisions are influenced by the mix of and the relative importance placed on different tiers of motives (Figure 10.3). Motives can be classified into three broad categories, with Core motives relating primarily to novelty, escape, relaxation and relationships. Mid-layer motives are more diverse and can include self-development, self-actualization and a range of other factors. Outer-layer motives include such features as social status, romance, nostalgia, isolation and the like.

While core motives represent the backbone or skeleton of all travel (Pearce and Lee, 2005), the extent to which they influence travel decisions depends on the relative importance individuals place on middle layer and outer layer motives. To a large extent, then, core motives are generic, while middle and outer layer motives become more specific. His framework identifies 14 different motive factors, which comprise more than six dozen specific motivate variables, as summarized in Table 10.2. His various studies have analysed the relationship at a factor level, more so than at an individual motive level.

Conclusions

Do we understand why people travel? The answer is an equivocal yes, for travel involves a series of personal decisions. Each one of us has a slightly different reason to travel and thus, it is impossible ever to come up with a definitive model, unless it contains a virtually unlimited set of inter-linked variables.

But, do we have a better theoretical understanding of travel? The answer here is an unequivocal yes! As this chapter illustrates, the last 40 years have seen a dramatic improvement in both the sophistication in understanding and theoretical development of ideas about why we travel. Early work in the 1970s tended to conflate ideas or borrow intuitively appealing but ultimately rather simplistic ideas that did not fit the complex nature of tourism. Moreover, these early models tended to place people in absolute categories that failed to appreciate that the factors that drove individual trips could vary from trip to trip. The early 1980s saw a well-spring of ideas that have currency today, especially the identification of the push-pull model and the beginning of efforts to isolate individual motives of escape, relaxation, relationship and the like. Work since then, culminating in the travel career pattern approach has added additional dimensions of both complexity and sophistication by recognising that we are driven to travel by a basket of motives and that the mix of motives and importance placed on each one varies by trip.

■ Discussion points / Exercises

1. Five different motivational models and theories were identified in this chapter. Identify common features that apply to all or most models.
2. Discuss how motivational studies have evolved over the past 40 years.
3. Can you map push-pull onto Pearce's travel career pattern model?

Points to consider for revision

- Models of increasing complexity have been developed to understand why people travel.
- Sociology and psychology are the dominant disciplines that have contributed to the understanding of the field.
- Motivational studies really began with Maslow, but for a variety of reasons, the model does not apply to tourism easily. Instead, it may reflect what people do, rather than why they travel.
- Push-pull theory represents the father of tourism motivation models.
- Crompton's nine motives encompass the main reasons why people travel.
- Pearce's evolving travel career pattern builds on Maslow's, Dann's and Crompton's works but importantly, represents the multi-faceted nature of motives and the different roles core, middle and outer layer motives play in pleasure travel.

■ Suggested additional readings

Crompton, J. L. (1979) Motivations for pleasure vacations. *Annals of Tourism Research*, 6(4), 164-182.

Dann, G. (1977) Anomie, ego enhancement and tourism. *Annals of Tourism Research*, 4(4), 184-194.

Pearce, P. & Lee, U. (2005) Developing the travel career approach to tourist motivation. *Journal of Travel Research*, 43, 226-237.

References

Cohen, E. (1979) A phenomenology of tourist experiences, *Sociology*, 13, 170-201.

Crompton, J. L. (1979) Motivations for pleasure vacations. *Annals of Tourism Research*, 6(4), 164-182.

Dann, G. (1977) Anomie, ego enhancement and tourism. *Annals of Tourism Research*, 4(4), 184-194.

Dann, G. (1981) Tourist motivation: An appraisal. *Annals of Tourism Research*, 8(2), 187-219.

Denning, S. (2012) What Maslow missed. *Forbes, March 29.* https://www.forbes.com/sites/stevedenning/2012/03/29/what-maslow-missed/#fb41ede661b5 <downloaded July 11, 2019>

Framke, F. (2002) The destination as a concept: A discussion of the business-related perspective versus the sociocultural approach in tourism theory. *Scandinavian Journal of Hospitality and Tourism*, 2(2), 92-108

Fyall, A., Legoherel, P., Frochot, I. & Wang, Y (2019) *Marketing for Tourism and Hospitality: Collaboration, technology, experiences.* Oxon: Routledge.

Graburn, N. (1983) The anthropology of tourism. *Annals of Tourism Research*, 10, 9-33.

Graburn, N. (2018) Secular pilgrimage: A general theory of tourism. In Gmelch, S. and Kaul (eds) *Tourists and Tourism: A general reader 3rd Ed.* Long Grove: Waveland, 17-28.

Gutman, J. (1982) A means-end chain model based on consumer categorization process. *Journal of Marketing*, 46(2), 60-72.

Hsu, C., Killion, L., Brown, G., Gross, M. & Huang, S. (2008) *Tourism Marketing: An Asia-Pacific perspective.* Milton: John Wiley and Sons Australia.

Huang, S. & Hsu, C. (2009) Travel motivation: Linking theory to practice. *International Journal of Tourism, Culture and Hospitality Research*, 3(4), 287-295.

Iso-Ahola, S. (1982). Toward a social psychological theory of tourism motivation: A rejoinder. *Annals of Tourism Research*, 9(2), 256-262.

Iso-Ahola, S. (1983) Towards a social psychology of recreational travel, *Leisure Studies*, 2(1), 45-56, DOI: 10.1080/02614368300390041

Jafari, J. (1987) Tourism models: The sociocultural aspects. *Tourism Management*, 8(2), 151-159.

Jamal, T. & Lee, J (2003) Integrating macro and micro approaches to tourist motivations: Toward an interdisciplinary theory. *Tourism Analysis*, 8(1), 47-59.

Jiang, S., Scott, N. & Ding, P. (2015) Using means-end chain theory to explore travel motivation: An examination of Chinese outbound tourists. *Journal of Vacation Marketing*, 21(1), 87-100.

Kaur, A. (2013) Maslow's need hierarchy theory: Applications and criticisms. *Global Journal of Management and Business Studies*, 3(10), 1061-1064.

Kay, P. (2003) Consumer motivation in a tourism context : continuing the work of Maslow, Rokeach, Vroom, Deci, Haley and others, in *ANZMAC 2003 : A celebrations of Ehrenberg and Bass: marketing discoveries, knowledge and contribution: Proceedings of the 2003 Australian and New Zealand Marketing Academy conference*, ANZMAC, Adelaide, pp. 600-614. https://dro.deakin.edu.au/eserv/DU:30036658/kay-consumermotivation-2003.pdf <downloaded July 11, 2019>

Kim, B., Kim, S. & King, B. (2016) The sacred and the profane: Identifying pilgrim traveler value orientations using means-end theory. *Tourism Management*, 56, 142-155.

Klenosky, D. (2002) The "pull" of tourism destinations: A means-end investigation. *Journal of Travel Research*. 40, 385-395.

Kotler, P., Bowens, J. & Makens, J. (2003) *Marketing for Hospitality and Tourism, 3rd edition.* Upper Saddle River: Prentice Hall.

Litvin, S. & Smith, W. (2016) A new perspective on Plog psychographic system. *Journal of Vacation Marketing*, 22(2), 89-97.

Lundberg, D. (1972) Why tourists travel. *Cornell Hotel and Restaurant Administration Quarterly*, 12(4), 64-70.

Mannell, R. & Iso-Ahola, S. (1987) Psychological nature of leisure and tourism experience. *Annals of Tourism Research*. 14(3), 314-329.

Maslow, A. H. (1943) A theory of human motivation. *Psychological Review*, 50(4), 370-96.

McDonald. S., Thyne, M. & McMorland, L-A. (2008). Means-end theory in tourism research. *Annals of Tourism Research*, 35(2), 596-599.

McKercher, B. & du Cros, H. (2008) A comparison of international travel between an emerging and a mature source market. *Asia Pacific Journal of Tourism Research*. 13(3), 265-280.

McLeod, S. (2018) Maslow's hierarchy of needs. *Simplypyschology.org*. https://www.simplypsychology.org/maslow.html <downloaded July 24, 2019>.

Mill R. C. and Morrison A. M. (2012) *The Tourism System: An introductory text*, Prentice Hall, NJ.

Moscardo, G. (2014) Do tourists travel for the discovery of 'self' or search for the 'other'? *Tourism Recreation Research*, 39(1), 2 – 10.

Moutinho, L. (1987) Consumer behaviour in tourism. *European Journal of Marketing*, 21(10), 3-44.

Nikjoo, A. & Ketabi, M. (2015) The role of push and pull factors in the way tourists choose their destination, *Anatolia*, 26(4), 588-597, DOI: 10.1080/13032917.2015.1041145

Page, S. (2019) *Tourism Management 6th Ed* Oxon: Routledge.

Pearce, P. (1992) Fundamentals of tourist motivation. In D. Pearce and R. Butler (Eds), *Tourism Research: Critiques and challenges* (pp. 85-105). London: Routledge and Kegan Paul.

Pearce, P. (2005) *Tourist Behaviour: Themes and Conceptual Schemes.* Clevedon: Channel View Publications.

Pearce, P. (2011) Travel motivation, benefits and constraints to destinations. In Wang, Y and Pizam A. (Eds) *Destination Marketing and Management.* CAB International, pp 39 – 52

Pearce, P. & Lee, U. (2005) Developing the travel career approach to tourist motivation. *Journal of Travel Research*, 43, 226-237.

Pearce, P. & Packer, J. (2013) Minds on the move: new links from psychology to tourism. *Annals of Tourism Research*, 40, 386-411.

Plog, S. (1974) Why destination areas rise and fall in popularity. *Cornell Hotel and Restaurant Administration Quarterly*, 14(4), 55-58.

Plog, S. (2001) Why destination areas rise and fall in popularity: An update of a Cornell Quarterly Classic. *Cornell Hotel and Restaurant Administration Quarterly*, 42(3), 13-24.

Prayag, G. & Ryan, C. (2011) The relationship between the 'push' and 'pull' factors of a tourist destination: the role of nationality – an analytical qualitative research approach, *Current Issues in Tourism*, 14(2):121-143, DOI: 10.1080/13683501003623802.

Ryan, C. (1998) The travel career ladder: An appraisal. *Annals of Tourism Research*, 25(4), 936-957.

Smith, S., Xiao, H., Nunkoo, R. & E. Tukamushaba (2013) Theory in hospitality, tourism and leisure studies. *Journal of Hospitality Marketing and Management*, 22, 875-894.

Song, H. & Bae, S. (2018) Understanding the travel motivation and patterns of international students in Korea: using the theory of travel career pattern, *Asia Pacific Journal of Tourism Research*, 23(2), 133-145, DOI: 10.1080/10941665.2017.1410193.

Tasci, A. & Ko, J. (2017) Travel needs revisited. *Journal of Vacation Marketing*, 23(1), 20-36

Uysal, M., Li, X., and Sirakaya-Turk, E. (2008) Push–pull dynamics in travel decisions. In Oh, H. (Ed) *Handbook of Hospitality Marketing Management*. London: Routledge. pp 412-440.

Woodall, G. (2013) Understanding consumer decision-making with means-end research. Rockbridge Associates. https://rockresearch.com/understanding-consumer-decision-making-with-means-end-research/ <downloaded Aug 1, 2019>

Yoo, C., Yoon, D & Park, E. (2018) Tourist motivation: an integral approach to destination choices. *Tourism Review*, 73(2), 169-185.

Yousaf, A. Amin, I. & Santos, J (2018) Tourists motivations to travel: A theoretical perspective on existing literature. *Tourism and Hospitality Management*, 24(1), 197-211.

11 Sociological and Anthropological Concepts in Tourism

At the end of this chapter, the reader will be able to:

- Understand tourism as a quest for contrived reality
- Evaluate how tourism is a quest for authenticity
- Define the environmental bubble and critique how it forms the basis of commercial tourism
- Evaluate Cohen's five roles of tourists
- Analyse Cohen's quest for the other
- Assess and be able to critique the concept of liminality
- Identify aspects of risk and how they affect travel behaviour

Introduction

Tourism is both a business and a social phenomenon. To understand how tourism works, then, one must understand both the business side of tourism as well as the human side. Ever since large scale tourism began, people have been examining it and trying to place it in a sociological perspective. These efforts have involved an examination of why people travel, the social reasons for travel and the type of tourist experience they are either getting or seeking. This chapter examines how the founding sociologists and anthropologists in the field looked at tourism and how they still shape our thinking.

Social perspectives on tourism

Three dominant themes in the sociology of tourism have emerged over the last 60 years. They are:

- Boorstin's essentially negative 1960s view that tourism represents contrived reality;
- MacCannell's 1970s more positive view that tourism represents a quest for authenticity; and
- Cohen's more neutral views that tourism represents a quest for one's centre.

In reviewing these themes, it is important to understand a few points. One is that most take a very American-centric view to the world. The second is to remember that these were written over a nearly 30-year period and that they are heavily influenced by the dominant social paradigm at the time of their writing.

Boorstin's contrived reality

In the early 1960s, sociologists were becoming increasingly concerned about the contrived and illusory nature of the human experience in American society. They were concerned that people were no longer experiencing reality in their lives; instead they were being presented a series of pseudo-events. As Boorstin (1964) described them, they were immersed in a thicket of unreality.

Tourism, especially large scale, mass tourism, was seen as being just another example of how American life had become overpowered by pseudo-events and contrived experiences. Boorstin felt that (in 1964, remember) the modern tourist was simply a passive onlooker who was isolated from the host environment and the local residents. As a result, tourists chose to disregard the real world around them, instead choosing the isolation of tourist ghettos. Both the tourist experience and the type of tourism development had to become more contrived, in which the tourist moved in a closed self-perpetuating environment that was isolated from the rest of the world. This was both caused by and reflective of the nature of modern (1964) American society, of which tourism was only a part.

There is much to what Boorstin discusses that is still relevant today. Mass tourists can only withstand a certain amount of strangeness. They like to witness strangeness of the host environment from the security of their own surroundings. Too much strangeness becomes threatening. In addition, he introduced the idea of the 'tourist ghetto' that is isolated from the real world. Destination areas like the Gold Coast, Disneyworld, Waikiki and many more are prime examples of tourist ghettos. On a smaller scale, many destination resorts are physically removed from the outside world. They have been constructed as discrete areas that, by design, inhibit any contact with the real world. The only types of contact that tourists are likely to have with locals are either in a servant-master relationship, in a commercial relationship, or through the window of a self-enclosed, air conditioned bus. No one would describe these places as being real except possibly as real tourist destinations.

But, as Cohen (1972) argues, Boorstin's concept has a number of fundamental weaknesses. To begin, he labels all tourists as being the same. He fails to recognize that there are a large number of tourists that do not fit the mass tourist model, and even among mass tourists significant differences exist. Moreover, his work is editorial and observational in nature, rather than scientifically rigorous.

Boorstin felt the modern tourist was simply a passive onlooker who was confined largely to tourist ghettos or enclaves and did not want to experience the destination in any meaningful way, nor was he or she able to do so. He felt places like Waikiki in Hawai'i placed the tourist in an enclave divorced from the real world. Some of his arguments are valid, but his core thesis fails to appreciate that this is exactly the type of experience many tourists seek, if you abide by Dann's ideas of anomie, Crompton's concept of relaxation and the core elements of Pearce's travel career pattern.

MacCannell (1973) – quest for authenticity

Some ten years later, Dean MacCannell (1973) came up with a revised view of tourism and tourist motivations. Indeed, he is rather dismissive of Boorstin's ideas. He argued that the modern tourist is not a victim of a contrived and illusory culture, but instead is on a quest for authenticity that involves paying homage to the symbols of modernity, in this case tourist attractions. He suggested that the modern mass tourism journey shares many similarities with religious pilgrimages. Tourism centres have become the modern equivalent of pilgrimage centres, and people go to these places for similar reasons that pilgrims go to the Vatican, Mecca and Lourdes. Whereas pilgrims attempted to visit a place where an event of religious importance actually occurred, tourists present themselves at places of social, historical, and cultural importance (MacCannell, 1973: 593). Likewise, while Boorstin suggested that attractions were of little significance, MacCannell adopted a completely different view and sees them as highly significant social symbols.

The quest for authentic tourist attractions, not contrived ones, becomes the central motivation for tourism. What is the difference between an authentic and contrived tourist attraction? Well, the difference is small and often perceptual. If the visitor goes to a destination and perceives it as being 'authentic' then it is. In a reflective piece published recently, he (MacCannell, 2018) commented that he "sees tourism as a ritual of modern peoples, in which they scour the world looking for some kind of *authenticity*, a quality which many think is lacking in modern social life; and in this intention they are aided by modern hosts who attempt to 'stage' it for them". On the other hand, if the visitor perceives an attraction as being contrived, then it is not regarded as being authentic.

This example also reflects MacCannell's idea of the concept of a 'front and back of house continuum'. Building on Goffman's (1959 as cited by MacCannell, 1973) idea of a front and back region, MacCannell identifies a six stage continuum (pg. 598), that includes:

- Stage 1: Goffman's front region; the kind of social space tourists attempt to overcome, or to get behind.
- Stage 2: a touristic front region that has been decorated to appear, in some of its particulars, like a back region... Functionally, this stage is entirely a front region, and it always has been, but it is cosmetically decorated with reminders of back-region activities: mementos, not taken seriously, called 'atmosphere'.
- Stage 3: a front region that is totally organized to look like a back region...This is a problematical stage because the better the simulation, the more difficult it is to distinguish it from stage 4.
- Stage 4: a back region that is open to outsiders... It is the open characteristic that distinguishes these especially touristic settings (stages 3 and 4) from other back regions. Here access to most non-touristic back regions is somewhat restricted.
- Stage 5: a back region that may be cleaned up or altered a bit because tourists are permitted an occasional glimpse in.
- Stage 6: Goffman's back region; the kind of social space that motivates touristic consciousness.

But, as the pressure by tourists to see 'real' places (or the back of the house) expands, destination areas are forced to create tourist attractions and present them as being real. In other words, to satisfy tourist demand, they have to stage authenticity. The creation of inauthentic 'real' tourist attractions is a consequence of tourism development.

MacCannell's work introduced the important concept of 'tourist space', which is different from tourist ghettos. Like Boorstin's work, his thesis is perceived to adopt a global view of tourism and failed to recognize that different types of tourists existed who are seeking different tourist experiences.

■ Cohen 1972 – strangeness vs. familiarity, or value adding?

Erik Cohen took Boorstin's initial concept and developed it one step further. Cohen's work examined tourism from the perspective of 'strangeness' versus 'familiarity' by looking at the extent, variety and degree of change tourists seek in their various social contacts. The degree to which strangeness or familiarity prevails in the tourist's role or activities determines the nature of the tourism experience as well as the effects he/she has on the host society.

While all tourists are to some extent strangers in the host community, the extent that the tourist's role is pre-determined will dictate the manner in which tourists interact and the images they will develop of one another. Tourism evolved from a basic change in our view of the world, from isolation to a global perspective. Initially few people travelled, primarily because of ignorance of other people's cultures which made people leery of seeking change. Because our view of the world changed, people became more interested in experiencing strange and novel situations but only to the point that this strangeness remained non-threatening.

People want to experience change from the security of their own environmental bubble. What is an environmental bubble? It is a bubble of familiarity that reduces the level of strangeness or threat to an acceptable level enabling people to participate. Think of an umbrella and raincoat. You can envelop yourself in a raincoat and use an umbrella to protect you from the rain, enabling you to go out and not have an unpleasant experience.

The idea is predicated on the assumption that people will want to participate in things only as long as the perceived risk is less than the perceived reward. If the reward is high and the risk low, then there is little need for a bubble. But, if the perceived risk is high, then things must be done to reduce it in order to facilitate participation. The environmental bubble exists throughout all aspects of the tourism system. Think of what hotels do to reduce perceived risk. Foreign television channels, multilingual staff, concierges who are familiar with the destination, different types of food on offer, especially at breakfast, in-room safes for security and many more actions are undertaken to be seen to reduce risk. Tour guides play the same roles by interpreting and mediating the destination experience. Airlines provide an environmental bubble through their safety announcements at the start of the flight and request that people keep their seatbelts fastened.

Most tourists are not adventuresome risk takers. Instead they are on an often, once a year break and are looking for relaxation, fun, spending time with family and novelty. They seek strange and exciting environments, only up to the point that they are not too strange and exciting. In other words, the thrill must be perceived as being greater than the risk. If the perceived risk is too great, then most people will choose not to participate or will alter their behaviour. This street market in Dhaka, Bangladesh is one such example where the perceived risk exceeded the perceived benefit of visiting. Few tourists feel comfortable enough to venture in. Alternately, this fruit market in Tehran, Iran proved to be very popular with tourists who felt the benefit of visiting exceeded the risk.

It is also important to appreciate that people have different levels of risk tolerance, and therefore, seek different levels of bubble provision. Here, Plog's categorisation is useful. The repressed Psychocentric/Authentic tourist cannot cope with much change. He or she will seek as much familiarity as possible when travelling. But, if placed outside their comfort zone, will require a massive environmental bubble in order to feel comfortable (fully escorted package tour, familiar food, known brands of hotels and restaurants, etc.). The various groups of Midcentric tourists are also risk averse, but not to the same extent. They are more willing to be adventuresome, but still desire a bubble of familiarity, such as branded hotels, availability of local food and perhaps some guided day tours or a clear itinerary they can follow. So while they want to see new sites, they still want to perceive them from the comfort of their own culture. The Allocentric/Venturer tourist, on the other hand is the least risk averse tourist and is most willing to accept the destination on its own merits. He or she eschews an environmental bubble.

Jaakson (2004) and others prefer the more neutral term 'tourist bubble.' It conveys the same idea but removes the discussion from insinuations that it is linked to the natural environment. He suggests the bubble has two dimensions – a physical emphasis of it as a type of place, and a psychological emphasis of it as the tourists' *eigenwelt* of motivations, attitudes and belief systems. Tourists thus place themselves in a secure physical place that is isolated from the rest of the destination and therefore only see a rather sanitized version of the place (Newman, 2002). Emotionally, as well they seek a secured, protected, and normalized environment, often in an artificial, segregated location devoted to consumption and play (Newman, 2002). It is for this reason that Jaakson (2004) has observed a pack mentality among many tourists. Andreu, Kozak, Avci and Cifter (2005) call this pattern the herding effect.

The likelihood of isolation is enhanced if substantial language or cultural differences exist between the hosts and the visitors. It is this controlling and standardizing of the tourist product, however, that often provides the images that visitors have of local arts and culture. Unfortunately, tourism destinations the world around look very similar. Countries lose their individuality as the area's geography and culture are packaged and reduced to bite sized elements for easy consumption. In a sense, mass tourism simply becomes a series of photographs and stamps on passports. Mass tourists themselves rarely walk away from a destination with any greater understanding of the area they are visiting.

Why has this happened? Urry (1990a and b) identified a number of reasons. First, most travellers are not adventuresome drifters who want to take the destination of their choice as it is presented to them. These people exist, but they are a minority. Most people do not want to feel that they are facing real risks and real discomfort in pursuit of their vacation. Second, the market dictates the types of services and amenities demanded. If the vast majority of travellers are looking for comfort and demanding some sort of environmental bubble, then the industry must accommodate them. But, third, people still want to feel like they are experiencing some type of novel experience and that they have been exposed to a different culture. It is just that they either do not want to or are incapable of experiencing this in any real or meaningful manner. As de Kadt stated (1978) most travellers are not anthropologists or sociologists. They are primarily consumers who are temporarily unemployed and who are seeking a break, an escape from their normal life. Finally, mass tourism is a commercial activity and as with any commercial activity, efficiencies can be achieved when the product is standardized, packaged, mass produced and easily consumed. Because of the nature of mass tourism, clients must be processed as efficiently and smoothly as possible. Therefore, the tourist's experience must be ordered, predictable and controllable, as much as possible. As such, attractions and facilities must be standardized for efficient consumption.

Cohen developed a four-fold typology of roles tourists play (Table 11.1). It is divided into two parts, those who follow an institutionalised path and those who he classifies as non-institutionalised tourists. The following section summarises Cohen's work (1972: 167-173).

Table 11.1: Cohen tourist typology

Institutionalized tourist	Organized mass tourist Individual mass tourist
Non-insitutionalized tourist	Explorer Drifter

Institutionalized tourists fall into one of two roles:

- **Organized mass tourists** are the least adventuresome and remain largely confined to their environmental bubble. The guided tour conducted in an air-conditioned bus represents the prototype of the organized mass tourist. The itinerary is fixed in advance and all stops are well prepared and guided. The tourist makes almost no decisions and stays almost exclusively in the micro-environment of the home country. Familiarity is at a maximum, while novelty is at a minimum. Mass tourism, and the commodification of the mass tourism product enables tourists to take in the novelty of the area without any physical or emotional discomfort. In a sense, they can observe it without experiencing it first-hand. One of the main purposes of mass tourism (remember the operative word is *mass*) is to see attractions, whether the attractions are genuine or contrived. Even if the attraction is genuine, it often has to be transferred and somehow controlled to make it more suitable for mass consumption.

- The **individual mass tourist** is similar to the organized mass tourist, except the tour is not entirely preplanned. However all the major arrangements are still made through a travel agent. Tourists experience the destination from within the environmental bubble of their home country and venture out of it only occasionally.

He then identifies two types of non-institutionalized tourist:

- The **explorer** who tries to get off the beaten track but is nevertheless comfortable in accommodations and reliable means of transport. This person dares to leave the environmental bubble more than the other two but is still careful to be able to step back when the going becomes too rough. Though novelty dominates, the tourist is not immersed completely in the host society.
- The **drifter** who ventures furthest away from the beaten track and from their customary way of life.

Value adding as a synonym for the environmental bubble

Cohen's idea is brilliant but he does not develop it further to explain the commercial nature of tourism, for the entire business model of tourism is devoted to environmental bubble creation. Only, industry uses the term, 'value adding'! The entire basis for the commercial delivery of tourism is all about bubble creation, right from the travel trade through transport providers, accommodation houses, in-destination transport and services and attractions. They exist to make travel more accessible by removing barriers and creating a sense of familiarity.

Think about what you are paying for. In reality, you are paying for the bubble. If you wanted, you could sleep on the street or stay in low cost hostels. But, most of us want the comfort and convenience of better quality accommodation. We could travel throughout a destination on public transit and explore sites on our own. But, commercially provided tours are popular because they remove the risk and highlight the thrill. We could buy and cook our food, but again restaurants are preferred for they remove the uncertainty of dining. The list goes on. What we pay for is less the tangible product on offer and more the augmented service provided that reflects the creating of a bubble.

There are added benefits as well. To begin, bubble creation broadens the market base, by reducing the skill level needed to participate, optimizing the thrill level and providing needed security to convince people that risk has been reduced to a sufficient level. Reisinger and Mavondo (2005: 213), reviewing the consumer behaviour literature, identify seven types of risk: financial (losing or wasting money), functional or performance (not performing, not meeting the customers' needs and requirements), physical (injury or illness), social (losing personal and social status), psychological (damaging self-image and/or reflecting poorly on personality), satisfaction (not delivering satisfaction), and time (not performing on time, taking too much time, and/or wasting time). Wickens (2002) argues the need for an environmental bubble relates primarily to risk reduction. Most tourists seek to balance the excitement of visiting a destination against the possible risks involved in doing so. If the risk exceeds the reward, tourists will need to enmesh themselves in a bubble of familiarity to reduce risk to an acceptable level.

The commercial delivery of tourism is all about environmental bubble creation as this image from Zagreb in Croatia illustrates. The tour product on offer included a tour guide, interpretation, bus transport and outside access to a number of historic sites in the city. Tourists could have undertaken this trip themselves for a fraction of the price, and indeed, could have walked much of the route for free. But they chose to purchase a day tour instead. Why would anyone pay for such a tour when you could do it for almost nothing? They are willing to pay for the value adding created by the tour company, or in sociological terms, the environmental bubble created by the operator. The guide offers some level of expertise, interpretation and significantly a safe, mediated experience. Tourists were taken to highlight sites and therefore do not feel they miss out on anything of importance. The coach ensures they keep on schedule. And, they did not have to worry about getting lost. In other words, they were willing to pay a price premium for the environmental bubble they were happy to join.

Moreover, satisfaction can be enhanced through the modification, standardization, and commodification of the product. Most tourists, and especially long haul tourists, may only visit a destination once or may only visit an attraction in a destination once. They want to have the best possible experience during this visit. The best way to do so for as many tourists as possible is to provide a highly commodified and standardized experience that seeks to offer a consistently high quality opportunity for all who visit. Within the cultural tourism space, for example, purpose-built cultural theme parks represent extreme environmental bubble formation that offer a safe, controlled and controllable environment (Craik, 1997), while Hughes (1998) also observes the arts industries have standardized their product in order to satisfy consumer demand and enable effective signification to the consumer.

Cohen (1979) and the 'Centre and the Other'

Cohen also built on MacCannell's ideas of tourism as a quest. His later work (Cohen, 1979) sees the individual's alienation from society as lying along a five scaled continuum. He tied his thesis into the contemporary sociological premise of 'the Centre' and the 'Other'. Basically, every society possesses a Centre, which is its ultimate moral value. The Centre represents the paramount symbols of a society, be they social, cultural or religious.

For most people, the Centre is located within the confines of that society. Leisure and recreation activities, and by extension, tourism offer a temporary release from the Centre that serves a number of beneficial purposes. If a person is closely tied to his or her Centre, then the type of tourism experience he or she will select may have little regard for authenticity in their tourist experience. Why? Because they are already connected to their Centre and thus don't need the reinforcement of a tourism experience.

But, most people in modern and post-modern societies are not centred, and indeed, are alienated. For them, tourism has a different meaning. For them tourism becomes an opportunity to look for their Centre, or to reconnect with the Centre they have become alienated from. This can happen sometimes just by seeking rest and relaxation. At other times, the quest becomes more complex.

Cohen identified five different roles tourists could play depending on their circumstances at the time. While his categories resemble Plog's to some extent, he argues that each trip is context specific. So, the same tourist may be on a quest in one trip, looking to re-create on another or simply want to sooth his or her soul on a third trip. Each one has different needs and each one perceives authenticity in a different manner. The five roles are:

- Recreational;
- Diversionary;
- Experiential;
- Experimental; and
- Existential.

Recreational tourists are basically content with their lives. They are like Plog's various Midcentric groups. For them, tourism is a move away from the Centre that eventually reinforces it. The goal of travel, as the label implies, is to re-create, to regenerate and to restore mental and physical well-being. For them, their vacation is simply an opportunity to recharge their batteries. It has no deeper meaning. Travel is not undertaken for self-realization or for self-learning. Essentially, these people are already well centred in their host culture, but need a break. As a consequence, they seek mass entertainment and often seek what some would call shallow, trivial and easy to consume tourism experiences. They seek recreation, want to be entertained and are willing to accept make believe or pseudo-events on their own terms.

Diversionary tourists are alienated from their Centre, but are not seeking other Centres. They resemble Plog's Psychocentric tourists at the time. Diversionary tourists, again as the name implies, are simply seeking a temporary escape from their boring and meaningless routine. Their goal is to heal the body and to sooth the spirit but not to re-create or to reconnect with their core values. Authenticity has no meaning for them. Instead they are simply seeking something that will entertain them for the short term. These people are most like the type of traveler Boorstin identified in his work. They are reflective of a broader societal malaise.

On the other hand, what happens if you are aware that you are alienated from your Centre. One response is to look for meanings in the life of others. This is where MacCannell's thesis comes into play. The quest for meaning and the quest for a Centre begins with a search, which in this case is the modern tourism experience. **Experiential** tourists want authentic experiences, and therefore the opportunity to reconnect, are available to those people who try to break the bonds of their everyday existence and begin to 'live.' Tourism, thus absorbs the social function of religion for experiential travelers, where they try to experience other cultures through their touristic travels. But, this is where the conundrum kicks in. The tourism industry stages authenticity that leads these types of tourists to believe that they have broken through the contrived front of the unauthentic. Still with me? But, they haven't really done that. Unfortunately, these types of tourists can only experience these things in a vicarious and ultimately shallow way. As such, the inauthentic may become authentic.

On the other hand, **Experimental** tourists go that next step by actively going on a quest for Centres in other societies. These people seek authentic life and lifestyle in other cultures, but do not fully commit to them. Instead, they sample a variety of things, but ultimately return to their old life. This type of traveler was typified by the hippies of the late 1960s and early 1970s who went looking for something different and better. Most of them sampled alternative lifestyles for a while, but most ultimately returned to their original lifestyle. To a lesser extent, the backpacker of today may be seeking similar experiences.

Lastly, **Existential** Tourists are those people who not only sought different Centres, but who fully committed themselves to it. They may be people who, for example, joined an ashram in India, or who liked visiting Australia so much that they decided to emigrate there. For them they seek and become immersed in the authenticity of the culture they visit.

Clearly, these five types of tourists are very different. The most commonly found types are the recreational and diversionary tourist, with the experiential tourist ranking third. As the level of commitment to seeking and adopting a different Centre increases, the number of tourists diminishes.

Tourism as a liminal experience

As discussed in Chapter 2, tourism involves a journey away, a stay at some distant location and a return journey home. But, it involves far more than the physical act of traveling, for it is also an activity that occurs outside of physical, social and psychological time and space. It thus involves both geographical and symbolic separation from one's home (Beckstead, 2010). As such, tourism is imbued with many meanings, where the act of becoming and then being a tourist, is associated with a temporary psychological and emotional metamorphosis of disassociation from the normal (to a great or lesser extent) (Carr, 2002a, b) and the assumption of a 'tourist' persona with its own roles, and associations. A number of sociologists and anthropologists have likened tourism to a liminal experience, where people experience communitas upon entering a liminoid landscape (Cohen, 1988).

A liminal state refers to a temporary transitional state between more permanent states where the person lacks social status and structure (Turner, 1974). The roots of the word can be traced to the Latin term 'limen' which refers to a state of being on a threshold or in an intermediate condition (Huang et al., 2018). It can be applied to individuals, groups or people or populations (Gisolf, 2013). As Turner (1974: 59, 60) notes, in liminality, profane social relations may be discontinued, formal rights and obligations are suspended and the social order may have been turned upside down. Instead, an anti-structure sets in with its own rules, rights and duties.

The anthropologist van Gennep (1906, as cited by Turner, 1974) was the first to coin the phrase when he examined rites of passage of people in traditional agrarian societies as boys moved symbolically from childhood to adulthood. He suggested that such rites involved separation from the ordinary, a period of marginality or liminality and then re-aggregation or reintegration where the person comes back to the normal society as a changed person (Graburn, 1983: 13). The person thus enters a betwixt and between state where, according to Wels et al. (2011), the person is understood to be 'no longer' and simultaneously 'not yet'. Typical examples include teenage years, where the person is no longer a child and simultaneously not yet an adult.

Turner was among the first to regard leisure, and by extension tourism, as being representative of a liminal state, noting "'leisure', then, presupposes 'work': it is a non-work, even an anti-work phase in the life of a person who also works" (Turner, 1974: 67). He saw a direct link to tourism for he felt the passage from one social state to another also often involved a parallel passage in space, or a geographical movement from one place to another. Importantly, as well, he noted liminal spaces provide two types of freedom: *freedom from* institutional obligations and *freedom to* play with ideas, with fantasies, with words (Turner, 1974: 68). It is for this reason that Foster and McCabe (2015: 48) observe that the liminal phase is simultaneously destructive and constructive. It is a time when social structures are inverted and through various ritual practices individuals are encouraged to question tacit knowledge. Pritchard and Morgan (2005: 764) add such spaces are intangible, elusive and obscure, and lie in a limbo-like space often beyond normal social and cultural constraints.

As McKercher and Lui note (2014), the core premises are that travel allows individuals to set aside conventional standards of conduct that would normally be repressed at home (Selanniemi, 2003), that being in a liminal touristic space provides relative anonymity and freedom from scrutiny (Kim and Jamal, 2007). Epstein and Kheimets (2001: 42) add "becoming a tourist, however briefly, means shedding part of one's old identity and normal behavior [and]...adopting a new, temporary identity that necessarily incorporates some elements that are the opposite of the habitual personality and behavior."

Typically, entry into and exit from a liminal state involves three stages (Nash and Smith, 1991). Individuals reside in a structured situation. They are then separated from this situation, where they are distanced from their routine social groupings. Then they enter the liminal phase in which the structured requirements of daily life disappear into an undifferentiated state. The third stage involves a reintegration in which the person passes back into his or her structured life.

Where does tourism it in? Well, tourism can represent the symbolic three staged process whereby by going away from a familiar place (separation) to a far place (liminality) and then returning to the familiar place (reintegration) you can reconnect to your Centre. The liminal state or far place in tourism represents another Centre that you temporarily visit before coming back to your Centre (Cohen, 1988). This 'Centre Out There' can play a key role in helping you reconnect to who you are. As a result, you can do things that you normally would not do: spontaneity, wholeness, social togetherness and participation in role reversal activities. Traditionally, men have rarely played a significant role in cooking. Cooking has always been (and to a large extent still is) considered to be women's work. But, tourism provides the opportunity for these roles to be reversed. The barbeque is the one place where it is socially acceptable for men to be the cooks.

In a similar fashion, men often assume a much greater child care role on vacation than they do at home.

Jafari (1987) popularized the idea of tourism as a liminal experience. He identified a five stage process (Figure 11.1) beginning with the Corporation phase which represents the individual's profane and ordinary life. Some factor or factors push the person to want to escape from his or her ordinary life and the person begins to mentally prepare for an escape.

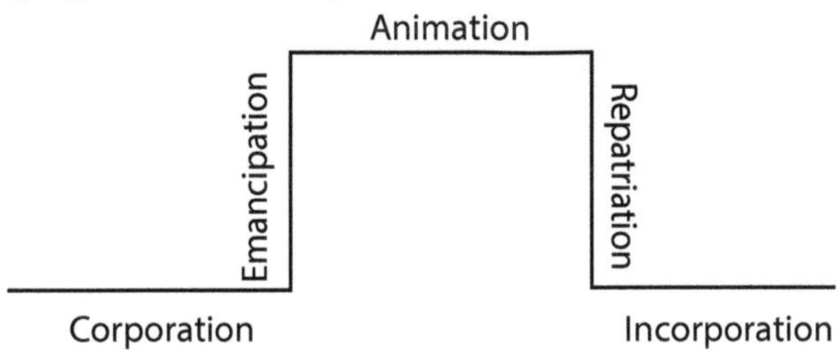

Figure 11.1: Jafari's liminal model of tourism

The Emancipation phase corresponds to the traditional separation phase in the three phase model. He identifies two stages in this phase. The first is the spatial notion where the person physically separates from the home region. The second, equally important stage, is one of declaration, where the individual declares he or she is no longer tied to the home situation. As Jafari (1987: 153) states these two stages work in consort for "separation signifies the *growing distance* between him and the ordinary, while declaration denotes the *shrinking distance* between him and the non-ordinary. Combined into one, the emancipation whole disconnects the cords of the ordinary womb and suddenly the tourist is born into the non-ordinary world." Cohen (1988) adds this stage is vital for the person is taken to an unfamiliar place peripheral to the ordinary place of abode and separated from the individual's ordinary social group. Here the process of leaving one's past behind and entering touristhood begins.

In the destination, the person enters a liminal state, or as Jafari suggests the Amination Phase, where the tourist enters a tourist culture and leaves behind the residual culture to a greater or lesser extent. Jafari (1987: 152) writes "as immersion into touristhood deepens, the tourist further internalizes the idea that he is no longer imprisoned by his former self but is transformed into a new person with a new identity (the tourist), that he is playing on a new stage (the magnet), and that he is living up to the rhythm of a new culture (the *tourist* culture)." Carr (2002a) suggests this process may be rapid or gradual depending on the nature and duration of the trip, travel party composition, the degree of spatial separation from home and also the degree of psychological separation desired.

Tourism involves entering liminal space, as shown in this safari in South Africa. Liminality in tourism involves at least three phases. The emancipation phase occurs when one leaves his or her normal world. This step involves both a physical and psychological escape. In the animation or liminal stage, the tourist enters a new world as a tourist, sharing that new, but temporary environment with other tourists. But alas, the trip must come to an end and the person returns to his or her ordinary life both physically and psychologically. While in the liminal state, the tourist has no status other than being a tourist and social norms of behaviour can be jettisoned to some extent. This break from social pressures enables the person to release a range of subterranean values and thus can return to the normal life as a much more complete person.

Liminal spaces are typically associated with freedom (Pritchard and Morgan, 2006) where antistructure rules and where the tourist temporarily abandons social status and home influence (Gisolf, 2013). While the person may physically leave the home environment, entry into the destination environment is incomplete (Ryan and Martin, 2001), heightening the sense of social marginality and opening up possibilities. Instead, one enters a netherworld of the tourist environment, populated by other tourists who are temporary residents of the destination and who function under their own sets of implied and semiotically induced norms and rules. Neither the rules of home or of the destination apply. As Jafari (1987) notes, there is a palpable sense of freedom, enabling the tourist to play any role of his or her choice.

This freedom has been labelled by Graburn (1983: 21) as the idea of inversion, whereby certain meanings and rules of ordinary behaviour are changed, held in abeyance or even reversed. Inversion does not mean revolution and entry into the opposite. Instead, it exists on a continuum, whereby tourists can choose which behaviours to change and the degree of change from their normal style. Inversion could occur at a spatial level (urban tourists going to rural locations); be seasonal (winter tourists becoming 'snowbirds'); or be class based (middle class people becoming backpackers).

But, touristic trips as temporary journeys must come to an end. The reintegration phase has been labeled as the Repatriation stage by Jafari (1987) Again, this stage involves two phases, Reversion and Submission. Reversion refers to the process of letting go as a tourist. It involves both a physiological process of leaving touristhood behind that culminates in the actual departure from the destination region. Submission is the process of reintegrating into one's profane life. After this stage, the tourist enters Incorporation, where he or she is subsumed into the ordinary life.

The extent to which a person can have a liminal experience, or simply enter a liminoid state depends on a number of factors. Indeed, many people argue that modal types of tourism do not necessarily represent a liminal transition from one stage to another and instead represent a temporary entry into liminoid spaces where the normal codes of everyday life are not valid and place and time lose their meaning (Cohen, 1985; Mustonen, 2006). Cohen (1985) notes the person may return refreshed to his or her mundane life, but the experience does not spiritually change the person's life. As Beckstead (2010) notes liminality in tourism is neither static nor a given. Instead, it is dynamic and negotiated by various social actors. People are far less likely to engage the liminoid during short duration holidays where they travel to culturally similar places and travel with a large group of closely associated companions (Gisolf, 2013). However, the longer the trip lasts, the more culturally distant the destination and the fewer the number of travel companions, the more likely one is able to peel back the onion of one's home cultural and leave a nub of their residual culture as their fully enter the touristic culture (Carr, 2002b).

There are many examples of liminal spaces in a tourism context. Pritchard and Morgan suggest hotels are liminal sites that exist out of time and out of place. They write (Pritchard and Morgan, 2006: 769,770) "to enter a hotel is to cross an imagined threshold into a liminal place which is strange, yet familiar, which offers freedom for some, but constraint, risk and unease for others. As potentially liminal and transgressive spaces they offer 'the spatial presence and practice outside of the norms of the prevailing (enforced) social specialization." Underthun and Jordhus-Lier (2018) also see hotels as a liminal space for transient workers who can work with different expectations and commitments than permanent staff. Huang et al. (2018) write about airports as being liminal spaces for passengers, for they are between time zones and countries and are metaphorically structurally invisible in a land of nowhere. Cohen (2018) makes a similar observation about backpacker enclaves where over time such places have become spaces of suspension in which backpackers are suspended between the local and global culture. Here a distinctive hedonistic lifestyle develops, marked by one's own ability to do one's own thing. Foster and McCabe (2015) feel residential activity camps satisfy the criteria of liminal spaces, while Mustonnen (2006) writes about volunteer tourism as a liminal activity where volunteers can engage projects to the extent they feel comfortable, but can leave any time they wish.

Hotels, like this one in Singapore, are liminal places. The guests are temporary residents where their status is unknown. All guests share the same experience, regardless of what they do in the profane world. The social order is turned upside down as all guests enjoy an elite status of being served. But stays are short and they must leave this unusual space and return to the normal world from whence they came.

Indeed, the idea of antistructure, suspension of social norms and freedom are essential elements of the liminal nature of tourism (Brooker and Joppe, 2014). These marginal places are privileged sites not normally afforded us where we can escape the oppressive places (Johnston, 2001). Huang et al. (2018) remind us this freedom equates to freedom from social norms and the existing social hierarchy we find ourselves in, and freedom to mix with different people and exist within a different social structure.

The benefits are well-recognized, for the opportunity to play gives access to valuable compensatory experiences (Cohen, 1988). Gisolf (2013) summarizing the literature identifies four main benefits of entering a liminal space. First, people can analyze their lives and background more clearly than they can at home and decide if the path they are one is the one to maintain. Second, social differences are deemphasized or ignored, resulting in a state of equality and solidarity among a group of tourists sharing the same experience. Third, the liberation of social constraints opens up the possibility of the more authentic self-emerging, at least temporarily. Finally, it may help people re-find themselves.

Matza and Sykes (1961: 716) suggest an even more profound benefit by arguing tourism provides an opportunity for the release of subterranean values, those values which are in conflict or competition with other deeply held societal values but which are still recognized and accepted by many, in a socially acceptable space and time. Buzzell (2005: 30) added later holidays "include behaviours that are by design out of the ordinary, expressions of misrule rather than rule. These occasions are moments when the dynamics of social control are altered, shifting from normative to acceptable non-normative acts." The ability to release these values in a safe and non-judgmental environment performs a number of important social roles. A holiday, therefore, is as much a psychological escape from the range of duties and obligations one might feel at home as it is a physical escape (Prideaux, Agrusa and Donlon Curran, 2004). Much has been written about the therapeutic benefits of such a break (Cohen, 1972; Urry, 1990a, 1990b), with Zhu (2012: 1498) observing the liminal space of tourism enables tourists to participate in temporary forms of transgression that enable their secret selves to be displayed.

Conclusion

Sociologists and anthropologists see tourism as an escape or quest, where the act of being in a different physical, social and psychological place provides a whole range of benefits to the individual that help her or him be better adjusted when returning home. The idea that tourism is a quest seems to form the basis for much of the thinking about this topic. However, it is worthwhile referring to the chapter on motivations as well to place these works in perspective. The need to escape either because of anomie, disequilibrium or seeking creates the initial impetus to travel. The destination provides the venue where the break can occur, or where the seeking can be satisfied.

Discussion points / Exercises

1. Cohen's typology based on alienation from one's centre and Plog's ideas overlap. Prepare a table showing how Cohen's and Plog's five groups are similar. Discuss.
2. Cohen's model may relate to Crompton's motives. Discuss.
3. Give examples of how tourism can release subterranean values.
4. How can tourism represent the different liminal stages?

Points to consider for revision

- The sociological perspective of tourism has evolved from seeing it as an essentially superficial and negative experience to a positive one, and now to one that is more neutral in its views.
- The environmental bubble is a core concept in tourism. It explains much of tourist behaviour and further explains the basis for the commercial delivery of tourism products.
- Cohen sees tourists as fitting four roles depending on the level of institutionalisation involved in the travel experience.
- Cohen also identified five types of tourist based on their level of alienation from their home culture.
- Liminality in tourism explains why tourists behave the way they do, for tourism provides a powerful release valve from the ordinary pressures one faces.
- Tourism provides the opportunity to release subterranean values.

Suggested additional readings

Cohen, E. (1972) Toward a sociology of international tourism. *Social Research*, 39(1), 164 .

Cohen, E. (1979) A phenomenology of tourist experiences, *Sociology*, 13, 170-201.

Jafari, J. (1987) Tourism models: the sociocultural aspects. *Tourism Management*, 8(2), 151-159.

References

Andreu, L., Kozak, M., Avci, N. & Cifter, N. (2005) Market segmentation by motivations to travel. *Journal of Travel and Tourism Marketing*, 19(1), 1-14.

Beckstead, Z. (2010) Liminality in acculturation and pilgrimage: When movement becomes meaningful. *Culture & Psychology*, 16(3), 383-393.

Boorstin, D. (1964) *The Image: A guide to pseudo-events in America.* New York: Harper and Row.

Brooker, E & Joppe, M. (2014) Developing a tourism innovation typology: Leveraging liminal insights. *Journal of Travel Research*, 53(4), 500-508

Buzzell, T. (2005) Holiday revelry and legal control of fireworks: A study of neutralization in two normative contexts. *Western Criminology Review*, 6, 30-42.

Carr, N. (2002a) The tourism-leisure behavioral continuum. *Annals of Tourism Research*, 29(4), 972-986.

Carr, N. (2002b) A comparative analysis of the behavior of domestic and international young tourists. *Tourism Management*, 23(3), 321-325.

Cohen, E. (1972) Toward a sociology of international tourism. *Social Research*, 39(1), 164.

Cohen, E. (1979) A phenomenology of tourist experiences, *Sociology*, 13, 170-201.

Cohen, E. (1985) Tourism as play, *Religion*, 15(3), 291-304.

Cohen, E. (1988) Traditions in the qualitative sociology of tourism. *Annals of Tourism Research*, 15, 29-46.

Cohen, E. (2018) Backpacker enclaves research: Achievements, critique and alternative approaches. *Tourism Recreation Research*, 43(1), 105-116. doi: 10.1080/02508281.2017.1388572

Craik, J. (1997) The culture of tourism sites, in C. Rojek and J.Urry (eds) *Touring Cultures: Transformations of Travel and Theory*, Routledge, London, pp 113-136

de Kadt, E. (1978) *Tourism: Passport to Development*, Oxford University Press, NY.

Epstein, A. & Kheimets, N. (2001) Looking for Pontius Pilate's footprints near the Western Wall: Russian Jewish tourists in Jerusalem. *Tourism, Culture & Communication*, 3, 37-56.

Foster, C. & McCabe, S. (2015) The role of liminality in residential activity camps. *Tourist Studies*, 15(1), 46–64.

Gisolf, M. (2013) Liminality in Tourism. *Tourism Theories*, https://www.tourismtheories.org/?p=700 <downloaded June 5, 2019>.

Graburn, N. (1983) The anthropology of tourism. *Annals of Tourism Research*, 10, 9-33.

Huang, W., Xiao, H. & Wang, S. (2018) Airports as liminal space. *Annals of Tourism Research*, 70, 1-13.

Hughes, H. L. (1998) Theatre in London and the interrelationship with tourism. *Tourism Management*, 19(5), 445-452.

Jaakson, R. (2004) Beyond the tourist bubble? Cruiseship passengers in port. *Annals of Tourism Research*, 31(1), 44-60.

Jafari, J. (1987) Tourism models: the sociocultural aspects. *Tourism Management*, 8(2), 151-159.

Johnston, L. (2001) (Other) bodies and tourism studies. *Annals of Tourism Research*, 28(1), 180-201.

Kim, H., & Jamal, T. (2007) Touristic quest for existential authenticity. *Annals of Tourism Research*, 34(1), 181-201.

MacCannell, D. (1973) Staged authenticity: Arrangements of social space in tourist settings. *American Journal of Sociology*, 79(3), 589-603.

MacCannell, D. (2018) The making of the tourist,*Via*, 13. http://journals.openedition.org/viatourism/2239.

Matza, D. & Sykes, S. (1961) Juvenile delinquency and subterranean values. *American Sociological Review*, 26, 712-719.

McKercher, B. & Lui, S. (2014) Becoming a tourist. *International Journal of Tourism Research*, 16, 399-406.

Mustonen, P. (2006) Volunteer tourism: Postmodern pilgrimage? *Journal of Tourism and Cultural Change*, 3(3), 160-177.

Nash, D. & Smith, V. (1991) Anthropology and tourism. *Annals of Tourism Research*, 18, 12-25.

Newman, H. (2002) Race and the tourist bubble in Downtown Atlanta. *Urban Affairs Review*, 37(3), 301-321.

Prideaux, B. Agrusa, J., Donlon, J. & Curran, C. (2004) Exotic or erotic-contrasting images for defining destinations, *Asia Pacific Journal of Tourism Research*, 9(1), 5-18.

Pritchard, A., & Morgan, N. (2006) Hotel Babylon? Exploring hotels as liminal sites of transition and transgression. *Tourism Management*, 27(5), 762-772.

Reisinger, Y. & F. Movando (2005) Travel anxiety and intentions to travel internationally: Implications of travel risk perception. *Journal of Travel Research*, 43(3), 212-225.

Ryan, C. & Martin, A. (2001) Tourists and strippers: Liminal theater. *Annals of Tourism Research*, 28(1), 140-163.

Selanniemi, T. (2003). On holiday in the liminoid playground: Place, time and self in tourism. In T. Bauer & B. McKercher (eds.), *Sex and Tourism: Journeys of romance, love and lust* (pp. 19-34). Binghamton: Haworth Press.

Turner, V. (1974) Liminal to liminoid in play, flow, and ritual: An essay in comparative symbology, *Rice University Studies*, 60(3), 53-92.

Underthun, A. & Jordhus-Lier, D. (2018) Liminality at work in Norwegian hotels. *Tourism Geographies*, 20(1), 11-28, DOI: 10.1080/14616688.2017.1314546

Urry, J. (1990a). The consumption of tourism. *Sociology*, 24, 23-35.

Urry, J. (1990b) *The Tourist Gaze.* London: SAGE Publications

Wels, H., van der Waal, K., Spiegel, A. & Kamsteeg, F. (2011) *Victor Turner and Liminality: An introduction. Anthropology Southern Africa*, 34(1-2), 1-4, DOI: 10.1080/23323256.2011.11500002

Wickens, E. (2002) The sacred and the profane: A tourist typology. *Annals of Tourism Research*, 29(3), 834-851.

Zhu, Y. (2012) Performing heritage: Rethinking authenticity in tourism. *Annals of Tourism Research*, 39(3), 1495-1513.

12 Tourism and the Individual

By the end of the chapter, the reader will be able to:

- Define propensity and intensity
- Analyse barriers to travel and know how to disaggregate interest as a discrete barrier
- Understand the role of the leisure paradox in influencing travel over one's life span
- Apply generational cohort theory to explain differences in travel by generations

Introduction

Tourism is a socially selective activity. While we may think that everybody has an opportunity to travel this is clearly not the case. Further, even among those who do travel, social, personal and demographic factors exert a significant effect on both propensity to and intensity of travel. This chapter examines a range of factors that influence tourism propensity and intensity among individuals. It begins with a definition of these terms and them explores barriers to participation, the traditional leisure paradox and how it affects tourism, and finally the role of generational cohort theory and tourism activity.

Propensity and intensity

Propensity relates to the participation rate, or the percent of individuals within a population who travel. It is defined as the share of the population that makes at least one (holiday) trip of a specified number of overnight stays during a given reference period (Eurostat, 2010). Like all things in tourism, though, this simple definition is not absolute, for propensity can be measured in a number of ways. Usually, the reference period is one calendar year or 12 months, depending on when the survey was completed. The length of stay to be considered can vary, with some European statistics looking at trips of four nights or longer (Eurostat, 2010), while others consider any overnight pleasure trip (Eurostat, 2017). The UNWTO, being an organisation that is interested in international travel calculates propensity on the basis of the number (or percentage) of international travelers as a proportion of a market's total population.

Intensity or frequency is a simpler concept for it relates to the average number of trips taken over the given period of time. Again, though, the denominator can vary using either the entire population, and, therefore, calculating non-travellers as part of the intensity figure, or by focusing only on those who travelled.

Non-travel

One of the myths of tourism, especially in a developed society, is that everybody travels or at least has the potential to travel. Yet, one does not need to look very far to see the inherent flaw in this assumption. Some people may not be physically robust enough to travel. Some people may simply not want to travel. Others may not have any travel companions. Others still may not be able to afford to travel. In Europe, about 62% of the population took either a domestic or international trip in 2016 (Eurostat, 2017). Propensity varied significantly by country of origin, though, ranging from a low of a 24% participation in Romania to a high of 91.4% in Finland. A study of Hong Kong (McKercher, 2009) showed propensity varied from the mid 60 percentile to the low 70 percentile depending on the state of the economy. Smith and Carmichael (2005) concluded that one in three Canadians do not travel.

Our conceptualization of why people do not travel grew out of leisure constraints theory. This theory postulates that everyone wants leisure but due to a variety of reasons not everyone can have it. It argues further that if constraints can be removed, then participation in leisure activity will follow. Crawford and Godbey (1987) and Crawford, Jackson and Godbey (1991) proposed that leisure constraints could be classified into one of three categories of intrapersonal, interpersonal, and structural barriers, and saw leisure participation as a function of the ability to negotiate each of the constraints.

Intrapersonal constraints represent psychological states or attributes that reflect preferences. They include such items as lack of interest, religion, reference group attitudes and perceived skill. Interpersonal constraints relate primarily to the lack of travel partners, while structural constraints lie somewhat beyond the individual's own control and can include time, cost, lack of opportunity, family commitments and the like. Leisure constraints theory argues that these three classes of constraints are hierarchical in nature, with intrapersonal issues needing to be resolved before interpersonal constraints can be considered. Interpersonal constraints must also be resolved before structural barriers can be addressed. However, in a more recent retrospective piece, Godbey et al. (2010) identified an intertwining relationship between the three dimensions, which challenges the linear hierarchical proposition associated with the model.

The model has been tested a number of times in a leisure context and found to be valid (Godbey et al., 2010). Jackson et al. (1993) evaluated this model from a tourism context, while it has also been used by a number of other researchers (Nyaupane and Andereck, 2008; Pennington-Gray and Kerstetter, 2002) with modest success. Intuitively it makes sense and some empirical evidence seems to support the existence of such barriers. For example, when considering intrapersonal constraints, Lee and Tideswell (2005) and McGuire (1984) found that societal pressures and in particular reference group approval from family and friends precluded many seniors from travelling. March (2000) came to similar conclusions examining propensity among Japanese women, where they are expected to be very travel active in their 20s, but once they marry are expected not to travel. Lack of travel partners has long been identified as a core intrapersonal constraint, especially among people with disabilities (Smith, 1987).

The impact of structural constraints has attracted most attention in the literature. Tourism Research Australia (2011a, 2011b) identifies income as the primary driver for both domestic and inbound tourism while Eurostat (2010) found a strong correlation between household income and overall travel propensity. The inability to speak foreign languages precludes some people, especially seniors from travelling abroad (Chen and Wu, 2009). Many others have identified work commitments as a factor that inhibits travel (Dickinson, 2014; Losada et al., 2016; Nickerson, 2000; Smith et al., 2011).

Reported constraints vary by age and life stage (Kattiyapornpong and Miller, 2009; Smith et al., 2011). Figure 12.1 shows the types of barriers that people at different life stages identified as reasons for not travelling. The intrapersonal constraint of social acceptability really only comes into play amongst the elderly, especially elderly singles and couples with low incomes who feel they are too old to travel. Instead, they feel that they have an obligation to stay at home and look after their grandchildren. Older working families, those in their 50s with children still living at home, identified structural constraints of family obligations and cost factors as inhibiting their ability to participate in tourism, while middle aged affluent couples and families cited work commitments as their main reason to not travel. Interestingly, Smith et al. (2011) also conducted a study of non-tourism and concluded that many people identified lack of money and time as factors that precluded travel, yet, their research found that most non-travellers can in fact afford to travel and have the time available to do so. But, as they note, something gets in the way as they make their purchase decisions.

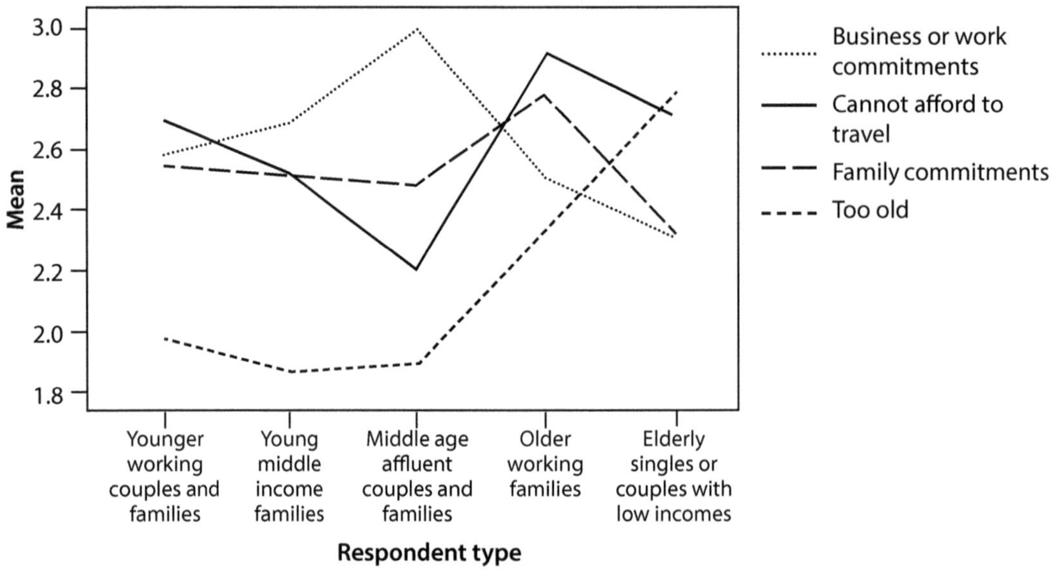

Figure 12.1: Impact of major barriers by cohorts. *Source*: McKercher, 2009.

Herein lies one of the dilemmas of importing theory from one discipline and applying it to tourism without necessarily considering the unique nature of tourism. If one frames research within the confines of leisure constraints, then it is easy to identify leisure constraints as factors that inhibit travel. But, is the use of this framework really unveiling the truth, for Kattiyanornpong and Miller (2009) argue us that travel constraints are quite different from general leisure constraints. Others note traditional leisure constraints theory may not be particularly relevant within a tourism context (Chen and Wu, 2009).

To begin constraints are not absolute (McGuire, Dottavio and O'Leary, 1986) and instead their impact varies depending on how people react to them. The same situation may act as a prohibitor that stops some people from travelling, a limitor that reduces participation or requires a modification of activities among others, or may be a non-issue for others still. For example, reference group pressures may stop some seniors from travelling, may modify their travel experiences by limiting them to travelling in seniors groups or with other family members, or may have no impact at all if the person is a keen tourist. The same can be said for interpersonal constraints where the growing body of solo backpacking literature, especially among women suggests it can be a liberating and life-changing event and not a prohibitor of travel (Bianchi, 2016; McNamara and Prideaux, 2010;

Wilson and Little, 2005). Likewise, income as a structural constraint may indeed stop some people from travelling but is more likely to limit others to taking a lower cost, short break, short duration holidays or may be a non-issue for some people.

More than that, though, lack of interest may play a far more important role than packaging it in a bundle of intrapersonal constraints suggests. Haukeland (1990) categorized travel along a 2 × 2 matrix and identified one group as 'Type A' non travellers who had no economic obstacles but simply preferred to stay at home. Litvin, Smith and Pitts (2012) found the stated reasons for non-participation, such as lack of money, contradicted with the demographic characteristic of respondents. They concluded that some element of social desirability bias prevented respondents from identifying the real root cause of their non-participation. Dolnicar, Lazarevski and Yanamandram (2013) and Dolnicar, Yanamandram and Cliff (2012) recognise that not all people want to go on vacation.

Lack of interest, more so than any other barrier, precludes many people from travelling. Simply stated, for some, travel is not a life priority or a high enough priority to engender action. Instead, different priorities dominate, including finding satisfying leisure activities close to home, as shown by these kayakers on the Rideau Canal in Ottawa, Canada. While we think travel is available to all, some people simply do not have the travel bug.

McKercher and Chen (2014) found a strong relationship between interest or the importance one places on travel and tourism propensity. Their study asked people to rate whether tourism was more, less or as important as a variety of other activities such as spending time with family and friends, saving money for long term goals or pursuing recreational interests to improve their skills. The results are summarised in Figure 12.2. People who identified tourism as being very important, felt it was equally as or more important than the other items measured, while those who felt travel was unimportant or very important to them ranked every other item as being of equal or greater importance than travel. The study concluded that intensity of and propensity to travel are related to the personal importance respondents place on tourism. They argued people who have the travel bug place the opportunity to travel as a high life priority, while those who do not see many other things in life as being of greater importance.

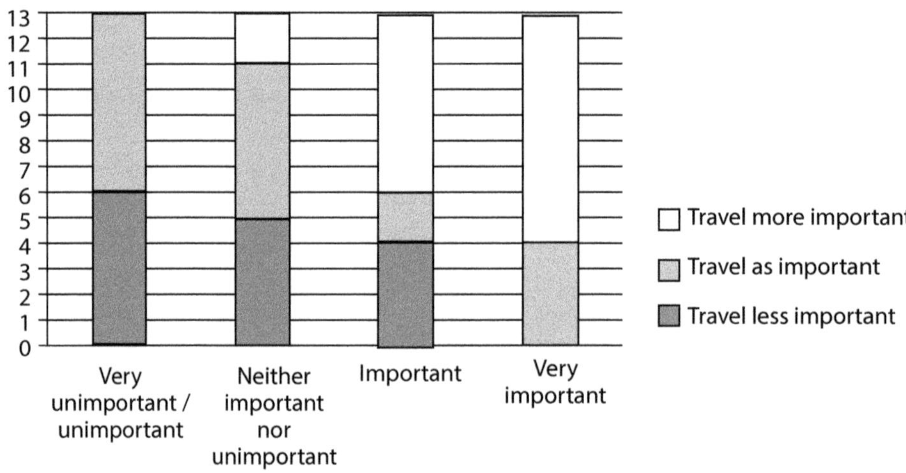

Figure 12.2: Tourism as a life priority. Source: McKercher and Chen, 2014.

An earlier study (McKercher, 2009) found a similar relationship between the importance one places on travel and the likelihood of taking a pleasure trip. It concluded a lack of interest created a significant barrier to participation. And, indeed, many people identified a range of intrapersonal and structural barriers as socially acceptable reasons to mask their lack of interest.

These findings along with aforementioned work by Haukeland, Smith and his colleagues, and Dolnicar and her team suggest we need to reconsider the leisure constraints model as it relates to tourism. In particular, the issue of lack of interest emerges as being the most critical early factor that influences travel participation. Again, leisure theory suggests lack of interest is just one of a bundle of constraints that can be easily overcome by removing barriers. Within a tourism context, though, it is difficult to overcome a deep-seated lack of interest in tourism, especially when other opportunities compete effectively for discretionary leisure time and income. Instead, lack of interest is of such profound importance that it deserves to be identified as a discrete barrier category in its own right. As such, it is proposed that the leisure constraints model be modified by adding a fourth foundation tier of lack of interest as shown in Figure 12.3. Thus, lack of interest represents the primary barrier to some people to participate in tourism and unless, and until, it can be overcome people will not travel or will travel infrequently. It is only once this constraint is addressed that the other types of intrapersonal, interpersonal and structural barriers can be considered.

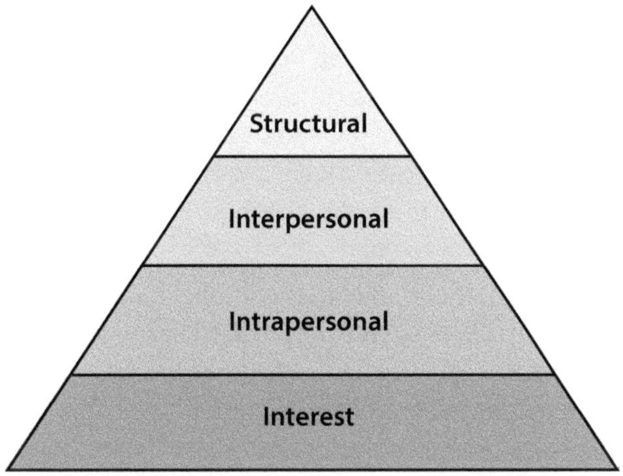

Figure 12.3: Tourism constraints model. *Source*: from McKercher and Chen, 2015.

The leisure paradox

Constraints theory has its merits, but life is rarely as simple as theory suggests, for a variety of intervening factors also affect one's ability to travel and the type of tourism activity pursued throughout one's life. Cooper et al. (1998) introduced us to the idea of the leisure paradox in tourism (Figure 12.4) where there is an inverse relationship between income and time availability. As Mill (2008) notes the paradox suggest that when people have money to engage in various leisure pursuits they lack time and when they have the time availability, income is at a premium. Added to this are different health, fitness and physical issues that mean both intensity and the type of travel will vary with age (Hong et al., 2005).

12

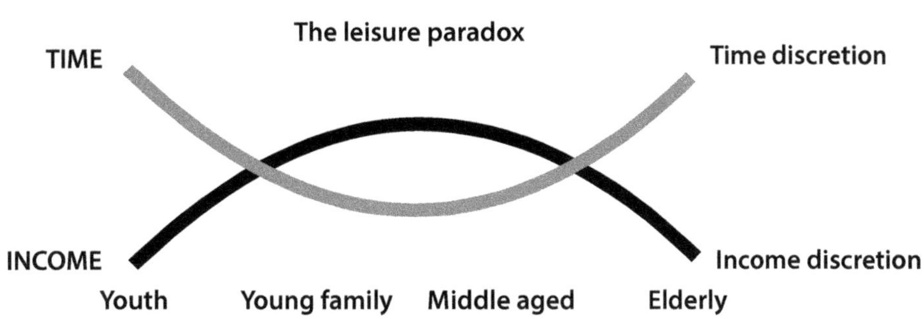

Figure 12.4: The leisure paradox. *Source*: Cooper et al., 1998.

Travel activities vary according to one's lifecycle stage. Singles, for example, are usually able to participate in a wider variety of travel experiences than are those people who have families, simply because they have fewer responsibilities. More than just simply the availability of time, life stage also shapes the motivation for and needs of a holiday. Prior to the onset of children, singles and couples are motivated to participate in an activity primarily for the intrinsic satisfaction that participation brings. Once children arrive, however, the primary motivation becomes one of providing benefits to the children and to the family more so than for the individual. By the same token, after children arrive, the social and financial realities of the couple or individual change. Travel is often curtailed as the family spends far more time at home.

Hong et al. (2005) found marrieds without children, full nesters with older children, and empty nesters spend significantly more than singles, whereas solitary survivors spend significantly less on leisure travel. Families with a higher number of members are more inclined to make a domestic trip than one abroad. However, the oldest cohorts are more inclined to participate in tourism than the youngest one, but they are less inclined to spend on tourism (Bernini and Cracolici, 2015). Backer and King (2017) saw an increase in visiting friends and relatives travel among the over 60s, while Huang and Petrick (2010) found older people travelled for longer periods of time and travelled greater distances. Boksberger et al. (2009) saw a greater tendency to seek more comfort as people age.

Fleischer and Pizam (2002) and Nickerson (2000) found that travel participation often peaks between the age of 55 and 65 when leisure time and income are at their peak and one's health is still strong. Children have grown, mortgages are paid and those in long term employment have plenty of leave time available. Demand remains strong until the early 70s when propensity declines. But the frequency with which seniors travel is lower than other age groups (Hung et al., 2007; Lung, 2017). Lung (2017) determined that age alone is not a factor, instead, he determined post retirement income is the key determinant. Fleischer and Pizam (2002) add, somewhat axiomatically that income and health are the major constraints that affect senior's travel propensity. Zimmer et al. (1996) add education, mobility problems and self-assessed health status, rather than absolute health condition, and interest in spending money on recreation are other factors that influence participation among seniors.

The changing profile of families, split families and the emergence of three generation families is changing travel patterns (Glover and Prideaux, 2008). Bernini and Cracolici (2015) cite a higher life expectancy, a healthier ageing population, smaller family size, a higher share of the population with a high level of education and migration as affecting tourism. Schänzel and Yeoman (2015), for example, noted families are becoming more vertical rather than statically horizontal, with grandparents enjoying more time with their grandchildren as they live longer. As a result, we are seeing an increase in multi-generational travel.

Education levels are closely related to travel propensity and the type of activity pursued (Bernini and Cracolici, 2015; Zimmer et al., 1996), in part because the better one's education, the greater the likelihood of having a higher income, and in part because people who are better educated are more interested in what is going on in the world (Plog, 1974). This in turn, often stimulates a desire to see different areas and to experience different cultures. A well educated person may seek status and learning opportunities from their travel, while a less highly educated person may seek simply escape.

Education also relates to the ability of the individual to want to break free of his or her own cultural confines. Better educated tourists are generally more willing to seek holidays in unfamiliar and culturally different places. Less well educated tourists, who have less of a world view, feel intimidated travelling in strange cultures and are more likely to confine their activities to culturally proximate destinations. Ecotourism, adventure travel, alternative tourism, nature based tourism, cultural tourism and much special interest tourism (festivals and the like) all appeal to more highly educated individuals.

Generational cohort theory

People resemble their generations more so than they resemble their parents. Indeed, in all societies, each generation experiences life differently (Alwin and McCammon, 2003). What this means is that factors that shape generations have a stronger impact on consumer behaviour than family influences. This observation translates into generational cohort theory. Generational cohort theory posits that important historical events and social changes encountered by a group of people who were born at approximately the same time, and experience these events during their critical development stages, shape their attitudes, values, beliefs and inclinations (Moss, 2016; Tang, Wang and Zhang, 2017). It is also felt to affect their consumption patterns (Li, Li and Hudson, 2013) for they are driven by a person's system of value priorities. Li et al. (2013) argue each generation tends to display patterns that are internally consistent but differ across generations.

It has been widely used in sociology, anthropology and social psychology (Hung et al., 2007), political science (Cheng and Foley, 2017; Gilleard, 2004) and consumer behaviour (Debevec et al., 2017). Limited research has also applied it within a tourism context (Benckendorff & Moscardo, 2013; Gardiner, King and Grace, 2012; Li, Li and Hudson, 2013; Pennington-Gray, Fridgen and Stynes, 2003).

The concept was developed first by Mannheim in 1923 (Mannheim, 1952). He argued that generations have much in common for they are similarly located, by being members of the same stage as the collective process, experiencing the same events and being born within the same historical and cultural region. This situation limits them to a specific range of potential experiences and thus predisposes them to certain characteristic modes of thought and experience. Alwin and McCammon (2003: 25) add that members of a birth cohort share a social history, experience their lifecycle at the same time, and importantly, share the experience of the cohort itself. Moreover, each generation is unique for each experiences life differently (Pendergast, 2009). Ryder (1965: 843) notes each birth cohort acquires coherence and continuity from the distinctive development of its constituents and from its own persistent macro analytic features. Successive cohorts are differentiated by the changing content of formal education, by peer-group socialization, and by idiosyncratic historical experience.

Generational cohort theory suggests your travel patterns are defined more by people born in your generation than your family members. Each generation is defined by key moments that occurred during their formative years that in turn influence their values, attitudes and behaviours. Many elderly people grew up during time of great hardship, be it the Great Depression or the Second World War, when a culture of pleasure travel had not been established. As such, they tend to travel less. People born later grew up during a time of relative affluence, when a culture of pleasure travel started to emerge, but because of technology limitations confined their travel largely to domestic tourism. To them, travel was still seen as a discretionary activity. More recent generations see travel as a right and not a privilege. For them, travel is seen to be part of their everyday expectations.

A few features are critical when considering the impact of generational cohorts on behaviour. To begin, attitudes and values crystalise during one's formative years from adolescence to the early 20s (Benckendorff and Moscardo, 2013; Cheng and Foley, 2017). Second, they remain largely unchanged when people reach adulthood and change relatively little thereafter (Hung, Gu and Yim, 2007; Inglehart, 2008). Thus how people think about the world may depend more on what was happening in the world when they were growing up than what is occurring in the present (Alwin and McCammon, 2003: 25) as individuals born during the same generational cohort may share the same inclinations and cognitive styles (Moss, 2016). This observation challenges the belief that people's values and beliefs change as they mature (Sessa et al., 2007).

Each generation has two essential components, a common location in historical time and distinct consciousness of that historical position (Tang et al., 2017). Sessa et al. (2007:47) add "individuals in the same age group at the same time (societal or historical) and space (location) will be limited 'to a specific range of potential experience' thereby predisposing them to a 'characteristic mode of thought and experience' and a characteristic 'historically relevant action." Generations, thus are as much social creations as they are biological ones (Sessa et al., 2007; Twenge and Campbell, 2008), with the length of the generation depending on the duration of events that shaped it during the person's formative years.

A wide array of factors can influence the values that are formed in a generation. Sessa et al. (2007) identified such issues as a traumatic or formative events such as war, a dramatic shift in demographics, an interval that connects a generation to great success or failure, the creation of sacred space that creates a collective memory or the presence of mentors and heroes that give a voice to a generation. Coomes (2004) adds history and popular culture can play a vital role in defining a generation, while the futurist O'Hara-Deveraux (2013) suggests technological changes, the emergence of an industrial revolution and transition from a largely rural agrarian economy to an urban society can also shape generations.

Generation cohort theory has received relatively little attention in the tourism literature, but what research has been conducted has tended to compare two generations rather than a number of generations. Nevertheless, it has the potential to have a profound impact on both demand and the delivery of tourism products. Benckendorff and Moscardo (2013) suggest that changing values of Baby Boomers and members of Generation X and Y may result in the potential decline of ecotourism and the need to change how products are delivered. Ishida et al. (2017) and McKercher et al. (2020) have identified markedly different attitudes toward travel in general and international travel in particular by successive generations of Japanese and Mainland Chinese tourists respectively. Here members of older generations participate less, because it is speculated a culture of participation in tourism was never formed during their formative years. By contrast tourism is seen as a norm and something to be expected by members of younger generations.

Empirical studies by Oppermann (1995) and Li, Li and Hudson (2013) found members of younger generations were more experienced tourists than members of older generations at the same age, but that older consumers had more accumulated travel experience. Gardiner, Grace and King (2014) found substantial differences in attitudes to domestic tourism and travel behaviour among different generations of Australian residents. Baby boomers were more indoctrinated into the perceived hedonic outcomes of travel than Generation X and Y. Instead Generation Y's travel patterns are influenced by functional evaluations of value. Likewise, Li, Li and Hudson (2013) found that Baby Boomers and Generation X were similar in their past destination experiences but different from Generation Y. They also discovered a significant difference among four generations in terms of their outbound travel experiences, with older generations having traveled more outside of the U.S. than their younger counterparts. Pennington-Gray, Fridgen and Stynes (2003) made an interesting observation in their study covering two time periods. They found the behaviour pattern of each generation changed as it aged and inferred, but did not conclude outright, that preferences may be a function of experience as well as generational values.

Conclusion

A number of factors influence travel propensity and intensity among consumers. This chapter began by examining barriers to travel and illustrated how the traditional leisure-oriented theory is not applicable to tourism for it does not consider the profound role that lack of interest plays as an inhibitor of travel. People will not travel if it is not a life priority in the first place, or not a high enough priority to engender action. No amount of barrier removal will overcome such resistance. As such, interest needs to be disaggregated from the bundle of intrapersonal constraints and identified as its own discrete barrier. However, even if interest exists, the presence of the leisure paradox may have a significant effect on travel by individuals, for an inverse relationship exists between time availability and affordability. Those with the most time may be able to afford to travel least and therefore could be constrained in their behaviour. Those with the most income, on the other hand, may have to overcome the most obstacles to enable travel. Together, they reflect the trend toward short break trips, rather than taking one long duration holiday.

Finally, the chapter discussed the little examined role that generational cohort theory plays in tourism. Each generation is unique, and so, lessons learned from one generation may not be able to be translated into a different generation's travel patterns. What this means is that each generation will, potentially, evolve through its own unique travel career pattern and that the pattern will differ from generation to generation.

■ Discussion points / Exercises

1. Barriers to travel affect people differently. The same barrier may be a prohibitor, a limitor or have no impact. How can lack of money affect people in different ways?
2. Lack of interest is a critical feature stopping some people from travelling. How can this barrier be overcome? Can it be overcome?
3. What impact do children have on family travel?
4. Research on generational cohort theory suggests it has an effect on travel propensity, especially relating to whether a culture of pleasure travel was created in different generations. Interview people from three different generations to determine when they first travelled for pleasure, flew and, importantly, whether they felt long haul travel was something that was available to them.

Points to consider for revision

- Barriers are dynamic and need to be negotiated. Depending on the person, the same barrier may represent a prohibitor to travel, may necessitate moderation of travel or may have no impact on travel.
- Leisure constraints theory says barriers are hierarchical, as one must overcome intrapersonal, interpersonal and structural barriers in order to participate. However, tourism constraints theory identifies a fourth more fundamental level of lack of interest. This represents the single greatest barrier to travel for many. Many people who have little or no interest in travel need to voice socially acceptable answers (such as lack of time and money) to mask their lack of interest.
- The leisure paradox suggests an inverse relationship exists between time availability and income.
- Generational cohort theory suggests people resemble their generations more so that their families. In a tourism context, it relates primarily to whether a culture of pleasure travel was created for each generation.

Suggested additional readings

Dolnicar, S., Lazarevski, K. & Yanamandram, V. (2013) Quality of life and tourism: A conceptual framework and novel segmentation base. *Journal of Business Research,* 66(6) 724-729.

Litvin, S. W., Smith, W. W. & Pitts, R. E. (2012). Sedentary Behavior of the Non-travel Segment: A Research Note. *Journal of Travel Research, XX*(X), 1-6

McKercher, B., Lai, B, Yang, L. & Wang, Y. (2020) Travel by Chinese a Generational Cohort Perspective. *Asia Pacific Journal of Tourism Research.* 25(4), 341-354.

References

Alwin, D. F. & McCammon, R. J. (2003) Generations, cohorts, and social change, in J. T. Mortimer and M .J. Shanahan (Eds.) *Handbook of the Life Course,* Springer, Boston, 23-49

Backer, E., & King, B. (2017). VFR traveller demographics: The social tourism dimension. *Journal of Vacation Marketing, 23*(3), 191-204. https://doi.org/10.1177/1356766716665439

Benckendorff, P. & Moscardo, G. (2013) Generational cohorts and ecotourism. In Ballantyne, R. and Packer, J. (Eds) *International Handbook on Ecotourism.* Cheltenham: Edward Elgar publishing, pp 135-154.

Bernini, C. & Cracolici, M. (2015) Demographic change, tourism expenditure and life cycle behaviour. *Tourism Management,* 41, 191-205.

Bianchi, C. (2016) Solo holiday travellers: Motivators and drivers of satisfaction and dissatisfaction. *International Journal of Tourism Research,* 18, 197-208.

Boksberger P.E., Sund K.J. & Schuckert M.R. (2009) Between past, present and future – implications of socio-demographic changes in tourism. In Conrady R. & Buck M. (eds), *Trends and Issues in Global Tourism 2009.* Springer, Berlin, Heidelberg.

Boorstin, D. (1964) *The Image: A guide to pseudo-events in America.* New York: Harper and Row.

Chen, C. & Wu, C. (2009) How motivations, constraints, and demographic factors predict seniors' overseas travel propensity. *Asia Pacific Management Review,* 14(3), 301-312.

Cheng, M. & Foley, C. (2017) Understanding the distinctiveness of Chinese Post-80s tourists through an exploration of their formative experiences. *Current Issues in Tourism,* 21(11), 1312-1328. DOI: 10.1080/13683500.2017.1406462.

Coomes, M. (2004) Understanding the historical and cultural influences that shape generations. *New Directions for Student Services.* 106, 17-31.

Cooper, C., Fletcher, J., Gilbert, D., Shepherd, R. & Wanhill, S. (1998) *Tourism Principles and Practices,* 2nd ed., England: Addison-Wesley, Longman.

Crawford, D. & G. Godbey (1987) Reconceptualizing barriers to family leisure. *Leisure Sciences,* 9 (2), 119–127.

Crawford, D., Jackson, E. & Godbey, G. (1991) A hierarchical model of leisure constraints. *Leisure Sciences,* 13(4), 309 – 320.

Debevec, K., Schewe, Madden, T. & Diamond, W. (2013) Are today's Millennials splintering into a new generational cohort? Maybe! *Journal of Consumer Behavior*, 12, 20–31.

Dickinson, J. (2014) Time, tourism consumption and sustainable development. *International Journal of Tourism Research* 16(1), 11-21.

Dolnicar, S., Lazarevski, K. & Yanamandram, V. (2013) Quality of life and tourism: A conceptual framework and novel segmentation base. *Journal of Business Research*, 66(6), 724-729.

Dolnicar, S., Yanamandram, V. & Cliff, K. (2012) The contribution of vacations to quality of life. *Annals of Tourism Research*, 39(1), 59-83.

Eurostat (2010) *Tourism Statistics in the European Statistical System 2008 data.* https://ec.europa.eu/eurostat/documents/3888793/5848705/KS-RA-10-010-EN.PDF

Eurostat (2017) *Tourism Statistics - Participation in tourism.* https://ec.europa.eu/eurostat/statistics-explained/index.php?title=File:Participation_in_tourism_of_EU_residents_(aged_15_and_over),_2016.png <downloaded Jan 6, 2019>.

Fleischer, A. & Pizam, A. (2002) Tourism constraints among Israeli seniors. *Annals of Tourism Research*, 29(1), 106-123.

Gardiner, S., King, C. & Grace, D. (2012) Travel decision making: An empirical examination of generational values, attitudes, and intentions. *Journal of Travel Research*, 52(3), 310-324.

Gilleard, C. (2004) Cohorts and generations in the study of social change. *Social Theory and Health*, 2, 106-119.

Glover, P. & Prideaux, B. (2008) Using population projections to identify aspects of future tourism demand, in Joseph S. Chen (ed.) *Advances in Hospitality and Leisure (Advances in Hospitality and Leisure, Vol. 4)* Emerald Group Publishing Limited, pp.185-209.

Godbey, G., Crawford, D. & Shen, X. (2010) Assessing hierarchical leisure constraints theory after two decades, *Journal of Leisure Research*, 42(1), 111-134, DOI: 10.1080/00222216.2010.11950197.

Haukeland, J. V. (1990) Non-travelers: The flip side of motivation. *Annals of Tourism Research*, 17(2), 172-184.

Hong, G., Fan, J. Palmer, L. & Bhargava, V. (2005) Leisure travel expenditure patterns by family life cycle stages. *Journal of Travel & Tourism Marketing*, 18(2), 15-30.

Huang, Y. & Petrick, J. (2010) Generation Y's Travel Behaviours: A comparison with Baby Boomers and Generation X. In Benckendorff, P., Moscardo, G. and Pendergast D (Eds) *Tourism and Generation Y.* CABI, pp 27-37.

Hung, K., Gu, F. & Yim, C. (2007) A social institutional approach to identifying generation cohorts in China with a comparison with American consumers. *Journal of International Business Studies* 38, 836-852. doi.org/10.1057/palgrave.jibs.8400288.

Inglehart, R. (2008) Changing values among Western publics from 1970 to 2006. *West European Politics*, 31(1-2), 130-146, DOI: 10.1080/01402380701834747.

Ishida, Y., Miyaki, M., Fujisawa, Y. & Iwasaki, K. (2017) How does tourism differ among generations? Tourists from the United States and their willingness to visit Japan, *International Journal of Tourism Sciences*, 17(1), 49-60, DOI: 10.1080/15980634.2016.1274526.

Jackson, E. L., Crawford, D. W. & Godbey, G. (1993) Negotiation of leisure constraints. *Leisure Sciences*, 15(1), 1-11.

Kattiyapornpong, U. & Miller, K. (2009) Socio-demographic constraints to travel behavior, *International Journal of Culture, Tourism and Hospitality Research*, 3(1), 81-94. https://dro.deakin.edu.au/eserv/DU:30016691/kattiyap-sociodemographic-post-2009.pdf <downloaded Dec 2, 2019>.

Lee, S.H. & Tideswell, C. (2005) Understanding attitudes towards leisure travel and the constraints faced by senior Koreans, *Journal of Vacation Marketing*, 11(3), 249-263.

Li, X., Li, R. & Hudson, S (2013) The application of generational theory to tourism consumer behavior: An American perspective. *Tourism Management*, 23, 147-164.

Litvin, S. W., Smith, W. W. & Pitts, R. E. (2012). Sedentary behavior of the non-travel segment: a research note. *Journal of Travel Research*, 52(1), 131-136.

Losada, N., Alen, E., Dominguez, T. & Nicalou, J. (2016) Travel frequency of seniors tourists. *Tourism Management*, 53, 88-95.

Lung, R. (2017) Travel wanderlust: Age is not a constraint, but income might be. *Visa*. https://usa.visa.com/partner-with-us/visa-consulting-analytics/senior-travel-wanderlust.html. <downloaded Dec 3, 2019>.

Mannheim, K. (1952) The problem of generations, in P. Kecskemeti (Ed.) *Essays on the Sociology of Knowledge: Collected Works*, Routledge, New York, pp. 276-322.

March, R. (2000) The Japanese Travel Life Cycle. *Journal of Travel & Tourism Marketing*, 9(1-2), 185-200.

McGuire, F. (1984) A factor analytic study of leisure constraints in advanced adulthood. *Leisure Sciences*, 6, 313- 336.

McGuire, F. A., Dottavio, D. & O'Leary, J. T. (1986) Constraints to participation in outdoor recreation across the life span: A nationwide study of limitors and prohibitors. *The Gerontologist*, 26(5), 538-544.

McKercher, B. (2009) Non-travel by Hong Kong residents. *International Journal of Tourism Research*, 11(6), 507-519.

McKercher, B. & Chen, F. (2014) Travel as a life priority? *Asia Pacific Journal of Tourism Research*, 20(7), 715-729. DOI: 10.1080/10941665.2014.931289.

McKercher, B., Lai, B, Yang, L. & Wang, Y. (2020) Travel by Chinese: a generational cohort perspective. *Asia Pacific Journal of Tourism Research*. 25(4), 341-354.

McNamara, K. E. & Prideaux, B. (2010) A typology of solo independent women travellers. *International Journal of Tourism Research*, 12(3), 253-264.

Mill, R. (2008) The inter-relationships between leisure, recreation, tourism, and hospitality in Brotherton, B. & Wood, R. (eds) *The SAGE Handbook of Hospitality Management*, London: Sage. http://sk.sagepub.com/reference/download/hdbk_hospitalitymgmt/n4.pdf <downloaded Feb 8, 2019>

Moss, S. (2016) Generational cohort theory. *SICO-Tests* https://www.sicotests.com/psyarticle.asp?id=374 <downloaded Dec 2, 2019>.

Nickerson, N (2000) Travel and recreation outlook 2000: Focusing on demographics. *Montana Business Quarterly*, 38(1), 23-26.

Nyaupane, G. P. & Andereck, K. L. (2008). Understanding travel constraints: Application and extension of a leisure constraints model. *Journal of Travel Research*, 46(4), 433-439.

O'Hara-Deveraux, M. (2013) China's 5 Generations: Diverging Lifestyles. *Global Foresight.* http://www.globalforesight.dreamhosters.com/future-of-china-emerging-markets/chinas-5-generations-diverging-lifestyles/ <downloaded Dec 2, 2-19>.

Oppermann, M. (1995) A model of travel itineraries. *Journal of Travel Research*, 33(4), 57-61.

Pendergast, D. (2009) Generational theory and home economics: Future proofing the profession. *Family and Consumer Sciences Research Journal* 37(4):, 504-522.

Pennington-Gray, L. & Kerstetter, D. (2002) Testing a constraints model within the context of nature-based tourism. *Journal of Travel Research*, 40(4), 416–423.

Pennington-Gray, L., Fridgen, J. & Stynes, D. (2003) Cohort segmentation: An application to tourism. *Leisure Sciences*, 25(4), 341-361, DOI: 10.1080/714044495.

Plog, S. (1974) Why destination areas rise and fall in popularity. *Cornell Hotel and Restaurant Administration Quarterly*, 14(4), 55-58.

Ryder, N. (1965) The cohort as a concept in the study of social change. *American Sociological Review*, 30(6), 843-861. www.jstor.org/stable/2090964.

Schänzel, H.A. & Yeoman, I. (2015) Trends in family tourism. *Tourism Futures*, 1(2), 141-147.

Sessa, V., Kabacoff, R., Deal, J. & Brown, H. (2007) Generational differences in leader values and leadership behaviors. *The Psychologist-Manager Journal*, 10(1), 47–74.

Smith, R. (1987). Leisure of disabled tourists: barriers to participation. *Annals of Tourism Research*, 14(3), 376-389.

Smith, W. W. & Carmichael, B. A. (2005) A geographical analysis of the rates of non-travel across the regions of Canada. *Tourism Geographies*, 7(3), 257-271.

Smith, W., Fralinger, E. & Litvin, S. (2011) Segmenting the U.S.A. non-travel Market. *Enlightening Tourism. A Pathmaking Journal*, 1(2), 137-151.

Tang, N., Wang, Y. & Zhang, K. (2017) Values of Chinese generation cohorts: Do they matter in the workplace? *Organizational Behavior and Human decision Processes*, 143, 8-22.

TRA (2011a) What is driving Australian's travel choices? Canberra: Tourism Research Australia. /www.tra.gov.au/tra/2016/documents/What_is_driving_Australians_travel_choices_FINAL_-_2_June.pdf

TRA (2011b) Factors affecting the inbound tourism sector - the impact and implications of the Australian dollar. Canberra: Tourism Research Australia. www.tra.gov.au/tra/2016/documents/Factors_Affecting_the_Inbound_Tourism_Sector_FINAL_-_2_June.pdf

Twenge, J.M. & Campbell, W.K. (2008) Increases in positive self-views among high school students, *Psychological Science*, 19, 1082. DOI: 10.1111/j.1467-9280.2008.02204.x

Wilson, E. & Little, D. E. (2005) A 'relative escape'? The impact of constraints on women who travel solo, *Tourism Review International*, 9: 155-175.

Zimmer, Z., Brayley, R. & Searle. M. (1996) Whether to go and where to go: Identification of important influences on seniors' decision to travel. *Journal of Travel Research*, 33(3), 3-10.

Part 5 – Tourism Impacts and Sustainability

Tourism has alternately been portrayed as either a benign 'smokestack-free' activity that produces limited adverse impacts or as a major source of social, cultural and ecological impacts. Part 5 explores this issue in greater detail.

Chapter 13 examines the origins of many of the concerns over the social impacts of tourism and how much of these concerns have been placed within a conflict paradigm. It examines the two core origins of this concept, by Doxey (1975) and Budowksi (1976). The chapter concludes by arguing most impacts are, in reality, a function of place change and residents' ability or inability to cope with it.

Chapter 14 develops this theme more by examining how concerns over tourism's impacts led to the emergence of sustainable tourism. It discusses the dilemma for sustainable development, and in particular, how the same words can be used by different stakeholders to espouse contradictory ideologies. It then looks at various models of sustainability.

Chapter 15 discusses the issues of climate change, resilience and the challenges of transitioning to a carbon-neutral economy. In doing so, it progresses the idea of how we can move to a more sustainable tourism future.

Chapter 16 discusses the challenges of planning for a sustainable tourism sector. A number of models are discussed, as are the difficulties in applying a one size fits all approach to tourism planning.

13 Impacts – Conflict over Place Change

By the end of this chapter, the reader will be able to:

- Describe Doxey's Irridex
- Analyse how impacts are caused by place change
- Evaluate Budowksi's relationships between tourism and the environment
- Understand how conflict theory underlies early attempts to assess social and environmental impacts of tourism.

Introduction

Concerns about unsustainable tourism practices have been a subject of academic inquiry since the earliest days of tourism scholarship. Indeed, it seems that a majority of the papers published in the first editions of *Annals of Tourism Research* documented adverse social and cultural impacts of tourism. As McKercher and Prideaux (2014: 21) noted:

> "Wenkman (1975) documented the adverse environmental impacts of tourism on Hawaii. UNESCO (1976) published a literature review of the adverse social consequences of tourism. Rodenburg (1980) condemned large scale tourism in Bali and argued that small, community based tourism is preferred. Farrell (1979) documented adverse host-guest interactions. Jafari (1974) wrote a lengthy article documenting explicitly focusing on the costs, and not benefits of tourism."

Since then, a range of topics has emerged with the sophistication of the research evolving as our understanding of the causes of impacts and possible mitigation strategies have matured. Yet, to a large extent, many of the critical issues remain unresolved, in spite of the emergence of sustainable tourism as a dominant paradigm, as discussed in the next chapter.

Hundreds, if not thousands of academic papers have been published examining the impacts of tourism on host communities (Nunkoo, Smith and Ramkissoon, 2013). Deery, Jago and Fredline (2012: 65), though, note that much of this work is derivative, leading them to conclude "research into the social impacts of tourism appears to be in a state of 'arrested development,' [where] there is a sense that the advances in understanding the impacts of tourists on host communities is incremental at best, or potentially circular." The reasons are manifold. Far too much of this research is descriptive in nature, and is typified by a 'fill in the blank study – Understanding social impacts/community attitudes of tourism in ___________ destination.' Many papers as well use the same metrics to measure attitudes, and therefore, unsurprisingly, come to the same conclusions. A third issue is that much of this research is of the self-fulfilling prophecy kind, where if one structures the study to look for adverse impacts, they can be found. A fourth and more critical issue is the lack of theoretical basis for most of the research, something identified by Ap (1990) 30 years ago that is still relevant today.

This chapter explores some of the challenges in the conceptual discussion of tourism impacts. It begins with quest for theory and then reviews the foundational models by Doxey (1975) and Budowski (1976) that framed unsustainable practices within a conflict paradigm. It then looks at such issues as impacts as a function of place change.

Impact research - A quest for a unifying 'theory'

Deery et al. (2012) identified four stages in the development of social impact research. The first stage began with the identification of definitions and parameters of the debate. This occurred in the 1970s. Shortly thereafter, incipient models were developed, with Doxey's (1975) and Budowksi's (1976) models, discussed below, being the two most commonly cited ones. Stage three then looked at ways to operationalise these models, with Ap and Crompton's (1998) work being among the most respected studies, for it developed rigorous questionnaires and study methods. The fourth stage represented instrument refinement.

Three issues confront social impact research. The first is that that there is no such thing as a single community, when one measures community attitudes. Instead, geographic spaces are comprised of multiple communities that react to tourism in different ways. Some may feel they benefit from tourism, others are adversely affected, other still feel tourism has no impact and others yet have a complicated relation to tourism, seeing both benefits and costs. This observation leads to the second point in that impacts are not absolute. Instead, they are perceptual and reflected by attitudes or changes in attitudes (Gu and Ryan, 2008). The third issue relates to what to measure, for the list of potential impacts is only limited by one's imagination. For example, Easterling (2004) identified over 40 different impacts that could be summarised into technological, political, socio-cultural and natural. Deery et al. (2012) wrote an excellent piece on many of the other challenges facing impact research.

Perhaps, the biggest challenge, though, is the ongoing struggle to find some type of universal conceptual framework to explore impacts. Easterling (2004), Nunkoo and Gursoy (2019) and others summarised this dilemma by illustrating that research has been framed alternately in different theories as shown in Table 13.1. Each of these 'theories' would be considered as lower order theories by Smith and his colleagues. While Easterling (2004) notes they help broaden our understanding of tourism systems, none really explains it fully.

Table 13.1: 'Theoretical' approaches to impact studies

Equity Theory	Assesses perceived costs vs. benefits. Tourism is seen as beneficial when positive and negative consequences are in balance or when the positive aspects outweigh the negative.
Growth Machine Theory	Certain stakeholders (those who seek to gain) will support growth in order to maximize their personal economic returns. Others may not support growth or may actively oppose it.
Lifecycle Theory	Attitudes change as tourism evolves through its lifecycle.
Power Theory	Personal power, based on property, money, skills, knowledge and competence, affects one's ability to exploit exchanges. Non-power residents were more favorable toward tourism development than those with power.
Social Exchange Theory	Voluntary actions that are motivated by expected returns. Individuals or groups will engage in an exchange if they value that which is being exchanged, perceive that costs do not exceed benefits and that the exchange will be rewarding. Residents will engage in exchanges with tourists, as long as they 'profit' or as long as benefits exceed costs.
Stakeholder Theory	As key stakeholders in tourism, residents need to be identified, included and satisfied.
Sustainability	Triple bottom line of social, economic and ecological impacts.
Dependency Theory	Applicable mostly to developing countries where dependency originates from and, eventually, reinforces unequal power relationships between wealthy countries (i.e. the center) and poor countries (i.e. the periphery).
Social Impact Theory	The consequences to human populations of any public or private actions that alter the way in which people live, work, or relate to one another or cope.

Source: From Andereck et al., 2016; Easterling, 2004; Nunkoo & Gursoy, 2019

It must also be noted that as a result, attitude, perception and impacts are often conflated. As an example, social exchange theory is an excellent tool to assess residents support for tourism or otherwise, but does not necessarily prove effective in documenting impacts. Instead, by extension ecological, social and economic impacts resulting from host-guest exchange affect resident attitudes to tourism (Nunkoo and Gursoy, 2019). Yet, it is a tool used in a number of studies. The net result is that social impact research still remains rather fractured and descriptive. Perhaps, it is better to look to the future by reverting back to the origins of much of the research in tourism.

Impacts framed within a conflict paradigm

When examined more closely, much of the work on tourism impacts is framed explicitly or implicitly within conflict theory. Doxey's (1975) famous, though rarely read Irridex, positioned conflict as a cause of social impacts. Budowski (1976) asserted tourism had a volatile relationship with the environment, inevitably leading to conflict, and Jackson and Wong (1982) felt conflicting land use demands led to clashes between recreational user groups.

Conflict is goal interference attributed to another's behaviour. Goals, in turn are preferred social, psychological or physical outcomes of a behaviour that provide incentives for that behaviour (Jacob and Schreyer, 1980). Conflict may exist at two levels: direct, where the actions of others affect one's enjoyment; and indirect, where a general and more pervasive feeling of dislike or an unwillingness to appreciate others' views exist (Jackson and Wong, 1982). The likelihood of conflict existing is enhanced when a perception exists that people must fight over a fixed-pie asset, which produces winners and losers (Gramman and Burdge, 1991). It is also likely to occur when the perceived power balance between stakeholders shifts, empowering one and disempowering the other. Tourism is ideally suited to become a conflict inducer for it often represents a powerful, new stakeholder with different values to existing stakeholders (McKercher, 1993).

Conflict is dynamic, evolutionary or revolutionary in style and rarely if ever permanent. Yet, as will be discussed, much of the early work in this field assumed conflict was an end state. It is understandable, for tourism was undergoing rapid development at the time and systems that had been pushed into phase shift have not had the chance to revert to a steady state (Russell and Faulkner, 1999). However, as Perdue, Long and Kang (1999) note, social disruption is not permanent. Instead it is transitional where, eventually communities adapt to these changes. For example, perceived quality of life may decline initially but eventually stabilizes and then improves for both individuals and communities are resilient (Greider, Krannich,and Berry, 1991).

Conflict is defined as goal interference attributed to another's behaviour. Often consumers are seeking similar goals, as in this case of this whitewater kayaker and clients of a commercial white water rafting company on the Ottawa River in Canada. It is just that one is looking for a more solitary experience that challenges one's skill level and others are looking for a more pleasurable experience that requires far less skill. The key to this type of conflict is that both types of users must share the same fixed pie resource. Whitewater kayakers discovered this portion of the river first and only later did commercial rafting companies start to use this resource. On busy weekends, a seemingly endless stream of rafts navigates the rivers, meaning kayakers have to modify their behaviours to paddle only during the small time gaps before the next wave of rafts comes down the river. Tourism is often seen as a powerful source of conflict for it represents a new type of user that does not easily fit into either the biocentric or anthropocentric stakeholder group. On the one hand, nature based tourism shares much with biocentrism, for both want a clean environment. But on the other hand, it shares much with anthropocentrics who see the use value of environmental resources being of greater importance than their existence values. As a result, tourism is often a stakeholder that is not really trusted by either stakeholder group.

Doxey's Irridex

George Doxey's (1975) work examining the causes of social impacts resonates today, in part because he coined the term 'Irridex' which has become part of the tourism lexicon. This work is based on a conference paper that is less than five pages long, with scarcely four paragraphs devoted to the core argument. In it, he reports on two small case studies conducted in the Barbados and Niagara-on-the-Lake in Canada. The basic premise is that "in the mind of a 'resident' anyone not a native of the destination is an outsider and consequently is a possible source of friction. What is important is the extent that, in the minds of the residents, the outsider represents a challenge to the life style of the destination" (Doxey 1975: 195). The concluding paragraph of his work states:

> "unless local residents are psychologically capable of adjusting to change, irritation levels will continue to rise with long term disastrous effects to the total environment of the tourist area, and to the touristy industry" (Doxey 1975: 198).

He then developed his 'Irridex' on the understanding that given time, conflict between hosts and guests will lead the hosts to express their feelings progressively more overtly. The Irridex basically follows the lifecycle model developed a few years later by Butler (1980), but instead of adopting a physical carrying capacity perspective, it focusses on the social carrying capacity of the destination, and how tourism can inevitably change the character of a place for the worse. The four stages and the descriptions in the original paper of causes and solutions are shown in Table 13.2 below.

Table 13.2 : Doxey's stages of relationship between hosts and visitors

Stage	Conditions	Impact and resolution
Euphoria	Associated with the initial stages of development	Visitors and investors are welcome Usually little planning and few control mechanisms
Apathy	Visitor numbers increase Tourists taken for granted	Contact begins to become more formal Most planning will be concerned with marketing
Irritation	Saturation point is approached	Residents begin to show misgivings about the tourist industry Policy makers increase infrastructure rather than setting limits to growth
Antagonism	Tourism overwhelms places Saturation point exceeded Reputation of the destination suffers	Irritation overtly expressed verbally and physically Mutual politeness gives way to mutual antagonism Outsiders seen as the cause of all problems Remedial planning and marketing to address issues and generate new business

Source: Doxey, 1975.

As can be seen, Doxey, an economics professor, focused his work on the need to plan tourism effectively, rather than engaging in a deep discussion of the social impacts on host communities. Yet, this work has been referred to thousands of times in the literature and has been used as a lens to examine a range of impacts in a variety of settings. In fact, some authors have devoted more time to a discussion of his four stages than he did.

The question must be asked, though, is whether Doxey's model is complete, or whether a fifth stage can be added? As mentioned, conflict is rarely an end state and instead tends to resolve itself as either people adapt or are displaced. Doxey's work, like many other studies of social impacts, examined attitudes to tourism during the rapid development phase of the destination lifecycle. It is during this phase that the most significant changes to local communities are likely to occur and where the whole range of negative impacts are most likely to be felt. Doxey's model seems to apply to the first three stages of the life cycle, but is incomplete in addressing latter stages. Communities adjust to tourism as it becomes part of its social, cultural and economic fabric. In some cases, people may modify their behaviours to cope with the annoying aspects of tourism, but they begin also to appreciate what it means to the community in terms of improved livelihood, increased leisure and recreation opportunities, more and better shopping and dining opportunities and a range of other possible benefits.

■ Budowksi – Coexistence, symbiosis, conflict

Gerardo Budowski (1976) came to the same conclusion as Doxey when he examined the relationship between tourism and the environment, or more precisely human uses of natural environments. He argued that any human use of natural or near natural areas exerts a range of impacts. Tourism, through the introduction of non-traditional activities, uses and/or user groups is no different. He felt tourism could have three possible relationships with the environment. It may coexist during the early stages when the tourism industry and those promoting conservation have relatively little contact. He feels coexistence is a transitory phase especially as tourism activity increases and induces a range of mostly adverse impacts. The second type of relationship is typified by symbiosis in which tourism and conservation derive mutual benefits from each other. Natural assets continue to be conserved as far as possible or evolve towards an even more satisfactory condition while providing a range of benefits, including economic benefits to user groups. He also feels that the situation occurs rarely. Instead, the most likely relationship is one is of conflict especially when the presence of tourism is detrimental to the needs of nature.

Bukowski (976) essentially saw the relationship as existing along a continuum from coexistence through symbiosis towards conflict due to an increase in tourism and the shrinking of natural areas. Indeed, he feels that the explosion of tourism in the 1970s concerned with wildlife, wildlife areas and scenic beauty, based on natural resources is inimical with nature conservation and almost always leads to widespread degradation of the resource or a reduction in the assets of nature. Like Butler's (1980) lifecycle model, once a saturation point is reached or when a critical carrying capacity threshold is passed, rapid degradation seems inevitable.

He does hold out some hope, though, that tourism may not necessarily be detrimental to protected areas providing appropriate management practices are undertaken. He qualifies this point by stating, in the 1970s, that the policy of building hotels, restaurants and road systems in natural areas leads to their degradation. The solution is to build tourism infrastructure outside of natural areas, limit access and provide some kind of interpretation center at the entrance. He also illustrates that a symbiotic relationship can be fostered when quality tourism programs support conservation goals, when tourism contributes financially to park management and when an overall conservation ethos drives protected area management practices.

This work must be placed within the context of the emerging environmental movement of the 1970s when the anthropocentric and biocentric approaches to protected area management were being debated (Hendee, Stankey and Lucas, 1978). Here, it must be understood that natural areas, or wilderness areas are essentially human constructs that produce human values. There is no such thing as wilderness. It is a human concept. The preservation of 'wild' areas is not done solely for the benefit of the flora and fauna, but also for the important human benefits and values such preservation provides. In fact, to say some feature is unique and worthy of preservation is a human value judgement. As well, fundamental to the preservation ethos is that these areas should be 'used' to some extent.

Thus, conflict emerges over how these areas should be used. The two conflicting ideologies of anthropocentrism and biocentrism emerged, with each having its own set of stakeholders, user groups, values and beliefs. The anthropocentric perspective argues that the main goal for gazetting natural areas is for the use and enjoyment opportunities these areas provide. Supporters of this viewpoint argue that programs should be developed to encourage and facilitate direct use of natural areas. People who support this approach view wilderness from a sociological, human oriented perspective. The naturalness of wilderness is less important that its direct use benefits. On the other hand, biocentrics advocate the philosophy that unhindered operation of natural ecological processes is the fundamental goal of any natural area. While human use can occur, it must only do so in a manner that is subservient to the natural ecosystem's needs.

Clearly, these two ideologies are incompatible. Natural areas managed for anthropocentric goals place management priorities on the provision of recreational services. In doing so, substantial impacts might be noted to natural processes to facilitate greater use. In its extreme form, the natural environment would be manipulated to provide tourist benefits, as has been the case in many island resort communities. On the other hand, efforts by people to restrict access or hinder use would be rejected. Indeed, anthropocentrics argue againsts wilderness management plans that serve only the privileged few people who are fit, healthy and comfortable in wilderness areas. Biocentrics, on the other hand support management approaches that encourage natural processes to continue unfettered as much as possible. Recreational use is permitted only to the point that it does not become intrusive. As such, biocentrics support activities that control visitor numbers, exclude undesirable types of uses, inhibit use and disperse impacts.

Works by Doxey, Budowski and natural area advocates, along with the emerging influence of *Annals of Tourism Research* and such books as Mathieson and Wall's (1982) *Tourism: Economic, Physical and Social Impacts,* that both documented and modelled adverse social and cultural impacts, had a profound impact on tourism studies, laying the groundwork for deeper research on impacts. But, it also had an unintended consequence whereby tourism advocates began promoting a series of alternative tourism activities in response. The early 1980s saw the emergence of the anti-tourism movement advocating alternative tourism and a range of other forms of tourism, culminating in the explosion of ecotourism as a panacea (McCool, 2016). These supposedly new types of socially and morally responsible tourism were engaged in by a new type of tourist, who sought to reduce or at least minimise their impacts. A series of small case studies, often published by the World Tourism Organization (1981, 1983) heralded the benefits of tourism and asserted that this type of activity represented the new mainstream. This type of tourism promoted the idea that big is bad and small is good and that small scale tourism, done on a large enough basis can replace mass tourism (Dernoi, 1981).

An alternative explanation – impacts as a function of place change

In essence, what Doxey and Budowksi document is the impact of place change on attitudes to tourism, as well as perceived conflicts. Doxey's work overtly addresses this issue, while Budowski infers place change strongly as protected areas become used increasingly for anthropocentric purposes at the expense of biocentric ideals.

Historically space and place have been theorized as bounded geographic entities (Tapsell and Tunstall, 2008). As Tuan (1975) notes, the social meaning people give to geographic space transform it into place and give it meaning. In other words, there is an emotional tie between people and place (Mahon, 2007), where sense of place is defined by place identity, place dependence, and place attachment (Jorgensen and Stedman, 2001). Place identity involves dimensions of self that define the individual's personal identity through a complex pattern of conscious and unconscious ideas, beliefs, preferences, feelings, values, goals and behavioural tendencies (Proshansky, 1978). Place dependence is the strength of the association between an individual and a specific place, while place attachment reflects a positive bond between groups or individuals and their environment. It relates to the idea of insidedness/outsidedness, where being an 'insider' is to feel safe and attached (Turner, Turner and Carroll, 2005) and have an emotional right to occupy a place (McCabe and Stokoe, 2004). Outsiders do not have that right.

McKercher, Wang and Park (2015) discuss each of the elements in detail. They argue that destinations are comprised of three types of place: tourism place, non-tourism place and shared place. Tourism place is signed and signaled as being for tourists, while tourists are welcome in shared space, providing they behave appropriately. Non-tourism place, as the name implies, is part of the community that is signed and signaled for use by local residents. Normally, as with most other systems, the relationship between these three types of place evolves toward some form of equilibrium, or coexistence, where each is well defined and both locals and tourists respect their boundaries.

Table 13.3: How tourism affects nine types of place

Place type	Indicators of negative place change	Indicators of positive place change
Generic community	Loss of traditional lifestyles; rapid cultural change; community character; moral decay; family breakdown	Improved quality of life; stronger ethnic identity; cultural exchange
Personal lifestyle	Sense of invasion; need to curtail activities; turn locals into attractions; tourism imposed on lifestyle	
Congestion	Traffic; shortage of parking spaces; crowding at public places	
Dislocation	Physical displacement	
Cost of living	Inflation; higher costs of living; higher housing costs	
Economic	Loss of shops; low paying jobs; job loss; economic dependency on tourism	More jobs; higher standard of living; reinvigorate local economy
Resource	Increased competition for natural and recreational resources; land disputes; urban sprawl	
Amenity	Loss of recreational diversity	More leisure and recreational opportunities; better public transport and other services
Structural	Unstable residential patterns; influx of new residents; changing demographic profile; second home development	

Source: McKercher, Wang and Park, 2015.

Their meta-analysis of over 100 social impact papers examined the relationship between attitudes to tourism, lifecycle stage and place change. As shown in Table 13.3, they identified nine types of tourism place. Adverse impacts had been documented in all nine, yet positive place change was also noted in three types of place. As they note, some forms of place change affect individuals directly (personal lifestyle, congestion, dislocation), others were felt indirectly at a community level (generic community, economic, structural), while some represented a hybrid of both (inflation, resource, amenity).

Impacts of tourism are most likely to occur when otherwise stable systems are disrupted. Social disruption theory is closely related to place change. This theory states that communities experiencing rapid growth typically enter a period of generalized crisis that may lead to a wide array of social problems as changes penetrate throughout communities and at individual levels (Park and Stokowski, 2009). Perceived quality of life may decline initially, but will eventually stabilize and then improve as both individuals and communities adapt for, as Greider, Krannich, and Berry (1991) note, they are resilient.

Many of the conflicts over tourism involve place or space change, where once non-tourism place evolves into shared or tourism place, or shared place becomes exclusive tourism place. In extreme cases, local residents are displaced as small businesses realise they can generate higher incomes by catering to the needs of tourists. The range of products changes to satisfy these new consumers. Here in Bangkok, Thailand, a former floating market has gradually evolved to become a tourist attraction catering exclusively to international visitors. The range of products has changed from fruit and seafood to tourist souvenirs. While some authors feel this type of transition inevitably leads to conflict, others are less sure, instead stating that residents adjust their behaviours over time and come to accept the new user group as being part of the cultural milieu of a destination.

Non-tourism place may become shared or even exclusively touristic place and in doing so displace local residents, while shared place may evolve into tourism place. Many examples abound. 'Place' within a touristic sense has multiple social, spatial and economic connotations. People want and need to feel attached, that they belong to places, and consequently look for signs that validate that sense of belonging. Impacts are most likely to occur when that sense of belonging is threatened, but over time, stasis is re-established and a new normal of tourist, shared and non-tourism place emerges (McKercher, Wang and Park, 2015).

Their study revealed impacts, both positive and negative represent a change in the community's perception of place identity, place dependence or place attachment. Positive impacts are noted usually through the creation of shared place. Negative impacts, on the other hand are associated with diminished place attachment, loss of distinctiveness, continuity, self-esteem or self-efficacy, that eventually make residents feel more like outsiders in their own community. Here, non-tourism place is seen to change into either shared to tourism place.

Importantly, there is a direct correlation between attitudes to tourism and the destination lifecycle stage. Figure 13.1 summarises the findings.

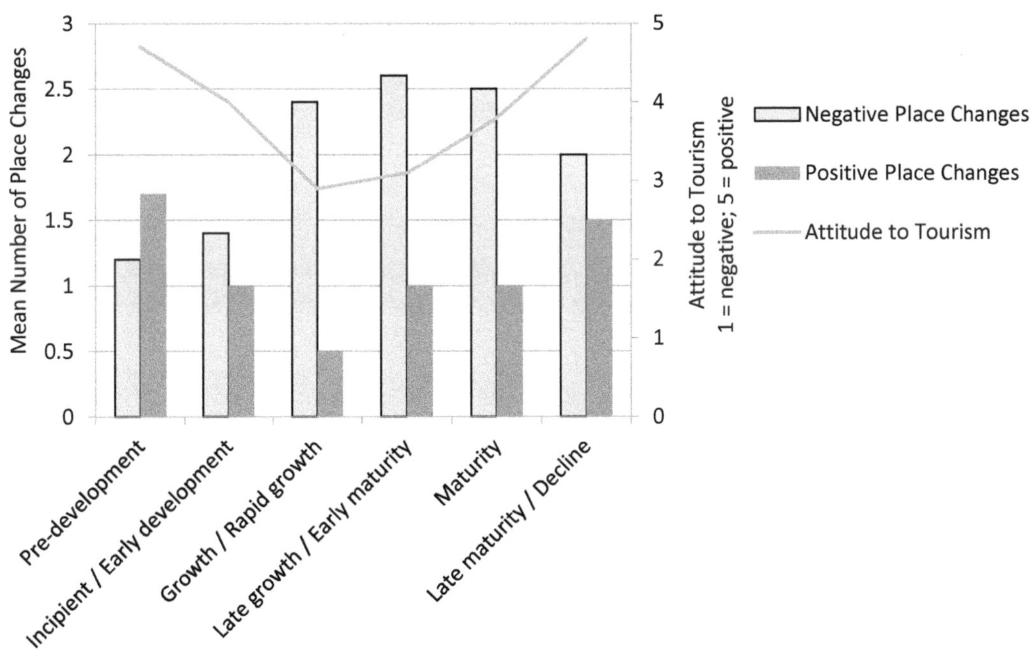

Figure 13.1: Social impacts as a function of place change. *Source*: McKercher, Wang and Park, 2015.

The line represents attitudes to tourism while the columns represent the perception of positive and negative impacts of tourism. As Doxey suggests, attitudes to tourism are generally positive in the early stages of development where little tourism occurs and where the perception exists that it will bring a number of positive changes to communities. But, attitudes change as tourism expands, and especially during the rapid growth stage, when a nadir is reached. Here the negative consequences of tourism are seen to far outweigh the positive benefits as place change is at its apex. But, a strange thing happens. While perceptions of negative impacts remain high, attitudes generally improve and the community begins to appreciate the positive impacts of tourism once more. What happened? The community has adjusted to the new normal and stasis has been regained.

Conclusion

The study of impacts of tourism, both positive and negative is an enduring theme in tourism scholarship. Indeed, impacts have been a core study area right from the earliest days of tourism research. While research has documented well the 'what' of impacts, it is still struggling to understand the 'whys' of impacts. Here a range of competing paradigms have been developed, each with its own merits but also each having its own limitations. This chapter tried to simplify much of the discussion by placing impacts within two dominant fields of study. The first is conflict literature, where the direct and indirect actions of one group of stakeholders can exert a range of impacts on another group of stakeholders. The second is the idea of space and place, whereby most adverse impacts can be summarised as a function of place change.

■ Discussion points / Exercises

1. Review a current tourism issue and determine what stage of Doxey's Irridex the destination is in. Is conflict inevitable and if not, what can be done to resolve the situation?
2. Impacts are a function of place change. Choose a destination of your choice and map out tourism, shared and non-tourism place. Look for areas where adverse impacts have been felt, and document if place change has occurred, and if so, which of the nine types of place change are evident.
3. Tourism becomes part of the lifeblood of most mature destinations, and a degree of stability emerges in the spatial structure of destinations. Indeed, in many cases, local residents are also heavy users of so-called tourism places. Identify and discuss how local residents have also made tourism place their own.

Points to consider for revision

- Early work on impacts of tourism was framed within a conflict paradigm.
- Multiple 'theories' have been developed that try to explore impacts, attitudes and perceptions. Each has its place, but none is ideal.
- Most impact are caused by place change, where tourism disrupts normally stable systems.

Suggested additional readings

Deery, M., Jago, L. & Fredline, L. (2012) Rethinking social impacts of tourism research: A new research agenda. *Tourism Management*, 33(1), 64-73.

McKercher, B., Wang, D. & Park, E. (2015) Social impacts as a function of place change. *Annals of Tourism Research*, 50, 52-66.

Budowski, G. (1976) Tourism and conservation: conflict, coexistence or symbiosis? *Parks*, 3(1), 27–31.

References

Andereck, K., Pachmayer, A. & Zhao, S. (2016) Resident attitude towards tourism – the state of knowledge. *Travel and Tourism Research Association. 10*. https://scholarworks.umass.edu/ttra/2012/Oral/10 <downloaded May 6, 2020>.

Ap, J. (1990) Residents' perceptions research on the social impacts of tourism. *Annals of Tourism Research*, 17, 610-616.

Ap, J. & Crompton, J. (1988) Developing and testing a tourism impact scale, *Journal of Travel Research*, 37(2), 120-130.

Budowski, G. (1976) Tourism and conservation: conflict, coexistence or symbiosis? *Parks*, 3(1), 27–31.

Butler R. W. (1980) The concept of a tourist area cycle of evolution: implications for management of resources. *Canadian Geographer*, 24(1), 5-12.

Deery, M., Jago, L. & Fredline, L. (2012) Rethinking social impacts of tourism research: A new research agenda. *Tourism Management*, 33(1), 64-73.

Dernoi, L. (1981) Alternative tourism: Towards a new style in North-South relations. *International Journal of Tourism Management*. 2(4), 253-264.

Doxey, G. V. (1975) A causation theory of visitor-resident irritants: methodology and research inferences. In *Travel and Tourism Research Associations Sixth Annual Conference Proceedings* (195-198), San Diego, September.

Easterling, D. (2004) The residents' perspective in tourism research. *Journal of Travel and Tourism Marketing*, 17(4), 45-62.

Farrell, B. (1979) Tourism's human conflicts: Cases from the Pacific *Annals of Tourism Research*, 6(2), 122-136.

Gramman, J. H. & Burdge, R. J. (1991) The effects of recreation goals on conflict resolution: The case of waterskiers and fishermen. *Journal of Leisure Research*, 13(1), 15-27.

Greider, T., Krannich, R. & Berry, H. (1991) Local identity, solidarity, and trust in changing rural communities. *Sociological Focus*, 24(4), 263-282.

Gu, H. & Ryan, C. (2008) Constructionism and culture in research: understandings of the fourth Buddhist festival, Wutaishan, China, *Tourism Management*, 31(2), 167-178.

Hendee, J., Stankey, G. & Lucas, R. (1978) *Wilderness Management*, Forest Service, US Dept. of Agriculture, Misc. Publication No 1365, Washington pp 15 - 22.

Jackson, E. L. & Wong, R. (1982) Perceived conflicts between urban cross country skiers and snowmobilers in Alberta. *Journal of Leisure Research*, 14(1), 47 - 62.

Jacob, G. R. & Schreyer, R. (1980) Conflict in outdoor recreation: A theoretical perspective. *Journal of Leisure Research*, 12(4), 368-380.

Jafari, J. (1974) The socio-economic costs of tourism to developing countries. *Annals of Tourism Research*, 1(7), 227-262.

Jorgensen, B. S. & Stedman, R. C. (2001) Sense of place as an attitude: Lakeshore owners attitudes toward their properties. *Journal of Environmental Psychology*, 21(3), 233-248.

Mahon, M. (2007) The changing faces of rural populations. *Journal of Rural Studies*, 23(3), 345-356.

Mathieson, A. & Wall, G. (1982) *Tourism: Economic, Physical and Social Impacts*, Longman, London.

McCabe, S. & Stokoe, E. H. (2004) Place and identity in tourist accounts. *Annals of Tourism Research*, 31(3), 601-622.

McCool S.F. (2016) The changing meanings of sustainable tourism. In S. McCool and K. Bosak (eds), *Reframing Sustainable Tourism. Environmental challenges and solutions*, Vol. 2. Springer, Dordrecht pp 13-32

McKercher, B. (1993) Some fundamental truths about tourism: Understanding tourism's social and environmental impacts *Journal of Sustainable Tourism*, 1(1), 6-16.

McKercher, B. & Prideaux, B. (2014) Academic myths of tourism. *Annals of Tourism Research*, 46, 16-28.

McKercher, B., Wang, D. & Park, E. (2015) Social impacts as a function of place change. *Annals of Tourism Research.* 50, 52-66. http://dx.doi.org/10.1016/j.annals.2014.11.002

Nunkoo, R. & Gursoy, D. (2019) Introduction to tourism impacts. In D. Gursoy & R. Nunkoo (Eds) *The Routledge Handbook of Tourism Impacts: Theoretical and Applied Perspectives.* Oxon: Routledge. Pp 1-20.

Nunkoo, R., Smith, S. L. J. & Ramkissoon, H. (2013) Residents' attitudes to tourism: A longitudinal study of 140 articles from 1984 to 2010. *Journal of Sustainable Tourism*, 21(1): 5-25.

Park, M. & Stokowski, P. (2009) Social disruption theory and crime in rural communities: Comparisons across three levels of tourism growth. *Tourism Management*, 30(6), 905-915.

Perdue, R., Long, P. & Kang, Y. (1999) Boomtown tourism and resident quality of life: The marketing of gaming to host community residents. *Journal of Business Research*, 44(3), 165-167.

Proshansky, H. M. (1978) The city and self-identity. *Environment and Behavior*, 10(2), 147-169.

Rodenburg, E. (1980) The effects of scale in economic development: Tourism in Bali. *Annals of Tourism Research*, 7(2), 177 – 196.

Russell, R. & Faulkner, B. (1999) Movers and shakers: Chaos makers in tourism development. *Tourism Management*, 20(4), 411-423.

Tapsell, S. & Tunstall, S. (2008) "I wish I'd never heard of Banbury": The relationship between 'place' and the health impacts from flooding. *Health & Place*, 14(2), 133-154.

Tuan, Y. F. (1975). Place: An experiential perspective. *Geographical Review*, 65(2), 151-165.

Turner, P., Turner, S. & Carroll, F. (2005) The tourist gaze: Towards contextualized visual environments. In P. Turner & E. Devenport (Eds.), *Spaces, Spatiality and Technology*. Netherlands: Springer, pp. 281-297.

UNESCO (1976) The effects of tourism on social-cultural values. *Annals of Tourism Research*, 4(2), 74-105

Wenkman R (1975) The Pacific tourist blight. *Annals of Tourism Research*, 3(2), 77.

WTO (1981) *Social and Cultural Impacts of Tourism Movements*. Madrid: World Tourism Organization.

WTO (1983) *Study of Tourism's Contribution to protecting the Environment*. Madrid: World Tourism Organization.

14 Sustainability

By the end of this chapter, the reader will be able to:

- Define the key concepts involved in sustainable development
- Critique the inherent contradiction in the term
- Analyse the core concepts of sustainable tourism
- Compare and contrast differing approaches used to achieve sustainable development and/or sustainable tourism.

Introduction

During the early 1980s, it became apparent that major global environmental changes were occurring suddenly and silently and were only beginning to be recognized as an emerging threat by the global scientific community. Scholars and an increasing number of citizens became aware that there was a growing element of uncertainty and risk from human activities on global environments The establishment of the Intergovernmental Panel on Climate Change (IPCC) in 1988 by the World Meteorological Organization (WMO) and the United Nations Environmental Programme (UNEP) and endorsed later by the General Assembly of the United Nations was a major step in focusing the global community's attention to the problem. At that time, moderate environmentalists were wrestling with the idea that we can live in a world where economic development and economic pressures for both developing and developed countries would continue. Short-term pressures to develop national economies (along with the social benefits that derive from it) were considered by policy makers to outweigh long-term potential impacts from that development process. In other words, we failed to ask the question of how can a sustainable economy be developed that integrates both industrial and ecological needs? More recently, advocates of degrowth suggest a more radical change to how the global economy functions. While it is unrealistic to suggest that the solution to our environmental woes would be to wind back the global economy, growing concerns about sea level rise, ecosystem collapse and the emergence of other global ecological crises challenge the idea that it is possible to promote economic growth while maintaining environmental protection.

This chapter explores a range of issues related to sustainable development in general and sustainable tourism in particular. It begins with an overview of the concept of sustainable development before moving onto a discussion of sustainable tourism. The chapter then discusses the range of models and concepts that have been developed to monitor asustainability.

The dilemma of sustainable development

The concept of sustainable development was formalised in 1987 in a report published by the World Commission on Environment and Development (WCED, 1987), and then crystalised in the 1992 Rio Summit on sustainable development (Obst, 2016; Torres-Delgado and Lopez-Palomeque, 2012). Sustainable development was defined as development that meets the needs of the present without compromising the ability of future generations to meet their own needs (WCED, 1987). This approach contains within it two key concepts: the concept of 'needs', in particular the essential needs of the world's poor, to which overriding priority should be given; and the idea of limitations imposed by the state of technology and social organization on the environment's ability to meet present and future needs. Sustainability has a number of key principles including:

- **Inter-generational equity** – meaning that the range of activities and the scope of ecological diversity available to future generations is at least as broad as that felt by current ones;
- **Intra-generational equity, social justice and poverty alleviation** – improving the well-being of all residents in a community, and not just benefiting the powerful or the rich;
- **Public participation** – which means that we all share a role to play and that communities need to collectively make decisions rather than having them imposed by external forces;
- **Environmental protection** as an integral component of economic development – economic development without environmental conservation is no longer acceptable;

- **Dealing cautiously with risk and uncertainty** – in situations where environmental impacts of activities are not known, the preferred option is to proceed cautiously or not at all, until the likely impacts can be determined.

But, almost from the beginning the term 'sustainable' was contested (Pearce et al., 1989; Pigram, 1990) and remains so to the present (Biely et al., 2018). As Liu (2003) argues, while the term sustainability is well established, it has been used loosely in the literature, leading Salas-Zapata and Ortiz-Munoz (2018) to conclude the lack of clarity about the concept is a recurring obstacle to its implementation. Carillo and Jorge (2017) add that multiple interpretations of the concept enable it to be moulded to fit the needs of different stakeholders with competing views.

Two competing and polar opposite approaches to sustainability and sustainable development have emerged. The weak sustainability model assumes sustainability can be achieved through a constant wealth approach. The alternate strong sustainability concept is based on the premise of constant natural capital. Both models accept that the world's total asset base is comprised of both natural (ecological and social) and manufactured capital (Islam et al., 2019).

The constant wealth or weak sustainability philosophy assumes that natural capital and manufactured capital are essentially substitutable and considers that there are no differences between the types of well-being they create (Pelanc and Dedeurwaerdere, 2015). It sees economic sustainability as dominating, with the belief that technological advances will resolve environmental issues (Biely et al., 2018). Thus, all that matters is that the total value of the aggregate stock of natural and manufactured capital should be maintained and ideally added to for the sake of future generations (Pelanc and Ballet, 2015). Hence, natural capital can be traded off providing it creates sufficient manufactured capital to compensate for the loss (Islam et al., 2019). This philosophy argues for a business as usual approach.

The constant natural capital or strong sustainability argument, on the other hand, posits that natural capital is non-substitutable, either totally or partially and needs to be maintained. As Islam et al. (2019: 160) note, the contribution of natural capital to the total wealth is unique and critical, so that it cannot be measured in monetary terms and straightforwardly compared with other types of capital, particularly in social cost–benefit analysis. Pelanc and Ballet (2015) also highlight that a distinction must be made between natural and manufactured capital. They point out that natural capital has the risk of irreversibility and thresholds. For example, the extinction of a species is irreversible, while above a certain threshold of pollutants, ecosystems can become overloaded and enter into decline. Importantly, the amount of manufactured capital can be increased or decreased, while natural capital can disappear if it has become stressed beyond the point of being able to replenish itself. Proponents of this view argue we need to do things differently so that the absolute stock of natural capital does not diminish.

Two competing views of sustainable development have emerged since the concept of sustainability was first mooted almost 40 years ago. On the one hand, weak sustainability argues that natural and manufactured capital are essentially substitutable, and as long as the net sum of both remains the same or increases, sustainable development can occur. This view means that natural resources can be traded off if they generate wealth. Unfortunately, it does not consider the adverse impacts development may have on the environment, as shown in the first picture of a heavily eroded beach in Vietnam. The ecological integrity of the beach was destroyed due to the encroachment of the resort onto the waterfront which resulted in erosion and the ultimate destruction of the resort. Strong sustainability, on the other hand argues that natural capital is non-substitutable. Development can occur and is encouraged to occur, providing that the net sum of natural capital does not diminish, as shown in the second image of beach hikers in Hong Kong. Both views represent the extreme ends of a development continuum. In reality, most tourism development will involve some loss of natural capital, and so the concept of critical natural capital has emerged as a preferred alternative path forward. It argues development can occur as long as the ecological functioning of natural ecosystems remains above a critical threshold.

In reality, definitions and the practices of sustainability exists along a continuum with the weak and strong positions representing the extreme ends. All consumption involves some use of natural resources, some of which are not renewable (Kuhlmann and Farringdon, 2010). As a consequence, the idea of critical natural capital emerged as a core tenet of sustainability. Here it is argued that critical natural capital highlights the need to maintain the ecological functioning of natural systems above certain thresholds of degradation to conserve the capacity of the natural capital asset to renew itself (Pelanc and Dedeurwaerdere, 2015). Moderately strong sustainability has evolved from a purist point of view to one that espouses the belief that sustainability can be seen as a series of thresholds that must not be crossed (Kuhlman and Farrington, 2010).

What are those thresholds? The concept of the triple or quadruple bottom line provides some answers. The triple bottom line is dual edged sword. Sardianou, Kostakis, Mitoula, Gkaragkani, Lalioti and Theodooropoulou (2016), summarising the literature, remind us that sustainable tourism development must focus on the four areas of economic, environmental, social and cultural sustainability. These themes adhere to Elkington's (1994) universally accepted triple bottom line approach to sustainability, or quadruple bottom line if the dimension of social impacts is disaggregated to include social and cultural dimensions. An alternative version of the quadruple bottom line adds the dimension of governance. As Bac (2012) noted, strong sustainability focused on natural resources has evolved to consider the social and economic aspects of the concept. Without economic viability, social and ecological sustainability are difficult to achieve.

Moscardo (2015) highlighted criticisms of the triple bottom line approach to sustainability in that it encourages people to think about the dimensions as forms of capital that are separate and equal, and that finding a balance amongst them is both desirable and easy. This approach is problematic on many levels. First, the timeframe to consider each dimension is variable, with economic sustainability measured in terms of daily, weekly or monthly cash flow, liquidity and profitability. Social sustainability is measured on a longer term, and ecological sustainability on an even longer term still. Second, the issue of what to measure is also problematic and is something that has vexed the UNWTO in its attempts to develop its own set of indicators (UNWTO, 2004). Again, economic sustainability has clearly defined and measurable indicators, while social and environmental sustainability can be measured in terms of thresholds or debatable limits of acceptable change. Third, and perhaps most importantly, it is difficult, if not impossible to separate values from facts (Carillo and Jorge, 2017).

Sustainable tourism

Sustainable tourism has been widely accepted at a conceptual level as a strategy able to balance tourism demand for built and natural resources with the long-term protection of these resources, although not all scholars agree with this view. At face value, tourism appears to be ideally suited to adopting sustainability as a guiding philosophy. At a conceptual level tourism's use of resources, both natural and cultural, should be non-consumptive, making them renewable. The reality however is that tourism is a fierce competitor for resources where the provision of cultural and ecotourism opportunities for tourists may mean that local residents are displaced, and pristine ecosystems redeveloped for tourist consumption. Moreover, since the needs of tourists are different than those of local residents, serving tourists may not suit the needs of local residents. Tourism is often imposed on local communities, especially rural and minority communities, at level and speed that causes great social disruption (McKercher, 1993a).

Government tourism departments, academics and small, nature-based tourism operators on the fringes of mainstream tourism were among the first stakeholders to adopt sustainability as a new way to think about tourism. Indeed, the natural link between tourism and the environment provided an obvious self-interest for tourism to protect its environment (McKercher, 1993b). It is for this reason that early policies, such as those produced in Australia, focused predominantly on ecologically sustainable development (ESDWG, 1991).

Following broad acceptance of sustainable tourism as a concept, a search commenced for models, theories and frameworks that could be employed to ensure that tourism was sustainable. Policies of this nature reflect what Bac (2012) noted as an initial focus on protecting natural resources. He adds that it was only after 2002, when the UN launched the ST-EP (Sustainable Tourism – Eliminating Poverty) programme, that attention began to shift to the social component. Torres-Delgado and Lopez-Palomeque (2012) documented the policy evolution of sustainable tourism from the Rio Summit, through the 1st World Conference for Sustainable Tourism held in Lanzarote (Spain) in 1995, the Committee for Sustainable Tourism (CST) established by the UN in 1999, to UNWTO's Global Ethical Code for Tourism. McCool (2016) and Ruhanen et al. (2015) suggest that this evolution represented the maturation of the field of sustainable tourism and closer integration with the broader concepts of sustainable development.

Since then, the field has advanced rapidly – so rapidly that it is impossible to track all the changes in sustainability in one brief chapter. Buckley (2012) suggested that about 5000 papers had been published on this topic by early 2010s. A Google Scholar search in early 2020 using the term "sustainable tourism" revealed 240,000 matches.

Significantly, we have moved beyond just looking at sustainable tourism in isolation and now realise that tourism is part of a larger debate about overall sustainability (Bramwell, Higham, Lane and Miller, 2017; Budeaunu, Miller, Moscardo and Ooi, 2016). Earlier efforts were coordinated by tourism departments with little or no input from other government agencies. Tourism departments or ministries often are seen as junior departments that have limited influence over other government agencies and even less ability to influence policy emanating from those departments (Muangasame and McKercher, 2015). It is impossible to move towards an effective sustainable tourism sector unless a whole of government approach towards sustainability is adopted. As Budeaunu et al. (2016) observed, desirable change in the tourism system is not possible without broader economic change.

Tourism represents a powerful stakeholder that has the potential to outcompete other stakeholders. One of the challenges facing sustainable tourism is that much of the resource base that it relies on is part of the social fabric of communities. Resources such as beaches, shopping districts, even car parking spaces are used by local residents. But, if tourism is allowed to develop in an unsustainable manner, the needs of residents can be compromised and they ultimately may be displaced. This economic transformation may benefit local business people, but comes at a cost to local residents. Here in Macau, the shops along main street leading to the Ruins of St Paul used to serve local residents' needs, but the influx of large numbers of tourists has led to the dislocation of traditional shops that served the local population and their replacement by shops that cater exclusively to the throngs of tourists who visit every day. Local residents must now go elsewhere.

The need to incorporate tourism into the broader sustainability debate received a boost with the publication of the United Nations Development Program (UNDP) 17 sustainable development goals. The UNWTO strongly endorsed this policy and developed its own policy document illustrating how and where tourism fits within these broad goals (UNWTO, 2013). The document states "tourism's role in achieving the 17 Sustainable Development Goals (SDGs) can be significantly strengthened when sustainable development becomes a shared responsibility and moves to the core of decision-making within the tourism sector" (UNWTO, 2013: 10). The UNWTO feels tourism can contribute to 14 goals directly or indirectly (UNWTO 2013: 16,17), while three goals are specific to tourism matters. They are:

- SDG 8 – promoting sustained, inclusive and sustainable economic growth, employment and decent work for all – where tourism as a services trade is a major employer than can provide decent jobs to youth and women and can foster policies through its value chain that support diversification;
- SDG 12 – ensure sustainable production and consumption patterns – where it recognises that tourism needs to move more in this direction, but that the tools to do so are already in place;
- SDG 17 – strengthen the means of implementation and revitalise the global partnership for sustainable development – in part because tourism is a cross sectoral activity that operates on an international, national, regional and local basis.

One could also argue that the greatest advancement and simultaneously the greatest challenge, is the recognition that sustainability involves a tripartite relationship between government, industry and the consumer (McKercher and Prideaux, 2011), as shown in Figure 14.1. Government can play a macro role through the development of policies and the enactment of legislation. Higgins-Desbiolles (2017) adds that governments must tackle the environmental limits to growth and climate change among other issues and then develop tourism within these limits.

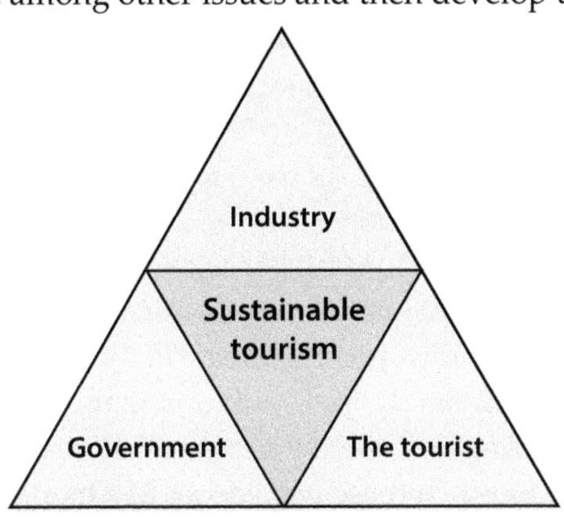

Figure 14.1: The tripartite solution to sustainable tourism

Government can enact sustainability legislation, while industry may be motivated to act in a more sustainable manner for economic reasons or as part of its broader corporate social responsibility mandate. The tourist, though, represents the third and most difficult component of this tri-partite sustainability relationship, for it involves changing behaviour at an individual level. A number of studies have indicated that while people are interested in environmental issues in general and climate change in particular, these considerations play no role in their travel decision making process. Indeed, the opposite may be true, for when the perceived benefits of participation are seen to be high, the perceived impacts of such activities are seen to be low. Moreover, tourists adopt a number of neutralisation actions – verbal justifications to rationalise behaviour that they know is inappropriate. 'Everyone else is doing it', 'I am just one of many people, my actions have no real impact' and 'I am a tourist, I came all this way to participate in this activity and no one is going to stop me' are common explanations used. And so, in spite of requests to behave, some tourists, individually and collectively, will act in a way that is socially, culturally and ecologically irresponsible, as shown in these two images of a tourist souveniring a piece of the Great Pyramid at Giza, Egypt and accumulated litter at the base of the Sphinx.

Industry can provide more environmentally friendly products and services, and can also engage in more sustainable business practices. There are two main reasons why it would choose to do so. The first is that it makes good business sense. The second is that business is increasingly recognising that it has a corporate social responsibility (CSR) that extends beyond simple business operations.

Changing consumer behaviour, is the third pillar of movement towards a more sustainable tourism sector. It is far more challenging for it involves convincing people individually to change their often long-established habits. Becken (2004) highlights the challenge that when people feel the personal benefits of engaging in an action are high, the perceived environmental costs are low. While tourists may see others as acting in an 'unsustainable' manner, they are unlikely to apply the same standard to their own actions. Anable, Lane and Kelay (2006) add further that people rarely make a link between their own actions and broader environmental problems. For example, most people are aware that driving a car produces greenhouse gases, but few people relate their own driving habits as making a direct contribution. McKercher and Prideaux (2011) also found that tourism ranked low on the list of personal environmental impact agendas, and moreover, because it represented a break from social norms, tourists could justify behaving differently than they did at home. And, thus, in spite of the fact that tourists espouse a desire to travel in a more sustainable manner, and in spite of a whole range of surveys that confirm this belief, most do not (anon, nd).

The ability to address the growth paradox remains unresolved. Mulls (2017) highlights that tourism is expected to continue to grow faster than the global economy as countries continue to move into the realm of the developed world, incomes grow and people have the ability to travel for the first time. One of the greatest challenges facing social scientists, in general, and tourism social scientists in particular, is that they (we) can raise issues but are largely impotent in resolving them, especially those issues that require engineering or technical solutions.

Models and frameworks

A number of models and frameworks have been suggested as vehicles to manage tourism in a 'sustainable' manner. Clarke (1997) for example developed a framework of approaches to tourism on a continuum that ranged from mass tourism to sustainable tourism where sustainable tourism is defined as small scale tourism. In the intervening years the scale of mass tourism has continued to grow, often engulfing efforts to restrict some areas for small-scale tourism. A number of concepts, many borrowed from other disciplines, have been employed by scholars on the basis that they offer pathways for assisting in the implementation of strategies aimed at achieving sustainability.

Tragedy of the commons

Hardin's (1968) essay on the tragedy of the commons drew attention to the over use of the common pool of resources that may lead to the exhaustion of the resources upon which they depend. As Gordon (1954) observed, common pooled resources such as fish do not yield an economic rent and are therefore difficult to manage. While criticized for failing to consider the role of institutional arrangements in resource management, Hardin's concerns about overuse of the common pool of resources can be applied to the tourism industry's use of resources such as rivers, beaches and iconic scenic views. Where the tourism industry enjoys the benefits of the resource without contributing to its maintenance, and in effect becoming a so-called 'free rider', it can be argued that tourism can be viewed as an extractive industry (Garrod and Fyall, 1998). Hardin's challenge has generated research including contributions about economic impacts in an accommodation setting (Pintassilgo and Silva, 2007), tourism landscapes (Healy, 1994), tourism policy (Briassoulis, 2002) and resource degradation (Cushman et al., 2004).

Carrying capacity

Carrying capacity is an attractive concept but as Fennell and Cooper (2020) note, measurement depends on a range of issues including indicators, location, ethical issues and who determines the physical limits and enforces decisions. In a thoughtful and exhaustive analysis of the origins and contemporary meanings of carrying capacity, Sayre (2008) noted that the concept was originally applied to calculations of limits in engineering contexts. In the 1870s its application was widened to include living organisms and natural systems and more recently to discussions on human populations (Odum, 1953).

Drawing on concerns about the ability of the Earth to sustain an ever growing population, first articulated by Malthus (1798) in his *Essay of the Principle of Population* (initially published anonymously), ecologist William Vogt (1948) published *Road to Survival*, a popularist book that expressed his concerns about the ability of an ever increasing human population to find an adequate balance between the availability of arable land with the need for increasing food production. Both Vogt and Odum have been criticised for their failure to explore the role that technologies such as genetics can play in changing the relationships that determine carrying capacity and therefore make it difficult to model. In a recent critique Sayer (2008: 132) states, "that the concept of carrying capacity has limits does not mean that the limits it purports to specify are nonexistent or meaningless – far from it. The point, rather, is that such limits are rarely static or quantifiable, let alone predictable and controllable."

More recently, the concept has found its way into debates on sustainable tourism. Echoing an earlier observation by Swarbrooke (1999), Fennell and Cooper

(2020: 121) state that in "implementing the system as an inter-organizational instrument, the carrying capacity concept has not been particularly useful". Earlier, Lindberg, McCool and Stankey (1997) stated that while carrying capacity has an intuitive appeal, it is not able to deal with the complexity that typifies many areas of tourism. Despite such concerns, carrying capacity has gained some support, particularly in areas where there is a desire to practice sustainable tourism.

In some settings, such as built attractions, the concept can have some utility for controlling visitor flows and volumes. However, as the introduction of non-pharmaceutical interventions (NPI) such as social distancing during the COVID-19 pandemic have demonstrated, even these parameters can be subject to rapid and unforeseen modifications. In future decades, the employment of carrying capacity, even on a site-specific basis will be become increasingly more difficult as ecosystems undergo rapid change driven by a plethora of changing environmental conditions.

The precautionary principle

The precautionary principle is often referred to in issues related to environmental and health policy as a risk management strategy. But it remains an ill-defined idea. A product of the increasing interest in environmental policy in Europe that emerged in the late 1970s, it has been adopted in many international treaties and declarations. For example, the precautionary principle was embedded in the Treaty on European Union (1992) as the basis for European environmental law. However, the precautionary principle has generated considerable controversy particularly over the interpretation of 'precautionary'. As Foster, Vecchia and Repacholi (2000) note, precautionary has been used as the basis by the EU to ban agricultural products that have been genetically engineered, potentially reengineering the principle as a protectionist tool. Sunstein (2003: 1003) suggests that "taken in its strongest form the precautionary principle should be rejected, not because it leads to bad directions, but because it leads in no direction at all. The principle is literally paralyzing – forbidding inaction, stringent regulation and everything in between". From this perspective, every step including inaction can be seen as creating risk.

In a review of the precautionary principle Sandin (1999: 898) suggested that it can be expressed as "if there is (1) a threat, which is (2) uncertain, then (3) some kind of action (4) is mandatory". This approach recognizes four dimensions of the precautionary principle:

- The threat dimension
- The uncertainty dimension
- The action dimension
- The command dimension

In a sustainable tourism context where timeframes are generally regarded as being long-term, Fennell and Cooper (2020) state that the precautionary principle is a useful mechanism for encouraging stakeholders, including operators and communities to exercise precaution in actions that may have short and long-term implications. This may take the form of audits that measure long-term affects of policies and actions, and that assist in the maintenance of appropriate standards of practice.

Indices

The development of sustainability indicators was an attempt to place measurable thresholds on levels of tourism activity. Butler (1999) reminds us that indicators are critical in the move toward sustainability, for without indicators, the term 'sustainable' becomes little more than a meaningless hyperbole. However, a number of scholars have suggested we are overwhelmed by indicators, rather than having too few (Marzo-Navarro, Pedraja-Iglesias and Vinzon, 2015). Part of the challenge is that tourism touches on so many aspects of society that Jovicic (2014) suggests an infinite number of indicators could be developed. The extent of the challenge is highlighted by the UNWTO's (2004) 500 plus page publication '*Indicators of Sustainable Development for Tourism Destinations: A Guidebook*'.

Our efforts to be inclusive have created a classic situation of choice overload where the number of alternatives or choice options is greater than the person's ability to make effective and efficient decision. As Agyeiwaah et al. (2017: 27) summarising the literature note, choice overload depends on the complexity of the choices available, how well the options align with each other and the presence or lack of a clearly preferred option. Preparing vast lists may be an impressive intellectual exercise, but is largely ineffective with the net result being a tendency to avoid acting (Park and Jang, 2013) or to pick and choose options that best suit one's narrow personal interests (Miller and Twining-Ward, 2005).

Life Cycle Assessment (LCA)

Life Cycle Assessment (LCA) is a technique that is used to evaluate the environmental impacts of services and products using a through-life approach (Ilgin and Gupta, 2010). Unlike other measures of impact, LCA seeks to assess the environmental impacts that occur cumulatively through the lifecycle of a product, service or process. Life cycle includes initial raw material extraction, processing, transport, manufacture, use, and disposal including recycling or dumping. A typical LCA study include developing an inventory of all the materials and energy required in the value chain of the product, service or process followed by calculation of cumulative emissions.

The International Organisation for Standardisation (ISO) has approved LCA which is now used in assessments in the ISO 14040 series of standards. A number of scholars (De-Camillis, Raggi and Petti, 2010; Filimonau et al., 2014) have demonstrated how LCA can be applied in a tourism setting. However, as Filimonau, Dickinson and Robbins (2014) caution, there remain a number of problems with LCA that may reduce its accuracy. Problems include lack of regular updates of data bases that LCA analysis draws on and limitations of the methodologies used to collect data.

■ Ecological footprint

One measurement technique that has attracted the attention of scholars working in the sustainable tourism area is the ecological footprint. First suggested by Rees (1992), the ecological footprint is an accounting approach for estimating the quantity of resources used on a range of scales and industries including tourism. On a global scale, it can be used to estimate the amount of resources used by the global society and then compared to the ability of the Earth to renew the resources used. In 2014, the Global Footprint Network estimated that humanity was using resources 1.7 times faster than they can be renewed (Lin et al., 2018) with rich countries consuming a far higher proportion of available resources than poor countries. The inequality of resource use between rich and poor countries is also reflected in tourism use, particularly in destinations located in less developed countries. The problems highlighted by the ecological footprint concept are likely to be magnified in the future unless the overconsumption that is a characteristic of many economies is reined in. Given that part of this 'overconsumption' is generated by tourism demand, future strategies to reduce the extent of the global ecological footprint may require rethinking the structure, availability and distribution of tourism activity.

Hunter and Shaw (2005) argued that ecological footprint analysis should be widely used as an environmental indicator. Despite significant data measurement problems that have been encountered, the successful development of a global measure of this type will allow for cross country and cross sector comparisons and change management. Building on the observations of Hunter and Shaw (2005) and acknowledging that it does not capture all environmental impacts, Castellani and Sala (2012) applied and compared both ecological footprint and Life Cycle Assessment (LCA) in a tourism destination case study in northern Italy. They found that it was a useful screening tool for resource use on a tourism system wide scale, but was not suitable for the more detailed assessments possible with LCA. In the future, continued refinement of the data measurement is likely to lead to the ecological footprint being used to measure impacts on a tourism system scale, while LCA is likely to offer a more detailed and comprehensive measurement at activity scale.

Limits of acceptable change

First introduced as a strategy to overcome problems encountered in applying the 'carrying capacity' approach in wilderness areas, the limits of acceptable change framework has been used adapted for use in nature-based tourism situations (McCool, 2013). It is described by McCool (2013: 286) as a framework that can be "embedded within planning processes of protected area decision making, or can be used as a framework to assist decisions that are specific to visitor and tourism management". As McCool (2013) states, limits of acceptable change should not be seen as a method for calculating carrying capacity but rather a process that assists natural area managers resolve problems that arise when attempting to mediate between overlapping goals such as preserving natural heritage values while enabling access to protected area.

Conclusion

There are a large number of models and frameworks that can be used to assess environmental impacts. In most instances, no one model, concept or framework is likely to successfully address the multitude of issues that can arise when managers and decision makers need information to solve problems related to use of the environment for tourist purposes. There are usually opportunities to use multiple tools as demonstrated by Castellani and Sala (2012) above when they used the ecological footprint and life cycle assessment in a case study in northern Italy.

The challenge remains of how to move from the current less sustainable practice of tourism to a more sustainable future. In many ways, sustainability is an unattainable goal. Instead, we have to consider how to move towards being more sustainable or less unsustainable. The challenge will become even greater as emerging economies gain the financial ability to travel, leading to greater demand than ever before.

Discussion points / Exercises

1. Adherents of hard sustainability argue that some activities are irreversible, such as species extinction, and therefore we must adopt this approach. Proponents of soft sustainability argue that we cannot upset the global economic system without creating huge social problems. Which point of view do you endorse and why?
2. Is tourism a smokestack free industry as early proponents of sustainability argued?
3. Which model, limits to acceptable change, or carrying capacity is most appropriate for tourism?

4. We are overwhelmed with sustainability indictors. Develop a list of five core indicators that you feel are both reasonable to achieve and essential for sustainable tourism.
5. Changing tourists' behaviour is the greatest challenge in developing a more sustainable tourism sector. Identify why most tourists seem unwilling to change and what actions may be adopted to encourage change?

Points to consider for revision

- Sustainability is a contested issue.
- Sustainable tourism is an ideal goal for the sector which corresponds closely with many of the UNDP sustainable goals.
- A number of concepts and models have been developed to encourage, monitor and move toward a more sustainable tourism future.

Suggested additional readings

Agyeiwaah, E., McKercher. B. & Suntikul, W. (2017) Proposing a core set of achievable sustainable tourism indicators: A meta-analysis approach. *Tourism Management Perspectives*, 24, 26-33.

Driml, S. & Common, M. (1996) Ecological economics criteria for sustainable tourism: Application to the Great Barrier Reef and Wet Tropics World Heritage Areas, Australia, *Journal of Sustainable Tourism*, 4(1), 3-16.

Fennel, D. & Cooper, C. (2020) *Sustainable Tourism: Principles, Contexts and Practices.* Channel View Publications.

References

Anable, J., Lane, B. & T. Kelay (2006). *An Evidence Base Review of Public Attitudes to Climate Change and Transport Behaviour.* London: Department for Transport. www.fcrn.org.uk/sites/default/files/Evidence_of_public_attitudes_and_behaviour.pdf <downloaded Dec 9, 2011>.

Bac, D. (2012) The emergence of sustainable tourism: A literature review. *QUAESTUS Multidisciplinary Research Journal*, 131-140. http://www.quaestus.ro/wp-content/uploads/2012/03/bac4.pdf <downloaded Oct 17, 2017>.

Becken, S. (2004). How tourists and tourism experts perceive climate change and carbon-offsetting schemes. *Journal of Sustainable Tourism*, 12(4), 332-345

Biely, K., Maes, D. & Van Passel, S. (2018) The idea of weak sustainability is illegitimate *Environmental Development and Sustainability*, 20(1), 223-232.

Bramwell, B., Higham, J., Lane, B. & Miller, G. (2017) Twenty-five years of sustainable tourism and the Journal of Sustainable Tourism: looking back and moving forward." *Journal of Sustainable Tourism*, 25(1), 1-9.

Briassoulis, H. (2002) Sustainable tourism and question of the commons, *Annals of Tourism Research*, 29(4), 596-611.

Buckley, R.C. (2012) Sustainability Reporting and Certification in Tourism, *Tourism Recreation Research*, 37(1), 85.

Budeaunu, A., Miller, G., Moscardo, G. & Ooi, C. (2016) Sustainable tourism, progress, challenges and opportunities: An introduction. *Journal of Cleaner Production*, 111, 285-294.

Butler, R. W. (1999) Sustainable tourism: A state-of-the-art review. *Tourism Geographies: An International Journal of Tourism Space, Place and Environment*, 1(1), 7-25.

Carillo, M. & Jorge, J. (2017) Multidimensional Analysis of Regional Tourism Sustainability in Spain. *Ecological Economics*, 140. 89-98.

Castellani, V. & Sala, S. (2012) Ecological footprint and life cycle assessment in the sustainability assessment of tourism activities, *Ecological Indicators*, 16, 135-147.

Clarke, J. (1997) A framework of approaches to sustainable tourism, *Journal of Sustainable Tourism*, 5(3), 224-233.

Cushman, C., Field, B., Lass, D. & Stevens, T. (2004) External costs from increased islands visitations: results from the Southern Thai Islands', *Tourism Economics*, 10(2), 207-219.

De-Camillis, C., Raggi, A. & Petti, L. (2010) Tourism LCA: State-of-the art and perspectives, *International Journal of Life Cycle Assessment*, 15(2), 148-155.

Elkington, J. (1994) Towards the sustainable corporation: Win-win-win business strategies for sustainable development. *California Management Review*, 36(2), 90-100.

Ecologically Sustainable Development Working Groups (ESDWG) (1991) *Ecologically Sustainable Development- Final Report - Tourism.* Canberra: Commonwealth of Australia

Fennell, D.A. & Cooper, C. (2020) Sustainable Tourism, Channel View

Filimonau, V., Dickinson, J. & Robbins, D. (2014) The carbon impact of short-haul tourism: a case study of UK travel to Southern France using life cycle analysis, *Journal of Cleaner Production*, 64, 628-638.

Foster, K., Vecchia, P. & Repacholi, M. (2000) Science and the precautionary principle. *Science*, 28 (5468), 979-981.

Garrod, B. & Fyall, A. (1998) Beyond the rhetoric of sustainable tourism. *Tourism Management*, 19(3), 199-212.

Gordon, H. S. (1954). The economic theory of a common-property resource: The fishery, *Journal of Political Economy*, 62(2), 124-142. Doi: 10.2307/1825571.

Hardin, G. (1968) The tragedy of the commons, *Science*, 162, 1243-1248.

Healy, R. (1994) The "common pool" problem in tourism landscapes, *Annals of Tourism Research*, 21(3), 596-611.

Higgins-Desbiolles, F. (2017) 'Sustainable tourism' is not working – here's how we can change that. The Conversation. https://theconversation.com/sustainable-tourism-is-not-working-heres-how-we-can-change-that-76018 <downloaded Feb 10, 2020>

Hunter, C. & Shaw, J. (2007)The ecological footprint as a key indicator of sustainable tourism, *Tourism Management*, 28, 46-57.

Ilgin, A. & Gupta, S. (2010) Environmentally conscious manufacturing and product recovery: A review of the state of the art, *Journal of Environmental Management*, 91(3), 563-591

Islam, M., Yamaguchi, R., Sugiawan, Y. & Managi, S. (2019) Valuing natural capital and ecosystem services: A literature review. *Sustainability Science*, 14, 159-174.

Kuhlman, T. & Farrington, J. (2010) What is sustainability? *Sustainability*, 2(11), 3436-3448; https://doi.org/10.3390/su2113436 <downloaded May 15, 2020>.

Lin, D., Hanscom, L., Murthy, A., Galli, A., Evans, M., et al. (2018) Ecological footprint accounting for countries: Updates and results of the national footprint accounts, 2012–2018. *Resources*, 7, 58.

Lindberg, K., McCool, S. & Stankey, G. (1997) Rethinking carrying capacity, *Annals of Tourism Research*, 24(2), 461- 465.

Liu, Z. (2003) Sustainable tourism development: A critique. *Journal of Sustainable Tourism* 11(6), 459-475.

Malthus, T.R. (1798) *Essay on the Principle of Population As It Affects the Future Improvement of Society, with Remarks on the Speculations of Mr. Goodwin, M. Condorcet and Other Writers . London: J. Johnson in St Paul's Church-yard.* Retrieved 20 June 2015. via Internet Archive

Marzo-Navarro, M., Pedraja-Iglesias, M. &Vinzón, L. (2015) Sustainability indicators of rural tourism from the perspective of the residents, *Tourism Geographies*, 17(4), 586-602, 10.1080/14616688.2015.1062909.

McCool, S. (2013) Limits of acceptable change in tourism. In Holden, A. & Fennel, D. (eds), *Routledge Handbook of Tourism and the Environment*, Routledge, Oxon, pp 285-298.

McCool, S. (2016) The changing meanings of sustainable tourism. In McCool S. & Bosak K. (eds) *Reframing Sustainable Tourism. Environmental Challenges and Solutions*, Vol. 2, Springer, Dordrecht, pp 13-32.

McKercher, B. (1993a) Some fundamental truths about tourism: Understanding tourism's social and environmental impacts *Journal of Sustainable Tourism*, 1(1), 6-16.

McKercher, B. (1993b) The unrecognised threat to tourism: Can tourism survive sustainability *Tourism Management*, 14(2), 131-136.

McKercher, B. & B. Prideaux (2014) Academic myths of tourism. *Annals of Tourism Research.* 46, 16-28.

Miller, G. & Twining-Ward, L. (2005). *Monitoring for sustainable tourism transition: The challenge of developing and using indicators.* Cambridge, MA: CABI.

Moscardo G. (2015) Tourism and sustainability: Challenges, conflict and core knowledge. In: Moscardo G. and Benckendorff P. (eds) *Education for Sustainability in Tourism. CSR, Sustainability, Ethics & Governance*. Springer, Berlin, Heidelberg. https://link.springer.com/chapter/10.1007/978-3-662-47470-9_2 <downloaded Oct 17, 2019>.

Muangasame, K. & McKercher, B. (2015) The challenge of implementing sustainable tourism policy: A 360-degree assessment of Thailand's "7 greens sustainable tourism policy" *Journal of Sustainable Tourism*, 23(4), 497-516, 10.1080/09669582.2014.978789.

Obst, C. (2016) *Framing Sustainable Tourism.* Madrid: UNWTO http://cf.cdn.unwto.org/sites/all/files/pdf/mst_issue_paper_1.pdf <downloaded Oct 17, 2019>.

Odum, E.P. (1953) *Fundamentals of Ecology*, Philadelphia: Saunders.

Park, J. & Jang, S. (2013) Confused by too many choices? Choice overload in tourism. *Tourism Management*, 35, 1-12.

Pearce, A. Markandya & E. B. Barbier (1989) *Blueprint for a Green Economy*. London: Earthscan Publications.

Pelanc, J. & Ballet, J. (2015) Strong sustainability, critical natural capital and the capability approach. *Ecological Economics*, 112, 36-44.

Pelanc, J. & Dedeurwaerdere, T. (2015) *Weak Sustainability versus Strong Sustainability*. UN Brief for GSDR 2015. https://sustainabledevelopment.un.org/content/documents/6569122-Pelenc-Weak%20Sustainability%20versus%20Strong%20Sustainability.pdf <downloaded Oct 17, 2019>.

Pigram J.J. (1990) Sustainable tourism – policy consideration. *Journal of Tourism Studies*, 1(2), 2-9.

Pintassilgo, P. & Silva, J. (2007) 'Tragedy of the commons' in the tourism accommodation industry, *Tourism Economics*, 13(2), 209-224

Ruhanen, L., Weiler, B., Moyle, B. & McLennan, C. (2015) Trends and patterns in sustainable tourism research: a 25-year bibliometric analysis, *Journal of Sustainable Tourism*, 23(4), 517-535, DOI: 10.1080/09669582.2014.978790

Rees, W. E. (1992). Ecological footprints and appropriated carrying capacity: What urban economics leaves out. *Environment & Urbanization*, 4(2), 121–130. doi:10.1177/095624789200400212

Salas-Zapata, W. & Ortiz-Munoz, S. (2018) Analysis of meanings of the concept of sustainability. *Sustainable Development*, 27, 153-161.

Sandin, P. (1999) Dimensions of the precautionary principle, *Human and Ecological Risk Assessment: An International Journal*, 5(5), 889-907, DOI: 10.1080/10807039991289185

Sardianou, E., Kostakis, I., Mitoula, R., Gkaragkani, V., Lalioti, E. & Theodoropoulou, E. (2016) Understanding the entrepreneurs' behavioural intentions towards sustainable tourism: A case study from Greece. *Environment, Development and Sustainability*, 18, 857–879 DOI 10.1007/s10668-015-9681-7.

Sayre, N. (2008) The genesis, history, and limits of carrying capacity, *Annals of the Association of American Geographers*, 98(1), 120-134, DOI: 10.1080/00045600701734356

Sunstein, C. (2003) Beyond the precautionary principle, *University Pennsylvania Law Review*, 151(3), 1003-1-58.

Swarbrooke, J. (1999). *Sustainable Tourism Management*. Oxon: CAB International.

Torres-Delgado, A. & Lopez-Palomeque, F. (2012) The growth and spread of the concept of sustainable tourism: The contribution of institutional initiatives to tourism policy. *Tourism Management Perspectives*, 4, 1-10.

UNWTO (2004) *Indicators of Sustainable Development for Tourism Destinations: A guide book*. United Nations World Tourism Organization. Madrid, Spain

UNWTO (2013) *Sustainable Tourism for Development* UN World Tourism Organization: Madrid. https://www.e-unwto.org/doi/pdf/10.18111/9789284415496 <downloaded Oct 24, 2019>.

Vogt, W. (1948) *Road to Survival*, New York: William Sloane.

WCED (1987) Our *Common Future*. World Commission on Environment and Development. United Nations. https://sustainabledevelopment.un.org/content/documents/5987our-common-future.pdf <downloaded Oct 22, 2019>.

15 Climate Change, Resilience and Transition to a Carbon-Neutral Economy

By the end of this chapter, the reader will be able to:

- Describe the role the Intergovernmental Panel on Climate Change plays in discussing climate change
- Discuss whether tourism is a victim, winner or loser in climate change
- Analyse tourism's contribution to climate change
- Discuss and analyse the concept of resilience as it relates to tourism
- Identify various models for tourism and climate change and/or resilience

Introduction

The tourism industry is in the unique and unenviable position of being simultaneously a major contributor of greenhouse gases (GHG) and a significant victim of changing climate. Historically, research on the relationship between tourism and climate lagged significantly behind that of other studies on climate change (Pang, McKercher and Prideaux, 2012) and to some extent remains on the periphery of mainstream tourism research when measured by the volume of published research, the number of active tourism researchers and the number of climate change subjects taught in tertiary level tourism programs. Much of the research has focused on social science aspects of climate change, without necessarily delving into deeper hard science based models. This chapter reviews a range of models and frameworks that are used in climate studies in a tourism context, followed by models and frameworks that relate to resilience as a response to climate change. It concludes with a review of three theories and associated models of the future transition to a carbon neutral economy.

Climate change

Research in the sciences has produced an almost overwhelming amount of data that has been used to build models that show how the drivers of climate change will affect global temperatures, precipitation, wind and sea level rise. These models can be used to investigate change at a system level, with the most widely used being those that appear in the Intergovernmental Panel on Climate Change (IPPC) reports. These models will become increasingly important in coming decades as the tourist industry begins to grapple with the impact of climate change at enterprise, destination and global levels.

■ IPCC modelling

The IPCC process presents a different perspective on how models are able to be used to investigate tourism related issues. The key to the IPCC process is the review and synthesis of a large range of publications (both peer reviewed and from reputable grey literature sources) and the development of a range of climate models based on scenarios of future increases in atmospheric concentrations of CO_2. The IPPC was established in 1988 as a joint initiative of the World Meteorological Organization (WMO) and the United Nations Environment Program (UNEP) with the objective of providing policy makers with regular scientific assessments of the implications of climate change. Reports include identification of risks and suggestions for adaptation and mitigation options. As of 2020, the IPCC had 195-member countries. The First Assessment Report (FAR) was published in 1990 (IPCC, 1990) and updated in 1992. FAR and subsequent assessment reports quickly became authoritative sources for model-based predictions used by a large number of researchers in the sciences and social sciences, governments and the private sector. Apart from regular assessment reports, the IPCC also produces special reports on issues that are of concern to member countries and methodology reports that outline guidelines for developing greenhouse gas inventories. Examples of recent special reports that may be of interest to tourism researchers include: a Special Report on Emissions Scenarios (SRES) (Nakićenović et al., 2000); a report on Global Warming at 1.5C (SR15) (IPCC, 2018); a Special Report on Climate Change and Land (SRCCL) (IPCC, 2019a); and a Special Report on the Ocean and Cryosphere in a Changing Climate (SROCC) (IPCC, 2019b). The sixth Assessment report is due to be published in 2022.

The methodology used in compiling IPCC reports is based on the synthesis of peer reviewed publications as well as appropriately documented industry publications by a panel of acknowledged academic experts (IPCC, nd). It should be noted that the IPCC process is one based on review, not the commissioning of new research. Once a draft report is completed, it undergoes a two-stage review process. In the first review, the draft is circulated to acknowledged experts for comment. The number of reviewers may be several hundred (IPCC, nd). After revision, the second draft is given to governments, authors and expert reviewers for further comments. The final draft must be approved by the IPCC Panel. The involvement of governments in determining the final version of Assessment Reports has been questioned by a number of scholars. Trenberth (2001: 11) for example, noted that while the rationale for this process is based on the view that "scientists determine what can be said, but the governments determine how it can best be said", there is scope for governments to blunt report findings. In general, the negotiations that occur over wording ensures accuracy, balance, clarity of message, and relevance to the scientific community and policy makers. While the inclusion of governments in the review process has been criticised as having the potential to distort findings, there has been a high level of acceptance of the reports by both the academic and government communities.

The FAR (IPCC, 1990) provided researchers with the first comprehensive assessment of the likely impact of climate change. Highlights of the report included:

- Confirmation that greenhouse gases (GHGs) were enhancing the greenhouse effect causing the Earth's surface to warm;
- Under a business as normal approach the global mean temperature will rise about 0.3°C per decade with an uncertainty rage of 0.2° to 0.5°C per decade;
- The rate of change contained in IPCC predictions was uncertain due to the magnitude, timing and regional patterns of climate change and incomplete understanding of a range of drivers including the melt rate of polar ice caps, sources and sinks of GHGs, absorption level of the oceans and impacts of cloud cover;
- Publication of climate scenarios based on modelling that make extensive use of data to predict future outcomes.

The most recent Assessment Report (AR5) (IPCC, 2014) stated that without world-wide adoption of mitigating policies, the global mean temperature will rise to between 3.7°C to 4.8°C above pre-industrial levels by 2100. AR5 was written by 831 authors and has become an important reference point for climate change related research in many disciplines including tourism. AR5 was a key document in negotiations to reduce carbon emissions at the 2015 Paris UN Climate Change Conference.

Climate change and tourism

As Scott, Hall and Gössling (2012) noted, initial interest in the impact of climate on tourism activity can be traced back to the 1960s, followed by a surge of interest in the mid to late 1970s. There was a further resurgence of interest in the late 1980s. This work predated the first IPCC report. The first decade of the 20th century saw an upturn of interest in climate change by tourism scholars (Scott, Simpson and Sim, 2012) which reflected increasing collaboration between the tourism and climate communities as well as an increase in specialist climate change and tourism conferences, and climate related reporting by the United Nations World Tourism Organisation (UNWTO). The release of three IPCC assessment reports by 2001 also contributed to this growth in interest.

Early research recognised that impacts would be uneven, potentially benefiting some destinations while harming others. This 'Victim, Winner, Loser' theme has continued to dominate much of the research agenda to the present. For example, under the victim theme low-lying touristic islands may become inundated (Storlazzi, Elias and Berkowitz, 2015); coral reef dependent destinations will also come under pressure as coral bleaching reduces their appeal (Hoegh-Guldberg, 1999); golfing destinations that rely on plentiful water supplies may be affected by a decline in rainfall (Simpson et al., 2008); the ski industry is particularly at risk (Abegg et al., 2007); and the list goes on. But, there may be some winners too. Rebetez (2011) found a correlation between the number of nights spent in mountain resorts in the summer and hot temperatures at lower elevations; seasonality may shift and extend the seasons into traditional shoulder seasons (Amelung, Nichols and Viner, 2007); climate change will result in a redistribution of tourists from more to less climate-sensitive destinations (Scott, McBoyle and Schwartzentruber, 2004); and domestic travel may dominate, especially if carbon taxes make long haul air travel too expensive (Hamilton and Tol, 2007; Taylor and Ortiz, 2009), helping to revitalise many destinations.

Tourism in general, and aviation in particular, are recognised as major contributors of carbon emissions. The UNWTO (Simpson et al., 2008; UNWTO, 2008) estimates that tourism contributed about 5% of total global CO_2 emissions and up to 14% of all emissions when other greenhouse gases are considered. The aviation sector alone accounts for up to 40% of total emissions. Moreover, airborne emissions may be between 1.9 and 5.1 times more harmful than surface-based emissions (Gössling and Peeters, 2007). The European Environmental Agency (EEA, 2011), for example, indicates that air travel contributes an average of 130g of CO_2 per passenger kilometer, while the typical passenger car emits 124.5 g/pkm, rail emits 46 g/pkm and coaches 34 g/pkm. Different types of accommodation also have different carbon footprints. Full service hotels have the largest footprints, emitting an estimated 20.6 kg of CO_2 per guest night (UNWTO-ENEP, 2008).

Many types of tourism will be victims of climate change, while some other forms may benefit. Low altitude ski fields as well as those in lower latitudes are likely to see less snowfall and shorter ski seasons, placing many ski resorts and the ancillary goods and services they use at risk, ranging from equipment rental through to accommodation and food and beverage outlets. Likewise, rising sea levels threaten many beach resort areas. In some cases, impacts like beach erosion may be mitigated through engineering. In many other instances, though, entire resort areas are at risk of being inundated by rising sea levels. Winners may include nearby domestic tourism destinations, especially if carbon taxes make air travel prohibitively expensive. In other instances, effective seasons may be lengthened, creating opportunities to appeal to new markets.

It has become increasingly apparent that tourism researchers need to engage in more than simple descriptive works, and instead need to be conversant with the relevant scientific literature, including IPCC reports. A number of models have been developed to demonstrate the links between science and social science research, and they show how each side of the research community needs to build bridges to facilitate a deeper understanding of the feedback loops that are generated by human impacts on the environment and subsequent environmental impacts on humans. One example is the climate change model developed by Prideaux, Coghlan and McNamara (2010) and modified by Ramis and Prideaux (2013) to investigate how coral bleaching may affect tourism demand. Using data derived from a survey of visitors to Australia's Great Barrier Reef (GBR) and observations in AR4 that at temperatures of 2C and above coral bleaching is anticipated to become an annual event, Ramis and Prideaux (2013) found that tourists respond negatively to a decline in the quality of their GBR experience. The modified model outlined in Figure 15.1 includes a feedback loop that recognises the need for the scientific community to incorporate human impact into their modelling. Feedback loops are a key element of recent IPCC reporting.

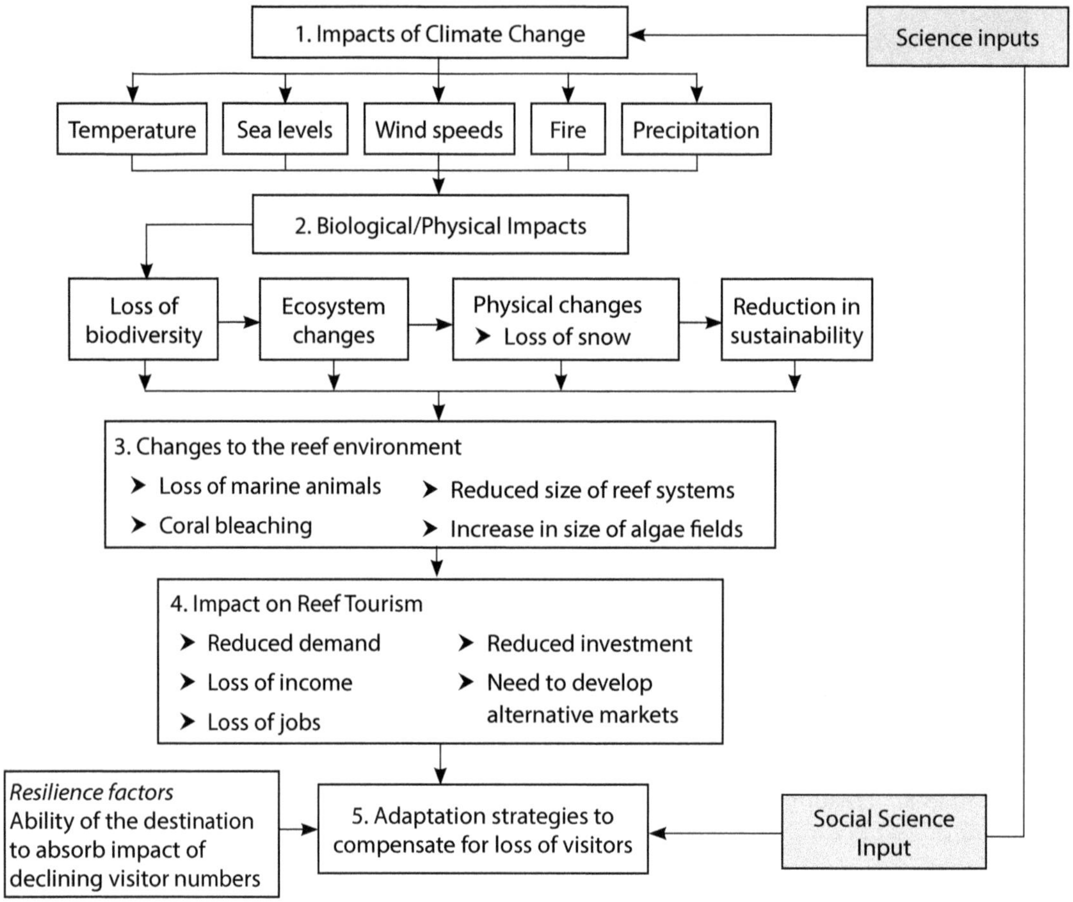

Figure 15.1: Climate change impact model applied to the Great Barrier Reef. Adapted from Ramis and Prideaux, 2013.

In a chapter that examined climate change risk on coastal island destinations and building on observations in AR5, Scott and Verkoeyen (2018) developed a climate change impact pathway that highlighted the impact of climate change on coastal tourism. While focusing on climate change, the model (Figure 15.2) also acknowledged the role of other factors (nine in all) on coastal tourism including technological change, demographic change and political instability. The importance of this model is that it highlights that climate change is one of many factors that can affect tourism flows in a destination setting.

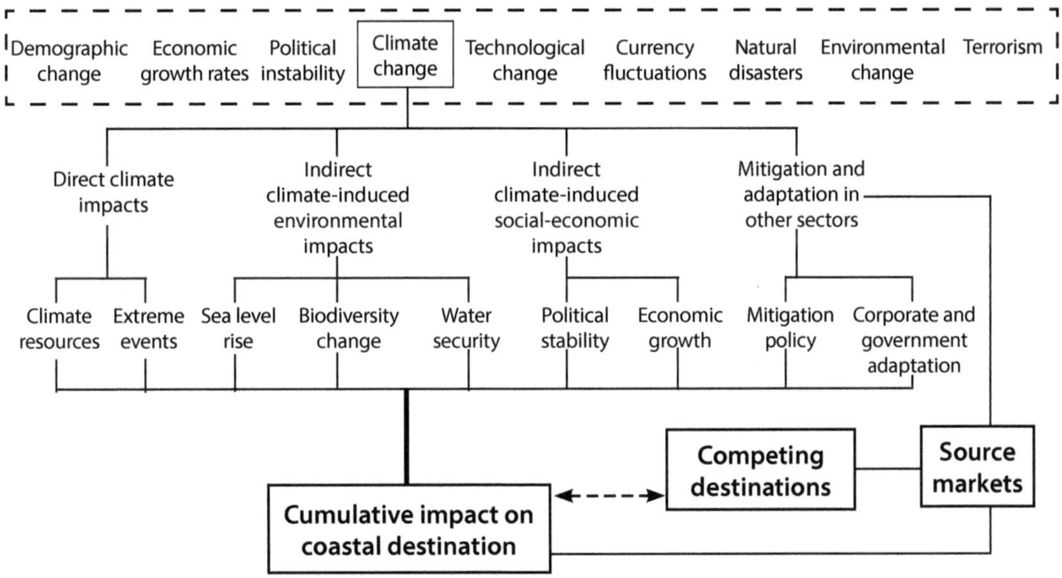

Figure 15.2: Climate change impact pathways on coastal tourism. Source: Scott and Verkoeyen, 2018.

On a destination scale the usefulness of indices such as the Tourism Climate Index (TCI) and Beach Comfort Index (BCI) allied with climate models of the type employed by the IPCC on a global scale, the United Kingdom Climate Projections models (UK Met Office, nd), and regional ecosystem models of the type produced by Williams, Bolitho and Fox (2003) can provide destinations with information that can used for future planning using scenarios. As the rate of global warming accelerates in the next few years (assuming that major emitters fail to agree on measures to curb GHG emissions), the use of the models similar to those outlined above will become critical in assisting the tourism industry respond to the changes that will occur.

Resilience

It is apparent that climate change will create significant disruption in the tourism industry in future decades. One of the key questions emerging from debates on climate change and the transition to a carbon-neutral production system is how will the tourism system cope? The ability to absorb and adapt to change to ensure ongoing economic and system functioning is an outcome of the level of social and economic resilience. Cochrane (2010: 173) described resilience as follows:

15

> "systems do not evolve in a linear fashion but according to a cycle, or loop; that the phases of the cycle roughly repeat, but the characteristics of each stage at different iterations are not necessarily identical; that the characteristics and the speed of recovery from a destabilizing event

> depend on the system's adaptive capacity; and that adaptive capacity depends on various forms of capital accumulated during previous phases and iterations."

From a slightly different perspective, Haimes (2009: 498) defined resilience as "the ability of the system to withstand a major disruption within acceptable degradation parameters and to recover within an acceptable time and composite costs and risks."

At destination level, resilience is a measure of the ability of the community and industries including tourism, to respond to, withstand and recover from adverse events such as climate change, the COVID-19 pandemic, earthquakes and tsunamis. If resilience is low there may be a need to change the structure of a destination's suite of tourism experiences or even move out of tourism altogether (Lew, 2014). Figure 15.3 illustrates one example of a tourism resilience model based on the Regional Tourism Sustainable Adaptation Framework proposed by Njoroge (2014), modified by Prideaux, Carmody and Pabel (2018) to include community values and further modified by the authors of this chapter to include community vision and alternative industries if tourism becomes unviable. The model proposes a four-stage approach to determining resilience where stage one involves assessment of a destination's tourism industry followed by risk identification, adaptive capacity and finally adaptation processes. If all or parts of a destination's tourism industry become unviable, the destination may need to look for alternative industries.

Transition to a carbon-neutral economy

The climate emergency currently confronting humanity is the result of a carbon-based linear production system that relies on cheap non-renewable energy and unsustainable use of resources. The greenhouse gas emissions of this production system are the main driver of global warming. The need to reduce and eventually eliminate greenhouse gas emissions through transitioning to a carbon-neutral production lies at the heart of solving the problems caused by climate change. The form that transition will take is unclear but will need to occur. As Mathews (2011: 869) argues, the current industrial capitalism system has outgrown the planet "destroying not only its own resource base but also the entire biosphere". He further argues that the next transition will, if it succeeds "enable global capitalism to co-exist with its natural biospheric limits; and if it fails, it will in all probability mean the end of our industrial civilization."

The tourism literature has yet to consider the implications of a transition of this type. This is a serious shortcoming given the disruptive nature of past economic transitions and how a future transition may impact on tourism. While there has been little discussion in the tourism literature on the implications of transition

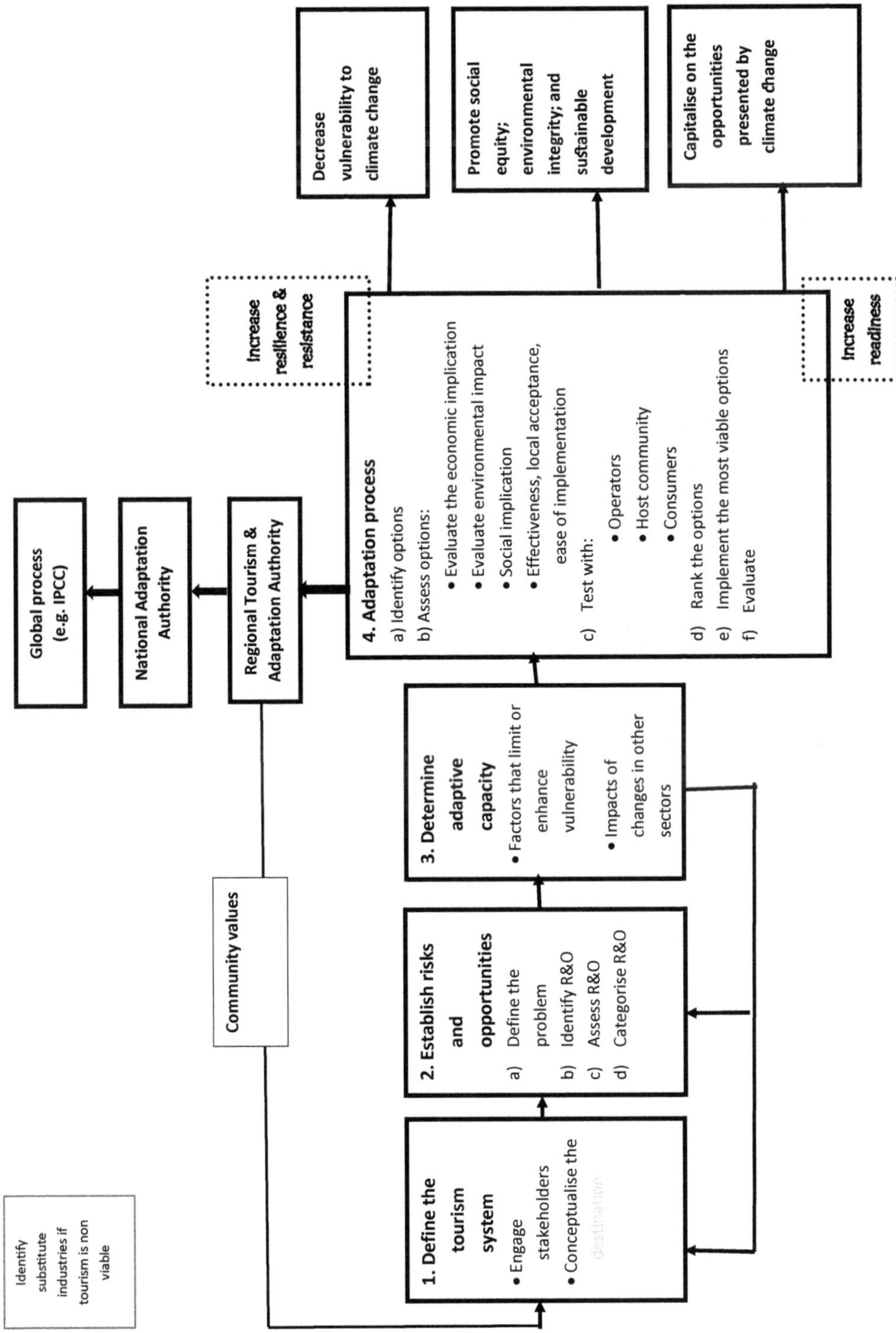

Figure 15.3: Modified version of the enhanced regional tourism sustainability framework, after Njoroge (2014) and Prideaux, et al., 2018.

15

there has been some discussion on models and theories that may be employed to achieve transition.

Transition to carbon neutrality will require radical changes to the linear economic production that underpins the current neoliberal economic model. Transition from the linear production system based on a 'take, make, use, dispose' approach to resource use (Gregson et al., 2015) will require a switch on a global scale to a new economy that will emphasise recycling, renewable energy and regenerative farming. The idea of global scale economic transition is not new. In the United Kingdom, the Second Agricultural Revolution (the first being the period when humans transitioned from hunting and gathering to cropping and pastoralism from 10,000 BC) transformed agricultural production (Thomas, 2005) and paved the way for the Industrial Revolution which lead to the introduction of new transport technologies (railways, steam powered ships, then oil-powered automobile and planes) and the emergence of the middle class. The global tourism system characterised by mass tourism was a later outcome. To date there has been a dearth of research that considers how the transition to carbon-neutrality will affect tourism. While the need to move to a carbon-neutral economic production system has been accepted by a number of multi-lateral organisations and countries, the timeframes for transition are measured in decades and may not occur before the 1.5°C threshold advocated by the IPCC (2018) is breached. This chapter briefly considers three theories for achieving the transition to a carbon-neutral production system. Unlike many of the models and theories previously considered in this book, the following models are economy-wide models that can be adapted to a tourism specific setting.

The circular economy model

While there are many schools of thought on the circular economy (CE), the underlying principle is better management of global resources (Pomponi and Moncaster, 2017). In broad terms the circular economy is based on an approach to economic production where the 'take' element of resource use is replaced by 'reuse' of resources through recycling, reuse, sharing and remanufacturing to create a closed-loop that minimises the use of new resources and reduces or eliminates pollution, waste and emission of carbon (Geissdoerfer et al., 2017). This approach recognises that the current system of resource use is a major generator of GHGs and responsible for unsustainable use of both renewable and non-renewable resources including land, mineral, agricultural land and ecosystems. A succinct definition of the circular economy is "an industrial economy that is restorative or regenerative by intention and design" (Ellen MacArthur Foundation, 2013:14). The Ellen MacArthur Foundation has emerged as a leading advocate for the circular economy approach and has been at the forefront in theory development. A more comprehensive definition describes the circular economy as a "regenerative system in which resource input and waste, emission, and energy leakage are minimised by slowing, closing, and narrowing material and energy loops. This can be achieved through long-lasting design, maintenance, repair, reuse, remanufacturing, refurbishing, and recycling" (Geissdoerfer et al., 2017: 759). To transition to an economic system of this type will require industrial processes and economic systems to move from an open loop system characterised by 'dispose' to a closed loop where resources are remade or recycled. It will also require extensive interventions by the public sector to ensure waste, reuse and energy reduction targets are achieved.

The Ellen MacArthur Foundation suggests the any future global production system should integrate agriculture with other forms of production using extensive recycling and powered by renewable energy. The impact of the CE model on tourism is not clear, a result of the lack of research into this potential model for economic organisation. If adoption of a CE model results in greater consumer surplus or reduced hours of work there is a strong possibility that tourism will be a beneficiary.

The need to shift to a carbon-neutral economy has been accepted by a growing number of European Union (EU) member states and adopted as a policy objective based on the European Circular Economy Package (European Commission, 2015). China has also adopted a policy of moving towards, efforts to encourage a shift towards a CE approach. These policies are outlined in the Circular Economy Promotion Law (Lieder and Rashis, 2016). In 2018 a coalition of over 40 firms, foundations, the World Economic Forum and the United Nations Environment Program (UNEP) launched the Platform for Accelerating the Circular Economy (PACE) as a public private partnership to advance the adoption of circular economy. PACE has three areas of specific activity: developing models for circular economy projects in developing and emerging countries; developing policy frameworks to overcome barriers to implementing the principles of the circuit economy; and promoting private-public partnerships to achieve these measures. The British Standards Institute released the first circular economy standards (BS8001:2017) in 2017 establishing a list of definitions and management frameworks for implementing the circular economy. At the enterprise level, circular business models have been developed to assist firms narrow (efficiency improvements), close (recycling) and/or dematerialise (use of software solutions) production cycles by adopting circular supply chains (Bocken et al., 2016).

There have been a number of critiques of the CE model. Hobson and Lynch (2016), for example, regard the prevailing view of the CE as a rebooting of capitalism variously described as 'cooperative capitalism', 'breakthrough capitalism' and 'naturalised capitalism'. This approach argues that the environment and capitalism can co-exist to ensure a level of sustainability that does not disrupt the status quo of politics and power (Hobson and Lynch, 2016). Other scholars view the idea of the CE as an alternative to the green growth model and for this reason an inadequate solution to ensuring long-term sustainability within the limits of the Earth's ecological systems (Schröder et al., 2019).

Degrowth

Degrowth is a philosophical world view that emerged from fears that the current linear production system will deplete non-renewable resources, accelerate industrialization, cause continuing deterioration of the global environment and lead to socio-economic collapse. Degrowth has been described as a critique of the current neoliberal economic development hegemony (Demaria et al., 2013).

The degrowth approach to transition calls for drastic reduction in consumption and production (Weiss and Cattaneo, 2017). Until recently, the debate about degrowth as a solution to climate change has been largely ignored by the tourism literature, however growing concern about overtourism (Fletcher et al., 2019; Milano, Novelli and Cheer, 2019) has stimulated interest in degrowth as a potential solution and reinvigorated interest in earlier work on degrowth by scholars including Andriotis (2014) and Scott, Gössling and Hall (2012). Hall (2009), for example, stated that degrowth may be used as strategy to overcome some of the impacts of climate change. However, as Fletcher et al. (2019: 1745) note, degrowth is "both social theory and social movement.... that has yet to engage systematically with the tourism industry".

The central idea of degrowth is an emphasis on: quality of life rather than quantity of consumption; structural transition to service-based activities by downscaling production and consumption; investment in ecological assets; and changes in worktime as a mechanism to achieve economic stabilisation (Jackson, 2009). One outcome of this process is an anticipated increase in human well-being, reduced consumption and greater leisure time. The latter outcome has significant implications for tourism. The flaw in the degrowth view of transition is that a model able to achieve the desired economic and political outcomes has yet to be developed (Weiss and Cattaneo, 2017). As Prideaux and Pabel (2021) observed, the degrowth literature has yet to address strategies needed to convince the general public that they will need to limit their consumption of touristic experiences.

The green economy

The concept of the green economy (GE) emerged out of the field of environmental economics (Borel-Saladin and Turok, 2013: 212) and assumes that it is not possible to separate environmental progress from economic growth and development. The GE has attracted interest as an alternative to the 'Brown Economy'. The UNEP (UNEP, nd) for example describes GE as "as low carbon, resource efficient and socially inclusive. In a green economy, growth in employment and income are driven by public and private investment into such economic activities, infrastructure and assets that allow reduced carbon emission and pollution, enhanced energy and resource efficiency, and prevention of the loss of biodiversity and ecosystem services". From this perspective, the GE is a holistic approach that requires integration of social inclusiveness and the environment into economic thinking, based on the establishment of a development path that understands that natural capital is a critical economic assert and public benefit. Law, De Lacy, Lipman and Jiang (2016) describe it in terms of its potential for dealing with the problems of climate change through transforming the current economic production system by adopting three foundations: low-carbon, resource efficiency and social inclusion. From this perspective, the green economy supports growth but with an emphasis on 'decoupling' it from negative social and environmental impacts.

Figure 15.4 illustrates a conceptual model of the green economy proposed by Law et al. (2016) based on an approach advocated by United Nations Department of Economic and Social Affairs (UN-DESA, 2012).

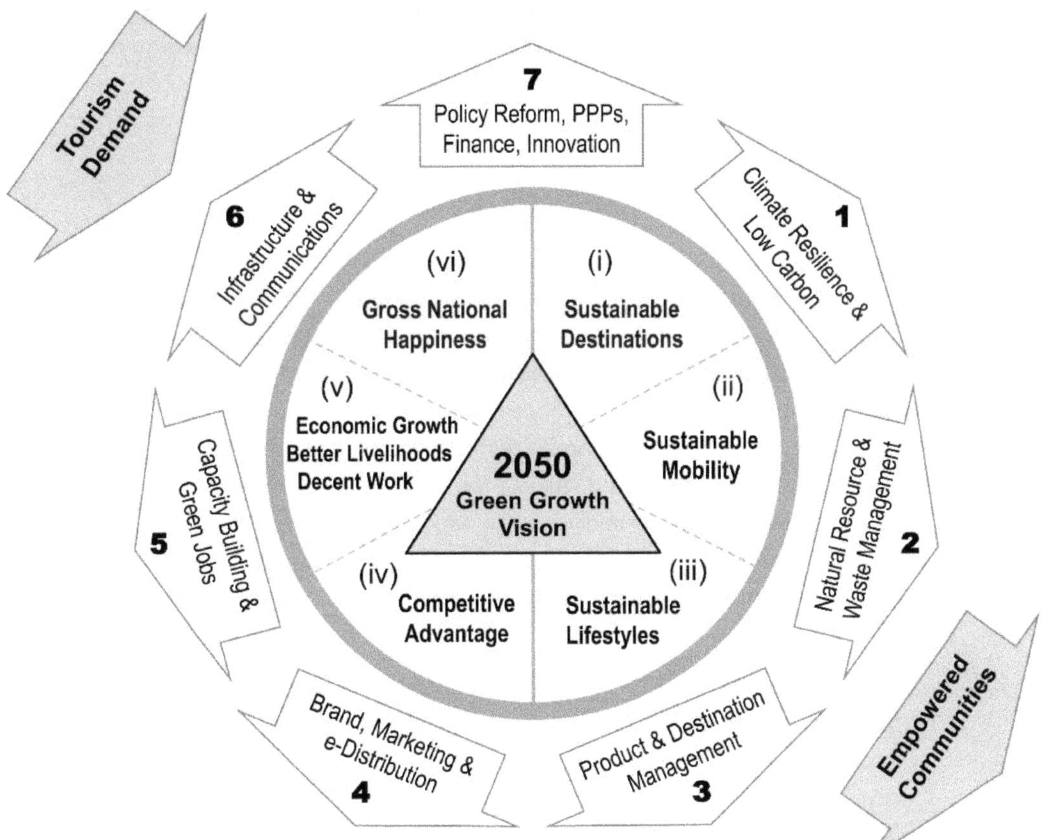

Figure 15.4: Green economy model for tourism destination stakeholder engagement. *Source*: Law, De Lacy, Lipman and Jiang, 2016.

The model was used to assess its potential to be employed as a strategy for tourism development in Bali. In this approach, nine issues were addressed: growth and economic development; environmental protection; low-carbon development; resilience; resource efficiency; ecological sustainability; human well-being; inclusivity; and equity. The conclusion reached by Law et al. (2016) was that successful implementation of the model will require sound strategy implementation and committed private and public leadership at all political levels plus citizen support over the long-term.

The model has been criticised in a number of studies based on vagueness (Schalensee, 2012), its continued emphasis on growth (green growth rather than brown growth) (Law et al., 2016), its conflict with the basic principles of sustainability and the difficulty of achieving behavioural change of the type required for consumers to support GE (Hall, 2013). Borel-Saladin and Turok (2013), for example, note that the green economy may not achieve all-round progress because it does not represent a sufficient departure from the current linear economy model. One key question is whether current environmental degradation and social inequity can be corrected by a modified version of the current economic system that created these problems in the first place (Borel-Saladin and Turok, 2013).

Conclusion

It is apparent that the current linear economic production system and the neoliberal model of political economy are not well suited to achieving outcomes that will lead to reduced GHG emissions, carbon neutrality and long-term sustainability of ecosystems. As the impacts of climate change become more obvious in terms of our daily life and future travel options, support for implementing alternative models of economic production is likely to grow. Until 2020 a reduction in tourism on the scale that occurred during the COVID-19 pandemic was unthinkable. For a sector that already is facing significant problems in successfully responding to climate change, the ability of many economies to continue to operate with reduced tourism flows poses further risks for the global tourism industry.

■ Discussion points / Exercises

1. Identify 10 different types of tourism that can be be a victim, winner or loser in climate change. Discuss possible mitigation strategies for victims and losers.
2. It has been argued that a high carbon tax would limit long haul travel and reduce what some say is unnecessary hedonistic travel. What impact would such a tax have on small island destinations that are reliant on tourism for a large part of their GDP?

3. Some people argue degrowth is not a realistic option and instead feel adopting a green economy approach is more suitable. Pick one and argue on behalf of the other concept.

Points to consider for revision

- Climate change is real and is already having a dramatic impact on tourism.
- The impacts of climate change on tourism are uneven, with some forms of tourism and some types of destinations being more at risk, while others are more resilient.
- Climate change may have a dramatic impact on tourism flows.
- Addressing climate change will involve a fundamental shift in the global economy.

Suggested additional readings

Schroder. P., Bengtsson, M., Cohen, M., Dewick, P. Hofstetter, J. & Sarkis, J. (2019) Degrowth within Aligning circular economy and strong sustainability narratives, *Resources, Conservation and Recycling*, 146, 190-191.

Ghisellini, P., Cialani, C. & Ulgiati, S. (2016) A review on circular economy: the expected transition to a balanced interplay of environmental and economic systems, *Journal of Cleaner Production*, 114, 11-32.

Ellen MacArthur Foundation (2015), *Growth Within – A Circular Economy Vision for a Competitive Europe*, Cowes, Isle of Wight, UK

References

Abegg, B., Agarwal, S. & deMontfalcon, A. (2007) Climate change impacts and adaptation in winter tourism. in Agarwal, S. (Ed) *Climate Change in the European Alps*, OECD, Paris, 26-60.

Amelung, B., Nicholls, S. & Viner, D. (2007) Implications of global climate change for tourism flows and seasonality. *Journal of Travel Research*, 45(3), 285-296.

Andriotis, K. (2014) Tourism development and the degrowth paradigm. *Turističko poslovanje*, 13, 37-45.

Bocken, N., de Pauw, I., Bakke, C. & van der Grinten, B. (2016) Product design and business model strategies for a circular economy, *Journal of Industrial and Production Engineering*, 33(5), 308-320.

Borel-Saladin, J. & Turok, I. (2013) The green economy: Incremental change or transformation? *Environmental Policy and Governance*, 23, 209–220.

Cochrane, J. (2010) The sphere of tourism resilience, *Tourism Recreation Research*, 35(2), 173-185.

Demaria, F., Schneider, F., Sekulova, F. & Martinez-Alier, J. (2013) What is degrowth? From an activist slogan to a social movement. *Environmental Values*, 22(2), 191-215.

EEA (2011) *Specific air pollutant emissions (TERM 028) - Assessment Published Jan 2011*. European Environmental Agency. http://www.eea.europa.eu/data-and-maps/indicators/specific-air-pollutant-emissions/specific-air-pollutant-emissions-assessment-3 <downloaded Nov 22, 2011>.

Ellen MacArthur Foundation (2013) *Towards the Circular Economy Vol.1*. Ellen MacArthur Foundation: Isle of Wight.

European Commission (2015) Closing the loop - An EU action plan for the Circular Economy, Com(2015) 614 communication from the commission to the European Parliament, the Council, the European Economic and Social Committee and the Committee of the Regions. *European Commission*, Brussels.

Fletcher, R., Mas, I., Blanco-Romero, A. & Blázquez-Salom, M. (2019) Tourism and degrowth: An emerging agenda for research and praxis. *Journal of Sustainable Tourism*, 27(12), 1745-1763.

Geissdoerfer, M., Savaget, P., Bocken, N. & Hultink, E.J. (2017) The circular economy a new sustainability paradigm?, *Journal of Cleaner Production*, 143, 757-768.

Gössling, S. & Peeters, P. (2007) 'It does not harm the environment!' An analysis of industry discourses on tourism, air travel and the environment, *Journal of Sustainable Tourism*, 15(4), 402-417.

Gregson, N., Crang, M., Fuller, S. & Holme, S. (2015) Interrogating the circular economy: the moral economy of resource recovery in the EU, *Economy and Society*, 44(2), 218-243.

Haimes, Y. (2009) On the definition of resilience in systems, *Risk Analysis*, 29(4), 498-501.

Hall, C. M. (2009) Degrowing tourism: Décroissance, sustainable consumption and steady-state tourism. *Anatolia*, 20(1), 46-61.

Hall, C. M. (2013) Framing behavioural approaches to understanding and governing sustainable tourism consumption: beyond neoliberalism, "nudging" and "green growth"?, *Journal of Sustainable Tourism*, 21(7), 1091-1109.

Hamilton, J. & Tol, R. (2007) The impact of climate change on tourism in Germany, the UK and Ireland: a simulation study. *Regional Environmental Change* 7(3), 161-172.

Hobson, K. & Lynch, N. (2016) Diversifying and de-growing the circular economy: Radical social transformation in a resource-scarce world, *Futures*, 15-25.

Hoegh-Guldberg, O. (1999) Climate change, coral bleaching and the future of the world's coral reefs. *Marine and Freshwater Research*, 50(8), 839-866.

IPCC (2018), *Global Warming of 1.5°C. An IPCC Special Report on the impacts of global warming of 1.5°C above pre-industrial levels and related global greenhouse gas emission pathways, in the context of strengthening the global response to the threat of climate change, sustainable development, and efforts to eradicate poverty*, Intergovernmental Panel on Climate Change (IPCC).

IPCC (1990) *First Assessment Report*, retrieved from https://www.ipcc.ch/report/ar1/wg1/

IPCC (nd) *Structure of the IPCC process*, retrieved from https://www.ipcc.ch/about/structure/

IPCC (2014) Field, C.B. et al. (eds.), *Climate Change 2014: Impacts, Adaptation, and Vulnerability. Part A: Global and Sectoral Aspects. Contribution of Working Group II to the Fifth Assessment Report (AR5) of the Intergovernmental Panel on Climate Change (IPCC)*, Cambridge University Press.

IPCC (2019a) *Special Report on Climate Change and Land* (SRCCL), IPCC, Retrieved from https://www.ipcc.ch/srccl/

IPCC (2019b) *Summary for Policymakers. IPCC Special Report on the Ocean and Cryosphere in a Changing Climate.* https://www.ipcc.ch/srocc/chapter/summary-for-policymakers/

IPCC. (2018) *Global warming at 1.5^{0}C.* IPCC. Retrieved from https://www.ipcc.ch/reports/ <on 23 September 2019>.

IPPC (nd) *The Intergovernmental Panel on Climate Change,* https://www.ipcc.ch/ <visited 10 June, 2020>.

Jackson, T. (2009) *Prosperity Without Growth. Economics for a finite planet.* London: Earthscan.

Law, A., De Lacy, T., Lipman, G. & Jiang, M. (2016) Transitioning to a green economy: the case of tourism in Bali, Indonesia, *Journal of Cleaner Production*, 111, 295-305.

Lew, A.A. (2014) Scale, change and resilience in community tourism planning. *Tourism Geographies*, 16(1), 14-22.

Lieder, M., & Rashid, A. (2016) Towards circular economy implementation: A comprehensive review in context of manufacturing industry. *Journal of Cleaner Production*, 115, 36-51.

Mathews, J (2011) Naturalizing capitalism: the next great transformation, *Futures*, 43(8), 868-879.

Milano, C., Novelli, M. & Cheer, J. (2019) Overtourism and degrowth: A social movements perspective. *Journal of Sustainable Tourism*, 27(12), 1857-1875.

Nakićenović, N. & Swart, R. (eds.) (2000) *Special Report on Emissions Scenarios: A special report of Working Group III of the Intergovernmental Panel on Climate Change*, Cambridge University Press.

Njoroge, J.M. (2014) An enhanced framework for regional tourism sustainable adaptation to climate change. *Tourism Management Perspectives*, 12, 23-30.

Pang, S., McKercher, B. & Prideaux, B. (2012) Climate change and tourism: An overview. *Asia Pacific Journal of Tourism Research*, 18(1/2), 4-20.

Pomponi, F. & Moncaster, A. (2017) Circular economy for the built environment: A research framework, *Journal of Cleaner production*, 142, 710-718.

Prideaux, B., Coghlan, A. & McNamara, K.E. (2010) Assessing the impacts of climate change on mountain tourism destination using the climate change impact model, *Tourism Recreation Research*, 35, 187-200.

Prideaux, B. & Pabel, A. (2021) Degrowth as a strategy for adjusting to the adverse impacts of climate change in a nature based destination, in Hall, CM., Lundmark, L. & Zhang, J. (eds), *Degrowth and Tourism: New Perspectives on Tourism Entrepreneurship, Destinations and Policy*, Routledge.

Ramis, M. & Prideaux, B. (2013) The importance of visitor perceptions in estimating how climate change will affect future tourists flows on the Great Barrier Reef, in Reddy, M. & Wilkes, K. (eds), *Tourism, Climate Change and Sustainability*, Routledge, London, pp. 173-188.

Rebetez, R. (2011) Relationship between tourism demand in the Swiss Alps and hot summer air temperatures associated with climate change. *Climate Change*, 108(1), 291-300.

Schröder, P., Bengtsson, M., Cohen, M., Dewick, P., Hofstetter, J & Sarkis, J. (2019) Degrowth within – Aligning circular economy and strong sustainability narratives, *Resources, Conservation and Recycling*, 146, 190-191.

Scott, D. & Verkoeyen, S. (2018) Assessing the climate change risk of a coastal-island destination, in Jones, A. & Philips, M. (eds), *Global Climate Change and Coastal Tourism; recognizing problems, managing solutions, future expectations*, CABI, Wallingford, pp. 62-73.

Scott, D., Gössling, S. & Hall, C. M. (2012) International tourism and climate change. *WIRES Climate Change*, 3(3), 213-232.

Scott, D., Hall, C. M. & Gössling, S. (2012) *Tourism and Climate Change*, Routledge, Oxon.

Scott, D., McBoyle, G. & M. Schwartzentruber (2004). Climate change and the distribution of climatic resources for tourism in North America. *Climate Research*, 27(2), 105-117.

Scott, D., Simpson, M.C. & Sim, R. (2012) The vulnerability of Caribbean coastal tourism to scenarios of climate change related sea level rise, *Journal of Sustainable Tourism*, 20(6), 883-898.

Simpson, M. C., Gössling, S., Scott, D., Hall, C. M. & Gladin, E. (2008) *Climate Change Adaptation and Mitigation in the Tourism Sector: Frameworks, Tools and Practices*. UNEP, University of Oxford, UNWTO, WMO: Paris, France.

Storlazzi, C., Elias, E. and Berkowitz, P. (2015) Many atolls may be uninhabitable within decades due to climate change. *Scientific Reports* 5, 14546 (2015). https://doi.org/10.1038/srep14546.

Taylor, R. & Ortiz, R. (2009) Impacts of climate change on domestic tourism in the UK: a panel data estimation. *Tourism Economics* 15 (4), 803–812.

Thomas, R. (2005) Zooarchaeology, improvement and the British agricultural revolution, *International Journal of Historical Archaeology*, 9(2), 71-88.

Trenberth K. (2001) Stronger evidence of human influence on climate: The 2001 IPCC Assessment, *Environment*, 43(4), 11-19.

UK Met Office (nd) Climate projections, retrieved from https://www.metoffice.gov.uk/research/approach/collaboration/ukcp/about.

UN-DESA (2012) *A Guidebook to the Green Economy: Issue 1*, United Nations Department of Economic and Social Affairs, retrieved from http://sustainabledevelopment.un.org/index.php?menu=1516.

United Nations Environment Program (nd) *Green Economy*, UNEP, RETRIEVED from https://www.unenvironment.org/regions/asia-and-pacific/regional-initiatives/supporting-resource-efficiency/green-econom.y

UNWTO (2008) *Climate Change and Tourism – responding to global challenges*. Madrid, Spain:

Weiss, M., & Cattaneo, C. (2017) Degrowth – Taking stock and reviewing an emerging academic paradigm. *Ecological Economics*, 137, 220-230.

Williams, S., Bolitho, E. & Fox, S. (2003) Climate change in Australian tropical rainforests: an impending environmental catastrophe, *Proceedings of the Royal Society London B*, 270, 1887-1892, DOI 10.1098/rspb.2003.2464.

16 Tourism Planning

By the end of this chapter, the reader will be able to:

- Describe the difference between planning and policy
- Identify and describe the scales of planning
- Explain the need to consider new approaches to planning in the future

Introduction

In the last few decades of the 20th Century, there was a shift from modernist forms of public administration, where government bureaucracies were given licence to 'define and act for the public good' (Dredge and Jamal, 2015: 287), to the current globalist/neoliberal perspective of governance, where governments have downsized bureaucracies and outsourced many functions. The outcome has seen many governments move into an arbiter/enabler role, shifting from public administration to public management. This change significantly influenced thinking about the role that governments play in planning. The consequence has been that while there is apparent public sector support for development that adheres to the principles of sustainability, the reality is that neoliberal style policy continues to place growth rather than sustainability at the heart of policy formulation. Previous chapters of this book have observed that maintening contemporary systems of economic production and consumption will create a future that will be significantly different from the present. Characteristics of this 'future' will include rising sea levels, degraded ecosystems and higher temperatures. Future planning challenges will include responding to the impact of climate change, the depletion of non-renewal resources, ecosystem decline and the need to transition into a new carbon-neutral economy. From a planning context these changes will require rethinking of how tourism planning is undertaken and the objectives that future planning may wish to achieve compared to the current objectives of planning, which are often based on a desire for continuing growth but with a cursory nod to a need for development that is 'sustainable'.

This chapter briefly reviews the role of planning, the relationship between planning and policy, and issues that planning will need to address in the future. Underlying this discussion is the view that a lack of planning, particularly at the strategic level but also at the destination and enterprise level, will result in ad hoc, unregulated, inefficient and reactive responses to calls to supply tourism infrastructure and services.

The literature on planning is vast and growing, with an increasing interest in specific forms and aspects of tourism. Readers are referred to Saarinen, Rogerson and Hall (2017), Dredge and Jamal (2015) and Rahmafitria, Pearce, Oktadiana and Putro (2020), Costa (2019) and Lew (2007) for detailed discussions on key trends in tourism planning. Hall (2015) provides a useful table (see Figure 16.1) that illustrates the relationships between planning traditions, key themes, global development milestones and humanity's global footprint. One outcome of the interest in planning has been the emergence of special interest areas in the planning literature. Examples include the role of planning in sustainable development, (Simão and Partidárion, 2012); power exercised through social representation in planning (Moscardo, 2011); community participation (Tosun, 2000); cultural heritage planning (du Cros, 2001); and destination development (Haughland, Ness, Gronseth and Aarstad, 2010). Commenting on the apparent gap between academic model building and their application in practice, Ruhanen (2004) observed that there is a need to move the 'wealth of academic knowledge' into the real world of planners and decision makers.

Planning traditions

Planning has a long tradition stretching back to the time when the first urban settlements emerged in the Middle East, and is now undertaken in a number of forms and at various scales. The forms of planning include public and private sector planning, development, infrastructure provision to support tourism activity, land and resource use, and promotion and marketing (Hall, 2005). Scales include sectoral, local, regional, national and international. Over time, planning has emerged as a systematic process that assists various private and public sector actors to engage in dialogue. This process acknowledges that there are usually overlapping and often competing interests that need to be considered in a policy environment that is becoming increasingly complex and contested, as factors such

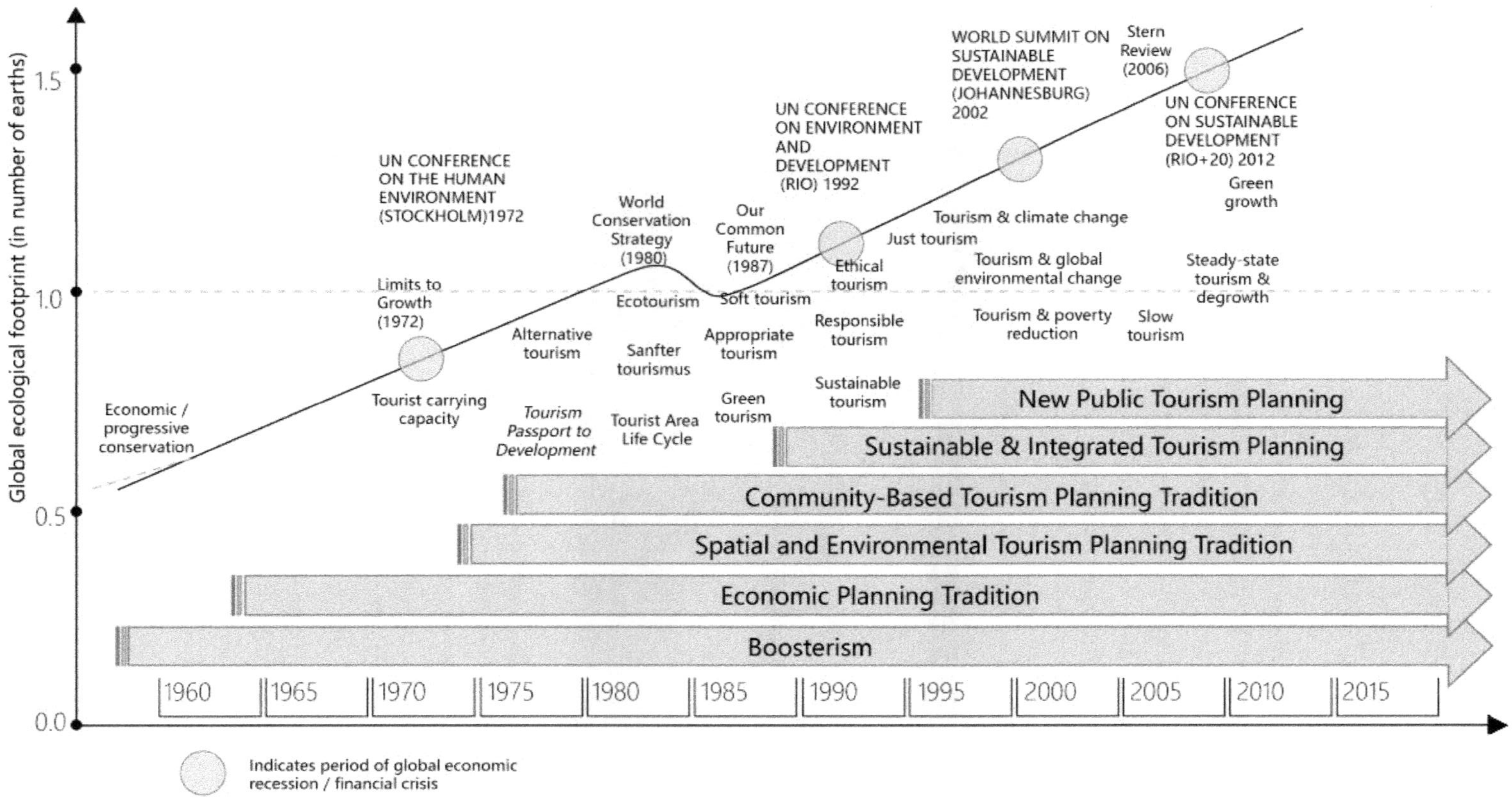

Figure 16.1: Relationship between planning traditions, key themes in tourism development, global development milestones, and humanity's global footprint (Hall, 2015).

16

as social justice, sustainability and impacts of climate change and other events are considered. The outcome of this process is generally expressed as a plan that incorporates strategies and targets designed to meet future objectives.

Tourism planning emerged as a specialised field in the 1960s in response to growing demand for tourism (Tosun and Jenkins, 1998). Governments began to support tourism planning at destination, regional and national levels, generally following a focus on investment incentives, industry organisation and spatial considerations (Yang and Wall, 2008). As Gunn (1988) noted, planning in this period was essentially an exercise in matching demand with tourism products. Planning failures highlighted the need for a more participatory approach (Harrill and Potts, 2003), supporting earlier calls by Gunn (1979) for participation by multiple stakeholders in tourism planning.

As the impacts of climate change accelerate in coming decades, planning will need to consider factors such as sea level rise, ecosystem loss and changes in temperatures and precipitation. In many coastal areas, communities will need to decide on flight or fight as sea levels rise. A decision to fight is likely to include erection of sea wall levees but there is no certainty about the level or timing of future inundations. Flight will require a withdrawal from some coastal areas and relocation in areas secure from predicted sea level rise. In both cases planners will require strong signals from government about future policy directions including possible subsidies and guidance on anticipated sea level rises, when investment decisions about tourism infrastructure including coastal hotels and airports are made. The beach facing tourist shopping, restaurant and accommodation pressure illustrated in this photo will be underwater at the 0.74m sea level scenario discussed in the latest IPCC (2018) report. The question coastal communities in particular face is how to adapt (fight or flight) and when. Addressing this problem will require rethinking and modification of many of the planning models discussed in this chapter and book.

In recent decades tourism planners have been confronted by the need to deal with the contradictions inherent in responding to increasing demand for tourism in a global context, where there are increasing concerns about the ability of the global environment to support growth. In practice, a gap has emerged between what scholars suggest should be done and the factors that should be considered, and what actually happens (Yang and Wall, 2008).

Although Hall (2000: 6) observed that "planning is an extremely ambiguous and difficult word to define", the need for a definition cannot be denied. Dredge and Jennings (2007: 22) for example, define planning as a "strategic activity comprising a number of stages that lead to the determination of a course of action to meet predetermined goals. Planning is concerned with the future; is devoted to acquiring knowledge and identifying appropriate courses of action; is about anticipating change, developing a strategic vision and facilitating political decision making; and is value-laden and political." Other definitions generally include similar relationships that should be embraced by planners. The specific methodology used in planning will depend on a large number of considerations, including the aim of a particular plan, the planning environment that planning occurs in, the objectives of the organisation sponsoring the plan, the level of consideration given to environmental, social and economic cost and benefits of the plan, and distribution of benefits between winners and losers of the planning process.

Planning is not a standalone activity and instead, takes place in a policy environment. However, planning and policy are dialectical concepts that have socially constructed meanings and are often defined by the context in which they are applied (Dredge and Jamal, 2015). That context may be 'fuzzy' or have a number of meanings depending on the context in which the terms are used. Dredge and Jenkins (2007: 7) for example define policy as being "a position, strategy, action or product adopted by government and arising from contests between different ideas, values and interests".

From a slightly different perspective, Hall (2000) observed that tourism policy is whatever governments choose to do or not do, including actions and inactions, decisions and non-decisions. In practice however, the ability of governments to balance community welfare and environmental conservation with the need for profit has created conflicts in policy making and planning (Rahmafitria, Pearce, Oktadiana and Putro, 2020). The need to balance tourism development with social aspects of community has also created other areas of tension. For example, in a neoliberal policy context the market is given primacy, while in a socialist context, the public sector sets the goals that policy and planning should achieve. Planning occurs in a political environment that determines how agendas are set, what issues are important, the roles that are assigned to the market and the state and the degree to which winners compensate losers. From these perspectives it is difficult to disengage planning from the political context that it occurs in and from the form of political economy that determines policy formation. While acknowledging the difficulty inherent in treating planning in isolation from policy and political economy, the discussion in this chapter will focus primarily on planning.

Planning approaches, typologies and frameworks

Getz (1986), in a study of over 150 tourism planning models identified a number of problems associated with tourism models including; limited analysis and evaluation of the costs and benefits of tourism development; the failure of many models to integrate into other development process; failure to take note of non-economic factors; and a narrow focus on specific development planning at project level. Planning (and policy) can be broadly categorised as either substantiative ideas that articulate the direction and underlying values of tourism policy and planning or procedural ideas that suggest how planning should be undertaken (Dredge and Jenkins, 2007). The typology suggested by Getz (1987, cited in Hall, 2005) provides an example of the substantive approach to planning. Getz identified four approaches:

- Boosterism
- Economic
- Physical/spatial
- Community.

Getz observed that these approaches were not mutually exclusive, and planning may incorporate elements of two or more of these approaches applied in tandem or sequentially. Hall (2000) later added sustainable planning in recognition of the growing interest in sustainability that had begun to occur in the literature and in global forums such as the United Nations Conference on Environment and Development (also referred to as the Rio de Janeiro Earth Summit). Soon after, Faulkner (2001) responding to the need to develop planning instruments that could be used for crisis management and suggested a six-stage community response plan for managing and responding to disasters in a tourism setting.

Commenting on the approaches suggested by Getz and Hall, Dredge and Jenkins (2007) cautioned that easily identifiable planning categories ran the danger of oversimplifying and obscuring the richness and diversity of the transcendent shifts that take place in social science and public administration ideology and the often-conflicting ideas that determine policy development and decision-making. As an alternative, Dredge and Jenkins (2007) suggested a typology based on six substantive approaches able to be identified in tourism planning (and policy formation):

- Industry development planning and policy – use of financial measures and tools such as incentives and tax breaks to strengthen industry capacity
- Market planning and policy – promote the development and expansion of markets with the aim of product development and diversification
- Spacio-physical destination planning and policy – spatial destination planning, modelling and relationships between attractions, services, origins, transport routes and marketing
- Conflict management planning and policy – measures to mitigate the social, cultural, environmental and economic effects of tourism on local destinations.
- Communicative tourism planning and policy- establishing collaborative dialogue where multiple stakeholders are involved
- Crisis-responsive planning and policy – addressing crises as they emerge and drawing where appropriate from the previous approaches.

Costa and Brandao, (2018) suggested a typology with planning models grouped into:

- Rational tourism models
- Systematic tourism models
- Production tourism models, and
- Strategic management tourism models.

The typology suggested by Dredge and Jenkins (2007) exhibits many similarities to those suggested by Getz (1987) and Hall (2000) with the exception of the crisis-response planning and policy approach. The contributions by both Hall (2000) and Faulkner (2001) led to significant advances in ideas about planning. For example, growing concerns about the ability of tourism planning to ensure that development is sustainable has generated considerable debate, and more recently has begun to merge with concerns over climate change and how tourism has contributed to it and suffered from it.

More recently still, the need to respond to the downturn in global tourism activity caused by the COVID-19 pandemic has generated an enormous uptick in interest in crisis and disaster management.

From a procedural perspective, spatial planning models based on a 'stagist' approach have been widely adopted by planning bodies. Popular models following the procedural approach include the Recreation Opportunity Spectrum (ROS), Zero Based Budgeting (ZBB), Management by Objectives (MBO) and Environmental Impact Assessments (EIA). Lew (2007) stated that many of these models have been subjected to considerable criticism because of underlying assumptions including unpredictable behaviour, limited knowledge and problem-solving capabilities, a failure to list all possible scenarios, and lack of objectivity in a closed system. In most cases planners have limited knowledge and time horizons and are subject to unpredictable events and influences. Examples of models based on sequential stages include Dredge and Jenkins (2008) who suggested eight stages and Lew (2007) who recommended 10 stages. Lew's stages are:

1. Formulation of goals and objectives
2. Identify a goal (to solve the problem)
3. Collect background data
4. Identify guidelines for assessing alternative plan scenarios
5. Identify alternative plan scenarios, including policies and guidelines to achieve the goals
6. Assess alternative plans scenarios using pre-defined assessment guidelines
7. Select the preferred alternative
8. Implement the plan
9. Monitor, evaluate, and revise the Implementation
10. Identify new problems and begin the process again. Development of alternative solutions or scenarios.

Bittar-Rodrigues and Prideaux (2018), provide a typical example of this type of 'stagist' approach but with a strong emphasis on community empowerment in a Community Based Tourism model. In their model (Figure 16.2) Column 2 identifies nine stages of actions, which are influenced by factors identified community issues (column 1), non-community issues (column 3) and external factors (column 4), such as national governance issues.

A new perspective

Many models have been applied on a 'one fit for all' solution to various issues including planning. In reality, one model is often insufficient to deal with the plethora of issues that may be encountered when looking for answers to tourism planning issues. Approaching the question of which model was the most appropriate for planning issues that were bedevilling a small remote destination, Prideaux and Thompson (2021) suggested an open architecture approach described as a process that allows various models and theories to be 'plugged in', on an 'as required' basis, as planning moves through a series of stages such as those identified by Lew (2007). The opposite to the open architecture approach is a 'closed architecture' approach where only one model is used to inform on planning issues. The open architecture approach facilitates flexibility, allowing models to be used in parallel or sequentially where required to solve various issues in the planning process. For example, supply considerations such as location and

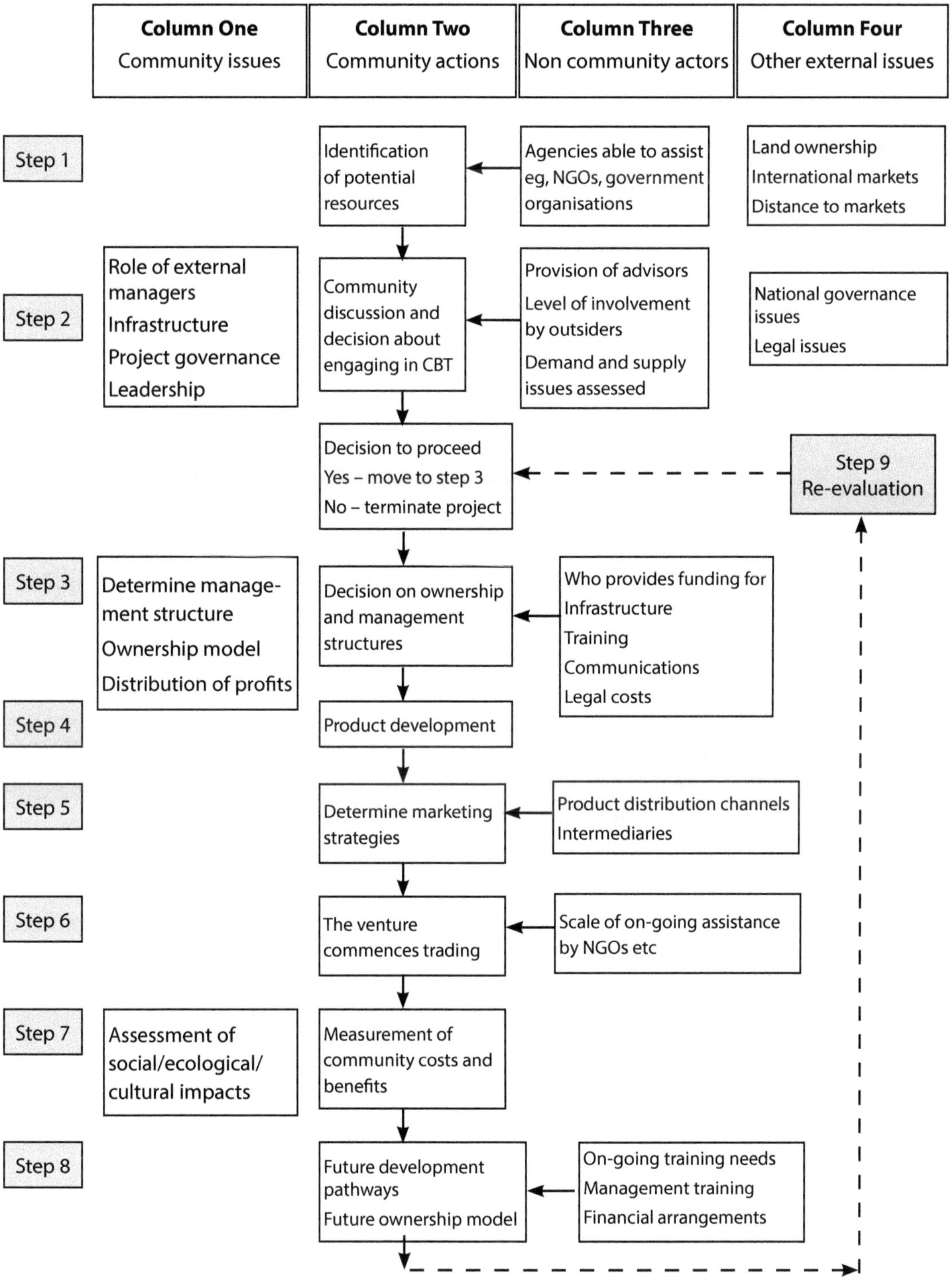

Figure 16.2 :CBT Implementation model. *Source*: Bittar-Rodrigues and Prideaux, 201.

resources, typically considered in stage 3 of the planning model suggested by Lew (2007), can be addressed by a variety of non-planning models including economic base, periphery-core and destination pull.

Conclusion

This chapter has highlighted a number of models, typologies and frameworks that have emerged in the tourism planning literature. The chapter has also observed that in the future, planning will need to take a path different from that of the present as the global economy responds to the problems generated by climate change. For example, government- rather than market-led intervention were required to deal with the economic problems caused by the COVID-19 pandemic, indicating a significant failure of the neoliberal approach to political economy. In the future it is reasonable to expect that governments and their bureaucracies will take a more interventionist role in public policy administration than has been the practice in recent decades, particularly as the need to transition to a carbon-neutral economy becomes more urgent.

On a different scale, future planning will also need to consider the implications of a rapidly urbanising global population. The United Nations Department of Economic and Social Affairs (2018) estimates that by 2050 about 68% of the global population will live in urban areas. Changes in technologies, including a possible switch to Autonomous Vehicles (Prideaux and Yin, 2019), increased use of new technologies including facial recognition and robots, Artificial Intelligence (AI), further growth of the sharing economy (with products such as Airbnb and Uber), will also call for new planning solutions.

On the climate change front, the impending loss of many coastal areas to sea level rise will generate a need to relocate some urban areas or parts thereof. This will require a significant re-evaluation of spatial and scale elements of tourism planning as well mobility factors. Emphasis will also need to be given to the changing moral and ethical aspects of tourism development, particularly in the Global South (Saarinen et al., 2017).

■ Discussion Points / Exercises

1. What are the major planning factors that should be considered when deciding on which planning method should be used?
2. What is the difference between the open architecture and closed architecture approach to planning?
3. What are the key differences between subjective and procedural approaches to planning?

Points to consider for revision

- Planning takes place in a policy environment, thus a range of non-spatial issues including community welfare, the environment and economic return need to be considered.
- The problems associated with climate change will require new planning approaches in future decades.
- Why will consideration need to be given to replacing neoliberalist approaches to planning in the future?

Additional readings

Costa, C. and Brandao, F. (2018) Tourism planning: evolution and future trends, in Cooper, C., Volo, S., Gatner, W. and Scott, N. (eds), *The SAGE Handbook of Tourism Management: Concepts and Disciplinary Approaches to Tourism*, Sage London, pp. 538-555.

Dredge, D. and Jamal, T. (2015) Progress in tourism planning and policy: A post-structural perspective on knowledge production, *Tourism Management*, 51, 285-297.

Rahmafitria, F., Pearce, P., Oktadiana, H. and Putro, H. (2020) Tourism planning and planning theory: Historical roots and contemporary alignment, *Tourism Management Perspectives*, 35, 100703.

References

Bittar Rodrigues, C. & Prideaux, B. (2018) A management model to assist local communities developing Community Based Tourism ventures: A case study from the Brazilian Amazon, *Journal of Ecotourism*, 17(1), 1-19.

Costa, C. (2019) Tourism planning: a perspective paper, *Tourism Review*, 75(1), 198-202.

Costa, C. & Brandao, F. (2018) Tourism planning: evolution and future trends, in Cooper, C., Volo, S., Gartner, W. & Scott, N. (eds), *The SAGE Handbook of Tourism Management: Concepts and Disciplinary Approaches to Tourism*, Sage London, pp. 538-555.

Dredge, D. & Jamal, T. (2015) Progress in tourism planning and policy: A post-structural perspective on knowledge production, *Tourism Management*, 51, 285-297.

Dredge, D. & Jenkins, J. (2007) *Tourism Planning and Policy*, Wiley, Milton, Queensland.

du Cros, (2001) A new model to assist in planning for sustainable cultural heritage tourism, *International Journal of Tourism Research*, 3(2), 165-179

Faulkner, B. (2001) Towards a framework for tourism disaster management, *Tourism Management*, 22, 135-147.

Getz, D. (1986) Models in tourism planning: Towards integration of theory and practice, *Tourism Management*, 7, 21-32.

Getz, D. (1987) Tourism planning and research: traditions, models and futures, paper presented at the Australian Travel Workshop, Bunbury, Western Australia, 5-6 November.

Gunn, C. (1979) *Tourism Planning*, Crane Rusak, New York.

Gunn, C. (1988) *Tourism Planning: Basics, Concepts, Cases*, Taylor and Francis, New York.

Hall, C. M. (2000) *Tourism Planning*, Prentice Hall, Harlow.

Hall, C. M. (2005) *Tourism: Rethinking the Social Science of Mobility*, Pearson, Harlow.

Hall, C. M. (2015) Economic greenwash: On the absurdity of tourism and green growth, in V. Reddy & K. Wilks (eds.), *Tourism in the Green Economy*, Earthscan, London, pp. 339-358.

Harrill, R. & Potts, T. (2003) Tourism planning in historic districts: Attitudes towards tourism development, *Journal of the American Planning Association*, 63(3), 233-244.

Haughland, S., Ness, H., Grønseth, B., & Aarstad, J. (2010) Development of tourism destinations: an integrated multilevel perspective, *Annals of Tourism Research*, 38(1), 286-290.

Lew, A. (2007) Invited commentary: Tourism planning and traditional urban planning theory – the planner as an agent of social change, *Leisure/Loisir*, 31(2), 383-391.

Moscardo, G. (2011) Exploring social representations of tourism planning: issues for governance, *Journal of Sustainable Tourism*, 19 (4-5), 423-436.

Prideaux, B. & Thompson, M. (2021) developing tourism in remote communities: an open architecture approach, *Advances in Hospitality and Leisure*, 16, 3-21. Doi 10.1108/S1745-354220200000016003.

Prideaux, B. & Yin, P. (2019) The disruptive potential of autonomous vehicles (AVs) on future low-carbon tourism mobility, *Asia Pacific Journal of Tourism Research*, 24(5), 459-467. https://doi.org/10.1080/10941665.2019.1588138.

Rahmafitria, F., Pearce, P., Oktadiana, H. & Putro, H. (2020) Tourism planning and planning theory: Historical roots and contemporary alignment, *Tourism Management Perspectives*, 35, 100703. https://doi.org/10.1016/j.tmp.2020.100703.

Ruhanen, L. (2004) Strategic planning for local tourism destinations: an analysis of tourism plans, *Tourism and Hospitality Planning & Development*, 1(3), 239-253.

Saarinen, J., Rogerson, C. & Hall, CM. (2017) Geographies of tourism development and planning, *Tourism Geographies*, 19(3), 307-317.

Simão, J. & Partidário, R. (2012) How does tourism planning contribute to sustainable development? *Sustainable Development*, 20, 372-385.

Tosun, C. (2000) Limits to community participation in the tourism development process in developing countries, *Tourism Management*, 21(6), 613-633.

Tosun, C. & Jenkins, C. (1998) The evolution of tourism planning in third world countries: A critique, *Progress in Tourism and Hospitality Research*, 4, 101-114.

United Nations Department of Economic and Social Affairs (2018) 68% of the world population to live in urban areas by 2050, says UN, United Nations Department of Economic and Social Affairs, retrieved from https://www.un.org/development/desa/en/news/population/2018-revision-of-world-urbanization-prospects.html.

Yang, L. & Wall, G. (2008) The evolution and status of tourism planning: Xishuangbanna, Yunnan, China, *Tourism and Hospitality Planning and Development*, 5(2), 165-182.

Epilogue

This book explored a range of theories, concepts, models and ideas that shape how we think about tourism, the way we do. In doing so, it revealed that tourism is a true multi-discipline. It is informed by such core disciplines as geography, anthropology, sociology, psychology, economics, leisure and demography, as well as by a multitude of other disciplines and fields of study as identified in Chapter 2. Historically, though, tourism studies has been beset by a high degree of silofication – a varied field of study examined strictly within the confines of individual disciplinary silos. Even when attempts have been made to be multi-disciplinary, the results have often been less than satisfactory, for usually one school of thought dominates, while others are placed in subservient roles. Add to this the force field of tourism, and it is not surprising that tourism studies have been labelled as fragmented and disjointed, typified by multiple communities of discourse with historically little cross-fertilization between communities.

Yet, at its most basic, we are talking about two simple things, regardless of the disciplinary focus, methodology adopted or personal bias of the author. They are: tourism as a phenomenon; and the tourist as the person participating in that phenomenon. The former relates to its structure, functioning, evolution, interaction with the broader community and its impacts; while the latter examines people who travel, why they travel, what they do and what impacts they exert. Indeed, different disciplines often examine the same event, but from different sides of the same coin. Destination evolution is a classic case in point. Butler (1980) adopted a supply side, geographer's perspective, Plog (1974) adopted a demand side, social psychologist's perspective, while the advent of complexity theory added a third dimension. The same situation occurs with motivation studies where sociologists and psychologists have vied for dominance. Geographers, anthropologists, economists and environmentalists have examine impacts. And the list goes on. Each approach has its merits and each has its limitations. None is perfect and certainly none is more 'right' or valid than others.

To date, though, the task of trying to present a holistic overview of tourism, what it is and how it works, has proven to be daunting. This book has attempted to do so by discussing the contributions made by many of the foundation scholars who legitimized tourism as a field of study and who explored this fascinating topic from the unique perspective of their own disciplines. How does it all fit together? Figure 1 lists about 50 major themes that formed the basis of the book and shows the interactions between and among them. Each of the themes is developed throughout the book, but because of the intricacy of interactions, it has been impossible to show the links within each chapter. Instead, the final task of integration rests with this Epilogue.

The chart highlights just what a multi-discipline tourism is and why it is such a fascinating field to examine. It also highlights the innate challenges we face when trying to teach it in a simplistic manner, for it is impossible to understand tourism with having at least a basic knowledge of the many area, fields and disciplines that have contributed to our collective knowledge, and how they interact.

We have written previously how we are now in the third generation of tourism studies. The first generation of tourism scholars studied it from within their own single disciplinary lens in order to inform that discipline. Few if any pure tourism programmes existed, and instead tourism scholars often worked on the fringes of their core disciplines, sometimes at the derision of their more 'serious' colleagues who scoffed at the validity of tourism as a research topic. These people had a strong core grounding and were then able to build a strong conceptual foundation that has informed the field for over 50 years.

The second generation came from industry or government, had a reasonably strong single disciplinary focus and were usually employed in burgeoning tourism departments, typically located in business faculties. Importantly, they usually completed their doctoral studies within a traditional discipline, as few of the nascent tourism programmes had reached the level of maturity to offer stand-alone doctorates. They were able to bring a combination of applied business skills that complemented their core disciplines. They are responsible for progressing the field to where it is today, for the work of the second generation of academics led to the proliferation of tourism journals and the ultimate acceptance of the field as a legitimate area of academic enquiry.

The third generation of tourism scholars emerged over the last 10 to 15 years and are now dominating the field as the first and second generations enter retirement. They take the field to new heights for they possess a unique skill set, having graduated from dedicated tourism programmes. But, the jury is still out to some extent, for they often lack a strong background in any single discipline to inform their studies. They may produce impressive empirical studies using sophisticated statistical techniques and have access to unprecedented big data sources. But a risk exists that their work will be atheoretical and add little real new knowledge to the field, for most have not had the chance to explore tourism as an integrated multi-disciplinary phenomenon. This book is written in part for them, in order to give them a chance to appreciate where our collective body knowledge came from and the various sources of these fonts of knowledge. In doing so, we hope that the reader can gain a deeper appreciation of tourism as a dynamic field of study.

Figure 1: Integration of Tourism

	Tourism as a multidiscipline	Tourism as a journey away and back	Heuristic vs technical definitions	Pleasure tourism as an extreme form of leisure	Tourism as a system	Individual tourism systems	Concept of a destination	Attractions	Tourist attractions system	Attractions taxonomy	Valuation of time	Distance decay	Market access	Effective tourism exclusion zones	Segment transformation	Itinerary models	Visa requirements	Bilateral air service agreements	Tourism as trade in services	Tourism as a democratic activity	Butler's lifecycle	Plog's lifecycle	Complexity theory	Prideaux's evolutionary model	Risk
Force field of knowledge	X		X	X	X		X	X	X	X		X	X	X		X				X	X	X	X	X	X
Tourism as a multidiscipline			X	X	X	X	X	X	X	X	X	X	X	X	X	X	X	X	X	X	X	X	X	X	X
Tourism as a journey away and back			X	X	X	X					X	X	X			X	X	X	X	X		X			
Heuristic vs. technical definitions				X			X				X	X	X							X		X			X
Pleasure tourism as an extreme form of leisure					X	X		X	X	X	X	X	X	X	X	X				X					X
Tourism as a system						X	X	X	X	X	X	X	X	X	X	X	X	X	X		X	X	X	X	X
Individual tourism systems							X	X	X	X	X	X	X	X	X	X	X			X			X		X
Concept of a destination								X	X	X				X		X					X	X	X	X	
Attractions									X	X	X	X	X		X	X		X	X		X	X	X	X	X
Tourist attractions system										X		X	X			X			X						X
Attractions taxonomy											X	X	X		X	X				X	X	X	X	X	X
Valuation of time												X	X	X	X	X	X			X	X	X	X	X	X
Distance decay													X	X	X	X	X	X		X			X	X	X
Market access														X	X	X	X	X				X	X	X	X

	Carrying capacity	Limits of acceptable change	Barriers to travel	Tourism as a life priority	The leisure paradox	Generational cohort theory	Boorstin's contrived reality	MacCanell's quest for authenticity	Cohen - strangeness vs familiarity	Cohen's tourist typology	Cohen's centre and the other	Tourism as a liminal experience	Dann's push pull model	Maslow's hierarchy of needs	Crompton's motives	Pearce's travel career pattern	Guttmann's means end chain	Doxey's irridex	Budowski relation with the environment	Tourism as a conflicting land use	Impacts as a function of place change	Sustainable development and tourism	Tripartite solution to sustainable tourism	Precautionary principle	Resilience
Force field of knowledge	X		X	X	X	X	X	X	X	X	X	X	X	X	X	X	X	X	X	X	X	X	X	X	X
Tourism as a multidiscipline	X		X	X	X	X	X	X	X	X	X	X	X	X	X	X	X	X			X	X			
Tourism as a journey away and back			X	X				X	X	X	X	X	X		X	X	X	X							
Heuristic vs. technical definitions			X	X	X	X	X	X	X	X	X	X	X	X	X	X	X			X	X				
Pleasure tourism as an extreme form of leisure	X	X	X	X	X	X	X	X	X	X	X	X	X	X	X	X	X	X	X	X					
Tourism as a system	X	X											X	X	X	X	X	X	X		X	X	X		X
Individual tourism systems	X		X	X	X	X	X	X	X	X	X	X	X	X	X	X	X	X	X				X	X	X
Concept of a destination	X												X						X	X	X	X			X
Attractions	X	X					X	X	X	X	X	X	X	X		X	X								
Tourist attractions system	X												X	X						X			X		
Attractions taxonomy	X			X	X		X	X	X	X	X		X	X		X				X			X		
Valuation of time			X	X	X	X							X	X	X	X	X			X	X				
Distance decay	X				X	X			X			X	X		X	X	X	X		X	X	X	X	X	
Market access	X						X	X	X	X	X	X	X	X	X	X	X	X		X	X	X	X	X	

Effective tourism exclusion zones															x	x	x						x	x	x										x	x	x	x			x	x								
Segment transformation																x	x	x				x	x	x	x	x		x	x	x	x							x			x	x								
Itinerary models																	x	x	x				x		x	x		x	x									x												
Visa requirements																		x	x	x		x			x	x		x	x		x				x	x	x	x		x	x	x	x	x			x	x	x	
Bilateral air service agreements																			x	x			x		x	x		x															x	x						
Tourism as trade in services																				x			x	x	x	x																						x		
Tourism as a democratic activity																					x	x	x	x	x	x		x	x	x	x	x	x	x	x	x	x	x	x	x	x	x								
Butler's lifecycle																						x	x	x		x	x	x						x	x	x	x	x		x	x	x	x	x	x	x	x	x	x	x
Plog's lifecycle																							x	x	x	x	x	x	x	x	x	x	x	x	x	x	x	x	x	x	x	x	x	x	x	x	x	x	x	x
Complexity theory																								x	x	x	x	x	x		x						x	x			x	x	x	x	x	x	x	x	x	x
Prideaux's evolutionary model																									x	x	x	x			x		x		x		x	x	x	x	x	x					x		x	x
Risk																										x	x	x	x	x	x	x	x	x	x	x	x	x	x	x	x	x	x	x	x	x	x		x	x
Carrying capacity																																											x	x	x	x	x	x	x	
Limits of acceptable change																																											x	x	x	x	x	x	x	
Barriers to travel																													x	x	x			x	x	x	x	x	x	x	x	x								
Tourism as a life priority																															x			x	x	x	x	x	x	x	x	x								
The leisure paradox																															x			x	x	x	x	x	x	x	x	x								
Generational cohort theory																																		x	x	x	x	x	x	x	x	x								
Boorstin's contrived reality																																	x	x	x	x	x	x					x	x						
MacCanell's quest for authenticity																																		x	x	x	x	x	x	x	x	x	x	x						
Cohen - strangeness vs familiarity																																			x	x	x	x	x	x	x	x	x	x		x		x	x	x
Cohen's tourist typology																																				x	x	x			x	x				x		x	x	x

	Tourism as a multidiscipline	Tourism as a journey away and back	Heuristic vs technical definitions	Pleasure tourism as an extreme form of leisure	Tourism as a system	Individual tourism systems	Concept of a destination	Attractions	Tourist attractions system	Attractions taxonomy	Valuation of time	Distance decay	Market access	Effective tourism exclusion zones	Segment transformation	Itinerary models	Visa requirements	Bilateral air service agreements	Tourism as trade in services	Tourism as a democratic activity	Butler's lifecycle	Plog's lifecycle	Complexity theory	Prideaux's evolutionary model	Risk
Cohen's centre and the other																									
Tourism as a liminal experience																									
Dann's push pull model																									
Maslow's hierarchy of needs																									
Crompton's motives																									
Pearce's travel career pattern																									
Gutman's means end chain																									
Doxey's irridex																									
Budowski relation with the environment																									
Tourism as a conflicting land use																									
Impacts as a function of space change																									
Sustainable development and tourism																									
Tripartite solution to sustainable tourism																									
Precautionary principle																									

	Carrying capacity	Limits of acceptable change	Barriers to travel	Tourism as a life priority	The leisure paradox	Generational cohort theory	Boorstin's contrived reality	MacCanell's quest for authenticity	Cohen - strangeness vs familiarity	Cohen's tourist typology	Cohen's centre and the other	Tourism as a liminal experience	Dann's push pull model	Maslow's hierarchy of needs	Crompton's motives	Pearce's travel career pattern	Guttmann's means end chain	Doxey's irridex	Budowski relation with the environment	Tourism as a conflicting land use	Impacts as a function of place change	Sustainable development and tourism	Tripartite solution to sustainable tourism	Precautionary principle	Resilience
Cohen's centre and the other												x	x	x	x	x	x						x	x	x
Tourism as a liminal experience													x	x	x	x	x								
Dann's push pull model														x	x	x	x				x	x	x	x	x
Maslow's hierarchy of needs															x	x	x		x	x					
Crompton's motives																x	x					x	x		
Pearce's travel career pattern																	x			x	x	x		x	
Gutman's means end chain																			x	x	x	x	x	x	x
Doxey's irridex																				x	x	x	x	x	x
Budowski relation with the environment																					x	x	x	x	x
Tourism as a conflicting land use																						x	x	x	x
Impacts as a function of space change																							x	x	x
Sustainable development and tourism																								x	x
Tripartite solution to sustainable tourism																									x
Precautionary principle																									x

Index

www.ingramcontent.com/pod-product-compliance
Ingram Content Group UK Ltd.
Pitfield, Milton Keynes, MK11 3LW, UK
UKHW051041120926
8556IPUK00030B/504

9 781911 635352